Adiabatic Thermodynamics of Fluids

From Hydrodynamics to General Relativity

Adiabatic Thermodynamics of Fluids

From Hydrodynamics to General Relativity

Christian Fronsdal

University of California, Los Angeles, USA

World Scientific

NEW JERSEY • LONDON • SINGAPORE • BEIJING • SHANGHAI • HONG KONG • TAIPEI • CHENNAI • TOKYO

Published by

World Scientific Publishing Co. Pte. Ltd.
5 Toh Tuck Link, Singapore 596224
USA office: 27 Warren Street, Suite 401-402, Hackensack, NJ 07601
UK office: 57 Shelton Street, Covent Garden, London WC2H 9HE

Library of Congress Cataloging-in-Publication Data
Names: Fronsdal, Christian, 1931– author.
Title: Adiabatic thermodynamics of fluids : from hydrodynamics to general relativity /
Christian Fronsdal, University of California, Los Angeles, USA.
Description: New Jersey : World Scientific, [2020] |
Includes bibliographical references and index.
Identifiers: LCCN 2019025705 | ISBN 9789811200670 (hardcover) |
ISBN 9789811200687 (ebook) | ISBN 9789811200694 (ebook other)
Subjects: LCSH: Thermodynamics. | Fluids--Thermal properties. |
Hydrodynamics. | Relativistic fluid dynamics.
Classification: LCC QC311.2 .F764 2020 | DDC 536/.7--dc23
LC record available at https://lccn.loc.gov/2019025705

British Library Cataloguing-in-Publication Data
A catalogue record for this book is available from the British Library.

For any available supplementary material, please visit
https://www.worldscientific.com/worldscibooks/10.1142/11288#t=suppl

Desk Editors: V. Vishnu Mohan/Kwong Lai Fun

Typeset by Stallion Press
Email: enquiries@stallionpress.com

This book is dedicated to Amelita, Tina-Marie, Gil and Tamara and their sons, Jenny and Lillian.

Preface

Thermodynamics is usually subdivided into a theory dealing with equilibrium and into one concerned with irreversible processes.

The quotation (Tisza, 1961) is testimony to the meaning that was and still is ascribed to the word "thermodynamics". Why is the extensive literature on adiabatic, **non-equilibrium thermodynamics** excluded from consideration? Why is the study of dissipative thermodynamics referred to as "non-equilibrium thermodynamics"? Laplace explained the speed of propagation of sound in air by postulating that the process is adiabatic; it is neither equilibrium nor dissipative. What about 200 years of atmospheric science? If it is regarded as an application of mere hydrodynamics then it is high time for an upgrade, for it is not possible to understand atmospheres without the concept of temperature.

This book is my attempt to understand thermodynamics, in the formulation as Eulerian field theory of temperature, pressure, densities and entropies, and completely integrated with hydrodynamics. Another hope was, and remains, to make classical thermodynamics more easily accessible to physicists with other specialties, people like myself who neglected to study the subject in school. To this end, the difficulties are emphasized rather than glossed over.

The title was chosen to announce that the subject is not the classical theory of equilibrium thermodynamics, but a dynamical theory (also classical) in which the variables are functions of position and time. Naturally, this implies an integration of hydrodynamics into thermodynamics. The attention is not limited to equilibria though, of course, the determination of equilibria is a major part of the study. Alternative titles such as "extended thermodynamics" were rejected because of close associations with particular trends and *dogmata*. I feel a close affinity to Prigogine (1949) who used the term "local thermodynamics" as opposed to "global (classical) thermodynamics". "Non-equilibrium Thermodynamics" would not be appropriate, for the important distinction is not between equilibrium and non-equilibrium configurations but between adiabatic and non-adiabatic processes.

Statements are heard everywhere to the effect that **entropy** is not defined for systems out of equilibrium. And yet adiabatic processes, defined as processes in which there is no change of entropy, are commonplace. Laplace (1825), departing from an older theory due to Newton, postulated that the entropy of each particle remains constant during this dynamic, non-equilibrium process. There may be some ambiguity in the concept of "fixed entropy". In this book it is interpreted in full accord with Gibbs' formulation of the minimum energy principle, but precisely. To focus on unary systems, for simplicity, **an adiabatic system is governed by a Lagrangian and the specific entropy density is a fixed, constant**. It was within such a system that Laplace's theory of adiabatic sound propagation was finally accepted, 50 years after publication.

A strong imperative characterizes the presentation adopted for this book: an exclusive preference for Lagrangian **variational principles**. This has proved to be an essential advantage in dealing with interesting applications, such as the theory of mixtures and the integration with General Relativity. But at the start of this project this was but a distant goal, since a variational formulation of hydrodynamics was then available only for the special case of irrotational flows. Some generality had to be sacrificed, provisionally. Thus, at the outset, it was clear that turbulence and related phenomena could be described only subsequent to further development. The required generalization of the variational principles to include general flows is introduced in Chapter 10. It is only then that interesting problems in hydrodynamics can be addressed.

As the work progressed and I had to study many fields that I had previously neglected, I came to realize that the potential benefits of action principles is much greater than I had anticipated. I learned that some, and perhaps all, those branches of physics that do not yet have an action principle formulation have very low **predictive power**; they describe, but they do not predict. This applies to astrophysics and atmospheric science, to electromagnetism of materials and to hydrodynamics. I hope to convince some readers of this, but the first evidence will not be presented until Chapters 5 and 6, with the new theory of mixtures.

In its original form, Gibbs' variational principle is a statement that characterizes the states of equilibrium from among all the configurations of a composite physical system as being states of lowest "energy". In this book the principle is applied within the context of an action principle, the Euler–Lagrange equations of which are the fundamental relations of thermodynamics. Gibbs' "energy" is the Hamiltonian of an action principle with a fully developed canonical structure. An adiabatic system is characterized by a Lagrangian and by a fixed entropy. The generalized action principle that is the core of the theory is limited to adiabatic processes, more precisely to processes where the entropy is fixed, see Section 2.3 and especially Section 2.4.

The approach to equilibrium via dissipation is always difficult, and in most of this book we aspire only to understand adiabatic dynamics. Dissipation is often characterized by a much longer time scale and is preferably studied as a sequence of adiabatic equilibria, with the object of discovering the final, absolute equilibrium and not the processes that lead to it. It is here that entropy becomes an essential part of our understanding and it is here that we shall take a far from confident attitude.

We read that "The proper definition of entropy for systems that are out of equilibrium is still an issue of debate". Yes, but actually this is a misstatement, for the entropy of an adiabatic system, at equilibrium or not, is a very clear concept, in thermodynamics if not in statistical mechanics. What is not well understood is the **distribution of entropy** between the components of heterogeneous systems, even at equilibrium, and the way that dissipation affects this distribution. It is for this reason that we prefer to reverse the usual presentation of thermodynamics, offered most convincingly by Callen, who approaches a problem by, first, making use of what is understood about the entropy, and only then introduces an adaptation of the principle of minimum energy to the type of process under consideration. The point is that the action principle covers not only equilibria, but all of adiabatic dynamics as well. We follow the lead of Prigogine (1970) in that we confidently apply adiabatic thermodynamics to systems that are adiabatic though out of equilibrium, in particular, to stationary or quasi-static phenomena. See the Wikipedia article on Adiabatic Process, dated 2015. An important caveat is that the problem of defining the entropy for non-adiabatic configurations has a quite different character when approached from the statistical point of view. See for example Lieb and Yngvason (2013).

The entire theory is based on a hydrodynamical action principle that is due to Lagrange (1760), resurrected by Lamb (1932) and brought to our attention by Fetter and Walecka (1980). This theory is explicitly Galilei invariant and so is all of this book. A large part of the book is dedicated to the study of mixtures. In spite of the very great emphasis on the work of Gibbs, who addressed himself especially to the study of heterogeneous systems, it breaks with Gibbs and with tradition in several respects. Gibbs and almost all others describe heterogeneous systems in terms of the total volume and molar fractions; here they will be replaced by densities, the natural thermodynamic variables, for I find that it greatly facilitates both clarity and convenience. The reader will be spared the Gibbs–Duhem relation and Euler's law, any discussion of intensive and extensive variables and "open" systems. Instead he will enjoy a much more dramatic presentation of the properties of mixtures in the two-dimensional density plane.

Localization (the transition from global to local thermodynamics) is far more natural in this formulation where densities replace extensive variables.

Of great consequence is the general attitude that will be taken with regard to binary fluids. It is a very common practice to treat a mixture of fluids as a single fluid with properties that interpolate between the properties of the components. This is sometimes a successful device, but in this book I insist on respecting the number of degrees of freedom. If a mixture occasionally behaves like a unary system, then the challenge is to understand how this comes about, as a result of interactions. The theory of mixtures in thermodynamics is approached in a manner, not entirely orthodox, that is implied by the reliance on action principles.

As in all textbooks on classical thermodynamics, the exposition relies very heavily on the wonderful example of the ideal gas. Most applications are subsequently upgraded to include the van der Waals fluids. The avowed plan is to describe the fundamentals in great generality but, even so, analogous systems such as magnetic phenomena and electrical circuits will not be discussed. The important roles of statistical mechanics and quantum mechanics will be alluded to when appropriate, but a full development of those subjects is beyond our scope. Applications that regrettably have not been included are electrolysis, electromagnetic fluids and wind tunnels. The main application is to the dynamics of fluids, fluid mixtures and phase transitions, in terms of concepts that are close to those of the experimenter.

A feature of this book is an unusual emphasis on the role of **`gravitation`** in thermodynamics. Gravitation enters as an important tool to help produce a useful separation of phases, as gas and liquid, and immiscible fluids. It enters as an essential complication in the observation of critical phenomena, it appears to be completely responsible for the existence of certain interesting convective phenomena, as Bénard cells, and it is a controversial element in the theory of atmospheres. And yet the fundamental role of gravity in determining the equilibrium state of an *isolated* atmosphere has never been subjected to experimental investigation, nor to sufficient theoretical study. This book includes a preliminary study of mixed atmospheres and a unification of General Relativity with Thermodynamics, provisionally limited to the case of irrotational flows. An extension to include general flows has been found recently. It is presented and developed in Chapters 10–13.

A major struggle has been to find a variational principle for hydrodynamics that allows for rotational motion and vortices. The problem found a solution in 2015. The non-relativistic theory is presented in Chapter 10. The special-relativistic version has been published on *arXiv* and *IJGMMP* and summarized in Chapter 12. An application to rotating planets is presented in Chapter 11. Plans for the future are presented on the last three chapters.

The need for Conservative Hydrodynamics in General Relativity is based, in the first place, on the constraint imposed on General Relativity by the Bianchi identity, and the associated Bianchi constraint on the source, as is explained in Chapter 13. The most precious guide is the work of Emmy Noether.

The discovery of Conservative Hydrodynamics, presented in the last chapters, was realized through an examination of popular applications of the Navier–Stokes equation. Within that context, the reliance on the concept of "energy" and, especially, "kinetic potential" contains a contradiction. Two kinds of velocity fields are involved, a gradient and a time derivative. The energy density is not $\rho/2(\dot{\vec{X}}+\vec{\bigtriangledown}\Phi)^2$ but $(\rho/2)(\dot{\vec{X}}^2+\vec{\bigtriangledown}\Phi)^2)$, and the **kinetic potential** is $\dot{\vec{X}}^2/2+\kappa\dot{\vec{X}}\cdot\vec{\bigtriangledown}\Phi+\vec{\bigtriangledown}\Phi^2/2$. The full theory is a field theory with two independent vector fields.

This explains why hydrodynamics is poorly represented in the first nine chapters of this book. What was missing was the creation of "Conservative Hydrodynamics", or Noetherian Hydrodynamics, free of the restriction to potential flows.

The successful creation of Conservative Hydrodynamics, during the last 3 years of the preparation of this book, is the culmination of a pursuit that has occupied me for the last 12 years. A more complete title for this book would have been *Adiabatic Thermodynamics and Conservative Hydrodynamics.*

About the Author

Christian Fronsdal received his "cand real" from the University of Bergen (Norway) in 1957 and his PhD from UCLA in 1959. He held a scholarship with CERN in Geneva for 2 years and was a Post-Doc at the University of Pennsylvania before starting a teaching career, teaching physics at UCLA for 60 years. He is known for his early work on Black Holes and wave equations for high spin particles. He participated in an important paper on deformation theory and developed the theory of gauge theories in anti-de Sitter space. Later, he made contributions to quantum groups until turning to hydrodynamics and thermodynamics (with connections to General Relativity). He is currently engaged in developing an action principle for these sciences. Other activity: Sailed from 57 deg S to 80 deg N (1974–), and he is interested to do further adventures with his sailboat.

Acknowledgments

It is my pleasure to thank my friends Daniel Sternheimer, Joseph Rudnick, Gary Williams, Tore Haug-Warberg, Alexander Zheltukhin, Frank Hansen, Hidenori Sonoda, Joel Lebovitz and Izumi Ojima for illuminating conversations. I also thank the Bogoliubov Institute in Dubna and my much regretted friend Volodya Kadeshevskij for hospitality. Special thanks are due to Elliott Lieb for hospitality at Princeton during a prolonged visit.

Contents

Preface vii

About the Author xiii

Acknowledgments xv

1. Introduction **1**

1.1 Elements of Hydrodynamics 1
The nature of a liquid: Density
Pressure
Energy
Equation of continuity

1.2 Hamiltonian and Lagrangian Formulations of Hydrodynamics 6
Action Principles: Lagrange multipliers
The Lagrangian of Fetter and Walecka

1.3 Temperature 11

1.4 The Internal Energy and the First Law 13
Heat and internal energy
Entropy: Carnot cycle
The first law and free energy

1.5 The Kinetic Potential, and a Warning 18

1.6 A Lagrangian in the Case of Dissipation 20

1.7 Hydrodynamics of Potential Flow 21

1.8 The "Lagrangian" Form of Hydrodynamics 25
The limitations of potential hydrodynamics

2. Action Principle for Global Thermodynamics **29**

2.1 The Internal Energy and the Free Energy 29
Fundamental relations: Legendre transformations
Normalization of the energy
2.2 The Ideal Gas 31
2.3 The Two Timescales of Thermodynamics 33
Adiabatic process
Dissipative process
2.4 Towards an Action for Thermodynamics 35
An action functional
Interpretation of the action principle
Free energy and Gibbs function
2.5 Lowest Energy and Maximal Entropy 40

3. Localization **43**

3.1 Hamiltonian Density and Adiabatic Equations of Motion 44
The hydrostatic condition
Thermodynamic pressure and the pressure on the walls
3.2 The Kinetic Energy 48
3.3 Lagrangian and Adiabatic Dynamics 50
3.4 The Second Law 52
3.5 Work and Pressure 54
Pressure and Lagrangian density
3.6 Important Lesson 59
Some questions
3.7 Local Thermodynamic Potentials 62
3.8 Conservation Laws and the Stress Tensor 64
The equalization of pressure
3.9 Minimal Energy and Maximal Entropy 68
The classical Lagrangian paradigm
Summary
3.10 The Third Law and the Normalization of Entropy 72
3.11 Galilei Invariance 73
3.12 The Space of Variations 76
Taking stock

4. Applications to Simple Systems **81**

4.1 Propagation of Sound in an Ideal Gas 81
Radial wave

4.2 The van der Waals Fluid 85
4.3 Saturation 89
Interpretation of the velocity potential
4.4 Dew Point, Bubble Point, Maxwell's Rule 92
Normalizing the free energy density
Calculating the dew point for Argon
4.5 The Critical Point 100
4.6 Sound Propagation at the Critical Point 102
4.7 Refinements of the Expression for the Internal Energy 107
The heat capacity
The speed of sound
4.8 The Cubic Equation of State 109
4.9 Stability 111

5. A Theory of Fluid Mixtures **113**

5.1 Variables and Potentials 115
Molar fractions and "free energy of mixing"
5.2 Entropy and Equations of Motion 119
Partial pressures
5.3 Mixture of Two Ideal Gases 124
5.4 Gibbs–Dalton Hypothesis for Ideal Gases 126
5.5 . . . and What Lies Behind It 130
5.6 Gibbs–Dalton Hypothesis for Some Real Gases 133
5.7 The Simplest Mixtures: Boltzmann Theory 136
5.8 Chemical Reactions: Dissociation 137
Gibbs' phase rule
Solving the equations
Minimal energy
5.9 Other Chemical Reactions 143
5.10 Taking Stock 146

6. Immiscible Fluids **149**

6.1 Ideal Gases with Mutual Interaction 149
Data
Conditions for equilibrium
The method of the common tangent
Symmetry
Direct search
Summary

6.2 Mixture of Two van der Waals Fluids 158
The spinodal of unaries
The spinodal of binaries
Calculation of the line of coexistence
6.3 Sample Applications: First Example 163
Binary van der Waals mixtures: Formulas
Coexistence
Points that are critical to the experimenter
6.4 Second Mixture of van der Waals Fluids 168
The square root interaction density for mixtures
Nitrogen and Argon
6.5 Helium and Xenon 176
The maximal (critical) temperature

7. Speed of Sound in Mixtures **179**

7.1 Sound Speeds in Gibbs–Dalton Mixtures 179
Equations of motion for ideal gases
The propagation matrix
Two types of mixtures
Summary
Type 1 mixture, "similar" gases
Type 2 mixture, "disparate" gases
7.2 Simple Model of Interactions 186
Variants
7.3 Two van der Waals Gases: Sound Speed 189

8. Superfluid Helium **193**

8.1 Experiments and Interpretation 193
The fountain experiment
The entropy
The equilibrium configuration
Second sound
Thermodynamic pressure
Critical behavior
8.2 First Analytic Models 203
Thermodynamic variables
Structure of the Lagrangian
The pressure
Entropy

Sound speeds
Equilibrium
The heat capacity
8.3 Dynamical Equations of the Model 210
The adiabatic index
Rotational flow
8.4 Superfluid Helium 213
Two fluids or two flows

9. Atmospheres 219

9.1 Introducing Gravitation 219
9.2 Mixed, Isentropic Atmospheres 221
9.3 Nitrogen and Oxygen 222
9.4 Air and Argon 225
The significance of a constant lapse rate
Van der Waals atmospheres
9.5 Air and Water: Introduction 227
Standard method
9.6 Phases in a Mixture of Air and Water 229
Interaction
Emden's theorem
The effect of entropic forces
The isentropic atmosphere
What physics is involved?
Reply to Maxwell
The U
The standard approach
Appendix: Polytropic Atmospheres 232

10. Conservative Hydrodynamics 239

10.1 The Variables of Hydrodynamics 240
10.2 Constraint and Euler–Lagrange Equations 242
10.3 Cylindrical Couette Flow 243
Laminar Couette flow
The question of sign, again
Rayleigh's mistake and what it tells us
10.4 Conservative Hydrodynamics has a Kinetic Potential 250
10.5 Stability of Laminar Couette Flow 251
Boundary conditions

Limits on Couette flow
A proposal
Main results
Discussion
Internal tension
A very special case, with application to superfluid He
Entropic forces
10.6 Conclusions 261
Future experiments
Methodology: Other applications
A theory of incompressible fluids?
10.7 The Energy–Momentum Tensor 266
Irrotational case
Angular momentum conservation
The general case
Interpretation
Appendix: Planar Couette Flow 270

11. Rotating Planets **273**

11.1 Introduction 274
A brief digression on the cylindrical Couette problem
11.2 A Compressible Fluid Rotating in a Fixed, Gravitational Field 278
The general case
Regularity at the poles
Overview of results
11.3 Earth and Mars 284
The quality of the fits, examples
Density profile, range of N
Rings, or not
11.4 Neptune and Uranus: Venus and Pluto 288
11.5 Saturn, Jupiter, The Sun, Haumea, Galaxies 289
11.6 Summary and Conclusions 292

12. Special Relativity **295**

12.1 The Relativistic, Antisymmetric Tensor Field 296
Photon mass
The covariant gauge
Conservative hydrodynamics

12.2 Non-relativistic Limit and Galilei Transformations 302
12.3 Dark Matter 305

13. General Relativity **307**

13.1 Astrophysics and Hydrodynamics 307
13.2 Introduction to General Relativity: The Metric 308
13.3 The Dynamical Metric 309
13.4 The Bianchi Identity 311
Implications of the Bianchi identity
Particles as sources: Gravity waves
13.5 Sources for G.R. Tolman's Formula 313
13.6 Conservative Sources for Relativity 314

Bibliography 317

Index 325

Chapter 1

Introduction

Thermodynamics is an extension of hydrodynamics by temperature.

It is difficult to believe that the work of Archimedes (287–212 BC) could have happened without an extensive prehistory of practical and theoretical work, but all we have is the archeological evidence of highly developed "pneumatic societies" (Heyerdahl, 1987). Historical hydro-thermodynamics began with Galileo Galilei (1638) and experiments by Torricelli (1644), with Newton (1730) and experiments by Hooke (1678) and others during the 17th century. It took off at the start of the 19th century with the work of Carnot (1824), Mariotte, Gay-Lussac, Joule and many others, interpreted by Laplace (1825), Fourier, Poisson, Thomson, Clausius (1864) and others. The first steps are concerned with the dynamics of fluids, principally gases, and consist of the identification of the main observables, density and pressure, and relations between them. Thermodynamics proper starts with the definition of temperature and reaches its modern development after a very hard struggle (Truesdell, 1980; Müller, 2007), with the introduction of the concept of entropy.

The story of the early stage of this developing science is ably presented in numerous textbooks. The following summary is therefore somewhat abbreviated, up to the point where we try to lay the foundations of adiabatic thermodynamics on the platform laid down by Gibbs, in Chapter 2.

1.1. Elements of Hydrodynamics

The nature of a liquid: Density

For an initial introduction to this subject, the book by Ramsey (1936) is recommended. It begins with a very lucid explanation of what it is that characterizes a **fluid**. For a complete treatment of classical hydrodynamics, the book by Lamb (1916) is famous, but difficult. For another extended text on hydrodynamics, see Lewis and Randall (1927). For an interesting and penetrating historical analysis of the early years, see Brush (1976). A fluid may be either a gas or a liquid. There is no precise distinction between the two, but the gas is the better intuitive model and the easier to understand. Consider a closed, rigid vessel filled with a gas, in the absence of discernible, macroscopic motion and in the approximation in which gravitation and other external influences are neglected. The most elementary concepts attached to this system are the volume V of the vessel (fully occupied by the gas) and the total mass M of the gas. The average **mass density** is M/V and we suppose that we can define a local mass density $\rho(x)$ at each point x in the gas. In the first round, we may assume that this density is uniform and constant in time, in which case $\rho(x,t) = M/V$ but in principle, it is a **density field** defined in the interior of the vessel, varying from point to point, and with time. Hydrodynamics is a theory of fields on space-time.

Throughout this chapter, we are concerned with **simple systems**, to the exclusion of mixtures, where additional dynamical variables are needed.

All quantities will be given with reference to the cgs system of units: centimeters, grams and seconds. The statement "$\rho = 3$" means that the density is $3\,\mathrm{g/cm^3}$. Molar densities will be used in Chapter 6 and in some later chapters.

There are enough **units** to cause confusion. For example, the basic unit of energy is either the "erg" or else the much larger **Joule**,

$$1\,\mathrm{erg} = 1\ \mathrm{g\ cm^2/s^2}, \quad 1\ \mathrm{Joule} = 10^7\ \mathrm{erg}.$$

The cgs unit of pressure is one "**Pascal**",

$$\mathrm{Pa} = 10\ \mathrm{erg/cm^3}$$

or the "mega-Pascal", $1\,\mathrm{MPa} = 1\,\mathrm{J/cm^3}$. The basic unit of energy density is $\mathrm{erg/cm^3}$. In a context that often adds energy density to pressure, it would be convenient to use either $\mathrm{erg/cm^3}$ or $\mathrm{J/cm^3}$ for both. But that is not customary. The "Joule" is often used for energy and "Joule/cm^3" for energy density but more commonly the "**bar**",

$$1\ \mathrm{bar} = 10^5\ \mathrm{Pa} = 10^{-1}\,\mathrm{MPa},$$

is used for pressure; it is essentially the same as one standard atmosphere; precisely 1 bar $= 0.986\,\mathrm{atm}$ or 1 atm $= 101325\,\mathrm{Pa}$, by definition. So, in that case, just divide the numerical value of the pressure (given in bars or atmospheres) by 10 and you have it in Joule/cm^3, then add it to other contributions to the energy density given in the same unit.

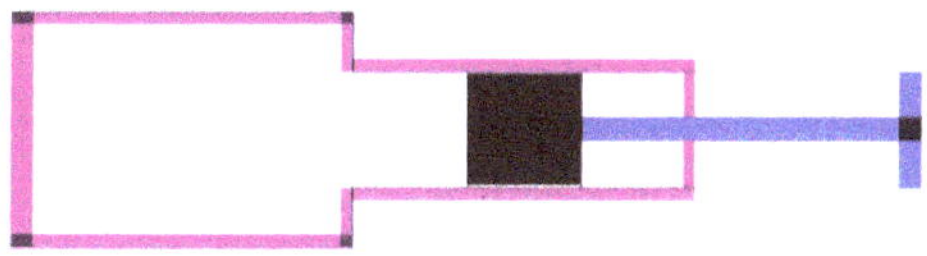

Fig. 1.1.1. Vessel with piston.

Pressure

Consider a vessel filled with a fluid, and furnished with a **piston** that can be moved to change the total volume, without changing the mass, thus affecting the density, supposed uniform. Move the piston very slowly so as not to agitate the fluid; this requires a force F, say (see Fig. 1.1.1).

The most elementary experiment of hydrodynamics measures the force F that is required to displace the piston. If the surface of the piston is normal to the displacement dx and has area A, then the change in volume is

$$dV = Adx.$$

The work done by the piston is Fdx and the **average pressure** against the piston is defined to be

$$P = F/A.$$

The **pressure** is expected to be uniform across the face of the piston. Usually, the pressure against any other portion of the walls of the vessel is not measured. But it can be measured, at the wall and in the interior, and it is a common experience that the pressure on the piston is transmitted through the fluid and that, in the case that there is no motion and no external forces, it is uniform. We postulate the existence of a field p over the extent of the fluid, eventually varying with position and with time, reducing to the value P at the boundary (see Gibbs (1878, p. 139) and Section 3.5 in Chapter 3).

The pressure on the walls can be understood in terms of classical particles bouncing off the walls. The notion of a pressure field in the bulk is the first axiom of hydrodynamics; it can be supported intuitively by placing imaginary walls that divide the fluid into separate compartments. At **equilibrium**, the two sides of such a wall are subject to forces of equal magnitude, in opposite directions. But these walls are not real, so to prevent macroscopic flow, it is necessary to suppose that the parts of the fluid exert forces on each other; it is thus not possible to maintain, as in the classical atomic picture of a collisionless gas, that the particles of the fluid do not interact with each other, but only with the walls, as is sometimes stated in elementary textbooks. This circumstance leads to certain difficulties in connection with the kinetic interpretation of ideal gases, to be discussed later. The resolution of the paradox is fundamental for an understanding of thermodynamics. See Clausius (1864).

The recognition that the atmosphere, as well as air in a confined space, exerts a pressure goes back, at least, to the experimenter Torricell (1644) and probably to Galilei (1638). It was hotly debated during the late 17th century. The result of experiments is known as Boyle's law,

$$PV = k,$$

where k is a constant that depends on the temperature. Later, it was found that k is proportional to T, where T is the absolute temperature (see Brush, 1976).

We have identified two of the three fields needed to describe a simple fluid in classical hydrodynamics. The third is a vector field $\vec{v}$, interpreted as being the instantaneous **velocity** of motion of a small portion of the liquid centered at a point. This fluid velocity is not interpreted in terms of the motion of individual atoms, nor is it associated with Brownian motion; it is entirely macroscopic and defined by the displacement of the fluid (see Clausius, 1864).

This completes the enumeration of the dynamical variables of a simple, hydrodynamical system and we are ready to begin the description of the simplest relations and the simplest experiments.

Energy

Theoretical hydrodynamics embodies a hallowed principle, conservation of energy. The energy expended by an external force F, by moving the piston in Fig. 1.1.1 inwards a distance dx (making $dV < 0$), is

$$F dx = -p\, dV.$$

This energy is conserved, somehow, and kept in reserve by the liquid. In fact, substantially, the same amount of energy is found to be returned to the piston if it is subsequently withdrawn. This is of course an idealization, but in most cases to be covered in this book it is true to very high accuracy if the motion of the piston is slow enough; the fluid behaves like an elastic body, more precisely as a spring, as is required for the good functioning of pneumatic operating systems.[1] This concept of mechanical work and energy is hydrodynamical. It is carried over to thermodynamics but in

[1]But see Section 3.6 in Chapter 3.

the wider context it is not conserved. The law of conservation of energy in thermodynamics depends on the introduction of the concepts of temperature and heat.

Equation of continuity

Experience shows that total mass is usually preserved. This law is expressed locally by the **equation of continuity**,

$$\dot{\rho} + \mathrm{div}(\rho\vec{v}) = 0. \tag{1.1.1}$$

To get a more concrete visualization of what this means, consider a portion Σ of the available volume, and integrate

$$\int_\Sigma d^3x\,(\dot{\rho} + \mathrm{div}(\rho\vec{v})) = 0.$$

By **Gauss' theorem**, the second term can be converted into an integral over the boundary $\partial\Sigma$ of the volume Σ, so that the last equation is the same as

$$\frac{d}{dt}\int_\Sigma d^3x\,\rho = \int_\Sigma d^3x\,\dot{\rho} = -\int_{\partial\Sigma} \rho\,\vec{v}\cdot d\vec{\sigma}; \tag{1.1.2}$$

that is, the increase of mass in Σ is accounted for by the flow of mass inwards through the boundary. The verification of the equation of continuity justifies the interpretation of $\rho\vec{v}$ as a **mass current** (see Fig. 1.1.2).

The equation of continuity is the very soul of hydrodynamics; it has equally important expressions in the related disciplines of quantum

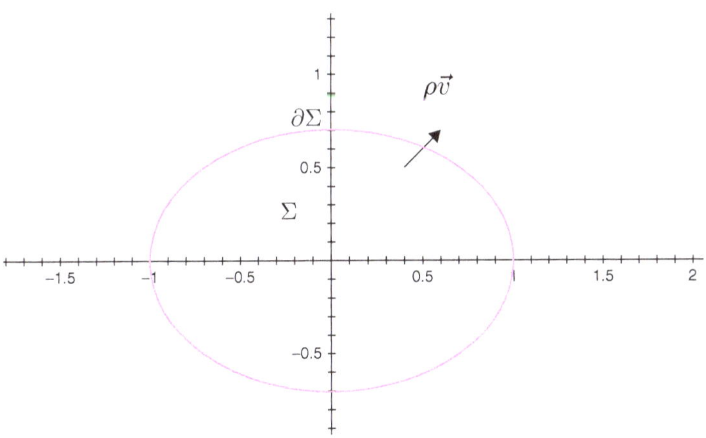

Fig. 1.1.2. The equation of continuity. Σ is a region (a domain) in space, $\partial\Sigma$ is the boundary of Σ and the arrow suggests a current $\rho\vec{v}$ flowing out of Σ (see Eq. (1.1.2)).

mechanics and quantum field theories where it is implied by the condition of **unitarity**.

In a "**Newtonian fluid**", each element of the fluid is assumed to accelerate in accordance with Newton's law, in the absence of other forces, in response to the pressure gradient,

$$\rho\frac{D\vec{v}}{Dt} = -\vec{\nabla}\,p. \tag{1.1.3}$$

This is the simplest case of an equation attributed to Bernoulli (1738) but already used by Newton (1687). The **total variation** with respect to time that appears here is partly due to the passage of time and in part caused by the motion of the fluid element through the displacement $d\vec{x} = dt\,\vec{v}$, which induces the change $\delta\psi = d\vec{x}\cdot\vec{\nabla}\psi$ in the value of any scalar field ψ. Thus,

$$D\vec{v} = dt\,\dot{\vec{v}} + dt\,(\vec{v}\cdot\vec{\nabla})\,\vec{v}.$$

Most fluids considered in this book are assumed to be Newtonian.

The next sections continue the formal development and lead us to the core of our subject. More intuitive aspects of hydrodynamics are discussed in Sections 1.7 and 1.8.

1.2. Hamiltonian and Lagrangian Formulations of Hydrodynamics

The Hamiltonian formulation of classical mechanics was invented after the equations of motion had been established and it may therefore be said that it is a reformulation. The same comment may be made with respect to the **Hamiltonian formulation of hydrodynamics**. But experience has shown and continues to show the value of the Hamiltonian and Lagrangian formulations in many fields of physics. Helmholtz and Maxwell are two of the many physicists who sought action principle formulations for every branch of physics.[2] One conclusion of our work is that hydrodynamics and thermodynamics are not exceptional and that a Lagrangian formulation is very valuable.

A remark by Poincaré expresses very precisely what I feel when confronted with a theory that is presented as a list of equations:[3]

> *We cannot content ourselves with formulas simply juxtaposed which agree only by a happy chance; it is necessary that these formulas come as it were to interpenetrate one another. The mind will not be*

[2]For references to some of the papers of this period, see the review paper by Bordoni (2014).
[3]See works on "extended thermodynamics", e.g., Müller (2007).

satisfied until it believes itself to grasp the reason of this agreement, to the point of having the illusion that it could have foreseen this.

The only context in which a collection of equations take on aspects of an organized entity with a soul is one that is governed by a principle of **least action**.

I find nothing more discouraging than a paper that begins by declaring that "the following n equations will be used in our calculation".

But the most pervasive argument in favor of action principles is their immense predictive power.

Action Principles: Lagrange multipliers

An action principle is a physical law or statement to the effect that something, "the action", tends to assume a minimal or maximal value. For a simple example, let ρ be the density of a fluid and let $W[\rho]$ be a functional that may be interpreted as an energy density, varying with position through its dependence on a density field ρ,

$$W[\rho](x) := W(\rho(x)).$$

The total energy is

$$E = \int d^3x\, W[\rho].$$

The problem is to determine the density that minimizes the total energy; this implies that this functional be stationary with respect to variations of the density, namely

$$\delta E = \int d^3x\, \delta W = \int d^3x\, \delta\rho \frac{dW}{d\rho} = 0. \tag{1.2.1}$$

This, as it stands, is not yet meaningful until we specify the class of variations of the density. We cannot allow all variations; to have any hope that there is a minimum energy, we have to be dealing with a fixed quantity (mass) of fluid,

$$M = \int d^3dx\, \rho.$$

This implies that the variation of the density is restricted by the condition

$$\delta M = \delta \int d^3x \rho = \int d^3x\, \delta\rho = 0.$$

In the absence of any restrictions on $\delta\rho$, Eq. (1.2.1) would imply that $dW/d\rho = 0$, a statement that is too strong; when the variation is restricted in this way it implies less, only that $dW/d\rho$ is constant,

$$\frac{dW}{d\rho} = c, \quad \text{say}.$$

Indeed, that is sufficient, since in that case

$$\int d^3x\, \delta\rho \frac{dW}{d\rho} = c\int d^3x\, \delta\rho = 0.$$

An efficient way to deal with this kind of situation is to modify the action by including a **Lagrange multiplier**. Let λ be a real number and consider the modified action

$$E_\lambda = \int d^3x\, (W(\rho) + \lambda\rho).$$

Now

$$\delta E_\lambda = \int d^3x\, \delta(W + \lambda\rho) = \int d^3x\, \delta\rho \left(\frac{dW}{d\rho} + \lambda\right) = 0.$$

This time, if we impose no restrictions on the variation, we get

$$\frac{\partial W}{\partial \rho} + \lambda = 0.$$

Since the value of λ is not fixed, this says only that W is a constant, which is the correct answer. This method is just a trick to allow us to forget the troublesome restriction to be imposed on the variation of the density, but it turns out to bring additional advantages.

The Lagrangian of Fetter and Walecka

A variational formulation of simple hydrodynamics may be found in a book by Fetter and Walecka (1980). The action, actually discovered already by Lagrange (1760), is[4]

$$A = \int dt \int d^3x\, \mathcal{L}, \quad \mathcal{L} = \rho(\dot\Phi - \lambda - \vec v^{\,2}/2) - W(\rho), \quad \vec v := -\vec\nabla\Phi. \quad (1.2.2)$$

The velocity field is restricted to be of potential type, the velocity vector field is the gradient of a scalar field Φ. This is a very strong restriction. Since Lagrange (1760) (or Euler?) first discovered this action principle, the search for a generalization has not borne fruit until quite recently. In this

[4] $\mathcal{L}$ is the Lagrangian density, "the Lagrangian" is the space integral $\int \delta^3x\mathcal{L}$ and A is "the action".

book, we shall work with irrotational flows, $\vec{\nabla} \wedge \vec{v} = 0$, in the first nine chapters.

With $\vec{v}$ identified as $-\vec{\nabla}\Phi$, the functional $A[\Phi, \rho]$ is an action and $\int d^3x\mathcal{L}$ is a genuine Lagrangian, a close analog of the familiar Lagrangian of particle mechanics, in which the "generalized coordinate" q is replaced by the field variable Φ. The conjugate momentum is

$$\frac{\partial A}{\partial \dot{\Phi}} = \rho.$$

It includes a complete Hamiltonian structure with

$$H = \int d^3x(\rho\dot{\Phi} - \mathcal{L}),$$

a first integral of the equations of motion; it will be shown later that, when boundary conditions are taken into account, as a consequence of the equations of motion,

$$\frac{d}{dt}H = 0.$$

The Lagrangian density in (1.2.2) includes the **kinetic energy density** $\rho\vec{v}^{\,2}/2$, as well as a Lagrange multiplier λ and a term $\rho\dot{\Phi}$. This last term will be present in all our work; it has a number of different personalities and interpretations.

In the first place, if the velocity vanishes, and $\dot{\Phi}$ is just a constant, then it serves as the required Lagrange multiplier associated with the mass. Hence, it can absorb the Lagrange multiplier λ and no explicit λ-term will be needed to allow for the fixing of the mass. But later this term will be related to the escape velocity and the energy of evaporation and more generally to the latent heat, and it will help us formulate the principle of maximal entropy. Most important, it relates the **Hamiltonian density** $\rho\lambda + \rho\vec{v}^{\,2}/2 + W$ to a **Lagrangian density**.

We must not forget that the action principle deals solely with the case that

$$\vec{v} = -\vec{\nabla}\,\Phi.$$

Only in that case is it legitimate to refer to the expression $\rho\vec{v}^2/2$ as an "energy". If the appearance of the kinetic energy in the Lagrangian and in the Hamiltonian seems completely natural, it is because of the evident analogy with particle mechanics. When the velocity can be expressed as the gradient of a scalar field, the velocity is said to be **irrotational** and **the flow** $\rho\vec{v}$ is said to be **of potential type**. That is a strong restriction on the velocity field, but we shall see that this provisional loss of generality still allows a significant field of applications. The restriction to potential flows will be lifted in Chapter 10. It will be shown that the appearance of the Lagrange multiplier as an adjunct to the **velocity potential** Φ is appropriate.

We have two scalar fields to vary, therefore two Euler–Lagrange equations. The variation of Φ and $\vec{v}$ gives

$$\delta A = \int dt \int d^3x \, (\rho \delta \dot{\Phi} - \vec{v} \cdot \delta \vec{v}) = 0.$$

A point that deserves to be strongly emphasized is that the variations of $\dot{\Phi}$ and $\vec{v}$ are not independent, and that both are determined by the variation $\delta\Phi$ of the dynamical field variable Φ, $\delta\vec{v} = -\vec{\nabla}\delta\Phi$. Consequently, to deduce the differential form of the last equation, it must be rewritten as

$$\delta A = -\int dt \int d^3x \, \delta\Phi(\dot{\rho} + \vec{\nabla} \cdot (\rho \vec{v})) = 0.$$

(We have made an integration by parts. "**Boundary terms**" are irrelevant in this context, but see Section 3.8 in Chapter 3.) Since $\delta\Phi$ is arbitrary, the **continuity equation** follows:

$$\dot{\rho} + \vec{\nabla} \cdot (\rho \vec{v}) = 0. \tag{1.2.3}$$

Independent variation of the density gives

$$\dot{\Phi} - \vec{v}^{\,2}/2 = \frac{dW}{d\rho}.$$

The Lagrange multiplier disappears when we take the gradient,[5]

$$\rho\dot{\vec{v}} + \rho\vec{\nabla}v^2/2 = -\vec{\nabla}\, p, \quad p := \rho\frac{\partial W}{\partial \rho} - W. \tag{1.2.4}$$

The combination $D\vec{v}/Dt$ is the acceleration of a fluid element; it appears here due to the identity

$$\vec{\nabla}\, \vec{v}^{\,2}/2 = (\vec{v} \cdot \vec{\nabla})\, \vec{v},$$

which is true by virtue of $\vec{v} = -\vec{\nabla}\Phi$. For a general vector field, we have instead the identity

$$\vec{v} \wedge (\vec{\nabla} \wedge \vec{v}) = \vec{\nabla}\, \vec{v}^2/2 - (\vec{v} \cdot \vec{\nabla})\vec{v}.$$

The left-hand side of Eq. (1.2.4) is the "momentum" of a fluid element, the mass times the acceleration. The equation thus expresses the idea discussed

[5]Equation (1.2.4) is known as Bernoulli's equation, more precisely the "differential form" of Bernoulli's equation. The preceding equation is Bernoulli equation in integrated form. Bernoulli is credited with the discovery that the gradient of the expression $\vec{v}^2/2$ is the centrifugal acceleration, but that is true only in this context, when the velocity field is a gradient.

in the subsection "pressure" of Section 1.1, that the pressure is transmitted through the gas. Any pressure gradient contributes to the accelerated displacement of the volume element and this affects the pressure on the neighboring volume elements. See Brush (1971) for interesting historical details.

The Hamiltonian is

$$H[\rho, \vec{v}] = \int d^3x \, (\rho \vec{v}^{\,2}/2 + W(\rho))$$

and the **Poisson bracket**

$$\{A, B\} = -\int d^3x \left(\frac{\partial A}{\partial v_i} \frac{\partial}{\partial x_i} \frac{\partial B}{\partial \rho} - A, B \right).$$

Here, A, B are functionals of $\rho, \vec{v}$ and Φ, with $\vec{v}$ defined as $-\vec{\nabla}\Phi$. The two fundamental equations (1.2.3) and (1.2.4) are the Hamiltonian equations of motion,

$$\dot{\rho} = \{\rho, H\}, \quad \dot{\vec{v}} = \{\vec{v}, H\}. \tag{1.2.5}$$

We shall return to these formulas below. The restriction to gradient flows will be lifted in Chapter 10.

Thermodynamics is usually understood as a study of equilibria. Here, we have taken hydrodynamics as our starting point, as an introduction to what we call adiabatic thermodynamics, thus we are aiming for a more encompassing theory. The next step is to introduce the temperature.

The Euler–Lagrange equations associated with the Fetter–Walecka Lagrangian are standard and often misused in wider contexts. A generalization of this action principle will be introduced in Chapter 10.

This very brief account of the foundations of hydrodynamics will allow us to introduce our main subject, thermodynamics. Some additional elements of hydrodynamics that will be needed are relegated to the end of this chapter.

1.3. Temperature

For a full appreciation of this topic, there can be no more enjoyable introduction than that of Saha and Srivastava (1935). For the historical development, see the very interesting book by Brush (1976).

The observation that a body, solid or fluid, can be hot or cold is familiar, as is the observation that, in a contact between two bodies, the difference tends to dissipate. It is observed that "heat" passes from the hot to the cold, never in the opposite sense. And if no change occurs, then both bodies must have **the same temperature**. This is the basis for the invention of thermometers (see Fig. 1.3.1); see the book just mentioned.

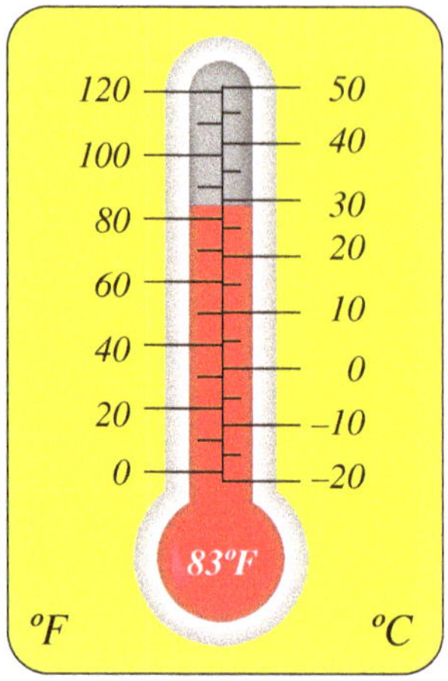

Fig. 1.3.1. Thermometer.

It is assumed that the temperature can be measured at each point of a fluid, so that there is a (scalar) field T, in principle, time-dependent, to be accounted for, in addition to density, pressure and the flow vector. This is the mark of thermodynamics, as compared to mere hydrodynamics.

One of the earliest laws that was derived from experiment was the **Gas Law**, according to which, **at equilibrium** (no motion and all fields constant in time),

$$p = \mathcal{R}\rho T, \quad \textbf{the ideal gas law},^{6} \tag{1.3.1}$$

where $\mathcal{R}$ is a constant and ρ is the molar density.[7] Since our densities shall be given in grams (not mols) per cm^3, we shall have

$$p = \frac{\mathcal{R}}{m}\rho T,$$

where m is the molecular weight. Sometimes, $\mathcal{R}_i$ will stand for $\mathcal{R}/m_i$ where m_i is the **molecular weight** of the ith component of a mixture. A gas for which this law is valid is often called an ideal gas. When this relation holds, when this or some other relation is used to express T in terms of ρ, then the temperature ceases to be an independent variable and the theory reduces in most respects to hydrodynamics. But temperature is vital to

[6]First deduced from experiments by Amontons (1699) but usually attributed to Lord Kelvin.

[7]If the densities are measured in mols per cm^3, then it is a universal constant,

$$\mathcal{R} = 0.8314 \times 10^8 (\text{cm/sec})^2/K = 8.314 \text{ Joule}/K.$$

The Joule is an energy unit equal to 10^7 erg.

any understanding of thermodynamics. By admitting the temperature as one of the dynamical variables, we are adding another dimension to the dynamics. The Euler–Lagrange equation that comes from variation of the temperature gives us the "**adiabatic condition**" that gives meaning to the terms adiabatic derivative and adiabatic thermodynamics. Bringing this equation to the front of the investigation is an important aspect of this treatise.

In Eq. (1.3.1), $T = T_C + 273$, where T_C is the temperature in degrees Celsius. The fact that the equation implies an absolute zero of temperature was noticed by Dalton (1802).

Example. Ordinary air at normal conditions (293 degrees Kelvin) has molecular weight 29 and a density of 0.0012 g/cm^3, thus

$$p = \frac{0.8314}{29} \times 10^8 \times 0.0012 \times 293 = 1.008 \times 10^6 \text{ ergs/cm}^3,$$

with the density $\rho = 1.2/29$ (mol/L). Alternatively,

$$p = 0.08314 \times 10^5 \times \frac{1.2}{29} \times 293 = 1.008 \text{ bars.}$$

1.4. The Internal Energy and the First Law

Heat and internal energy

The first treatise on thermodynamics was probably the work of Boyle, dating from 1660. But further development had to wait until Thomas Savery invented the steam engine in 1698 and Sadi Carnot (1796–1832) took up the study of its efficiency.

It was noticed that some fluids behave, when worked by a piston, as **elastic springs**. When work is done on the fluid, by compressing it, it appears to gain energy and to be able to return it later, for when the piston is withdrawn slowly, it is pushed out by the pressure of the fluid, the fluid produces **mechanical energy**.[8] When the principle of energy conservation was recognized, it became necessary to suppose that the fluid is capable of storing energy in some form, ready to release it when offered the opportunity. This led to the idea that there must be a function U, the "**internal energy**" depending on the volume and on a second variable, the "**heat**

[8]This is an oversimplification; much more will be said about this comparison later on.

content" Q, such that, for variations of the equilibrium configurations,

$$dU = dQ - pdV. \tag{1.4.1}$$

This conclusion is contradicted by observation. The equation defines dQ but there is no variable or *function* Q; dQ is not a perfect differential. The idea of "heat content" belongs to the older theory of heat; it is not a state variable of modern thermodynamics.

The idea of an energy functional expressed in terms of relevant dynamical variables is valid, but the "heat content" is not one of those dynamical variables.

The figure that more than anyone else dominates the earliest part of the history of thermodynamics is Sadi Carnot. When it was found that heat can be converted to mechanical energy, by the steam engine, he set out to determine the most efficient "heat engine" (Carnot, 1824). In a cyclical engine, you must supply heat at high pressure, then remove it at low pressure. The long and painful process that led from this observation to the second law of thermodynamics is recounted in most books on thermodynamics, most entertainingly by Truesdell (1980).

Entropy: Carnot cycle

In a nutshell, the conclusion is that there is no function $U(V,Q)$ that satisfies Eq. (1.4.1); that equation needs an **integration factor**: instead, there is a variable S, related to dQ by $dQ = TdS$ at constant V, and a function $U(V,S)$ such that

$$dU(V,S) = TdS - pdV. \tag{1.4.2}$$

The variable S is called **the entropy** and the energy U is called **the internal energy**. The factors T and p are the partial derivatives of this function,

$$\frac{\partial U}{\partial V}\bigg|_S = -p, \tag{1.4.3}$$

$$\frac{\partial U}{\partial S}\bigg|_V = T. \tag{1.4.4}$$

Another way to state this result is that the differential $dU+pdV$ is not exact. It needs a **multiplier**. After a time, it was realized (Clausius, 1864) that the multiplier is $1/T$: $(dU + pdV)/T = dS$.

The experimental evidence for this statement is very direct.

Consider an ideal engine that can exchange heat with a reservoir and that can do mechanical work by pushing on a piston. Suppose its operation is **cyclic** and that it starts each cycle with volume V_1, pressure p_1 and temperature T_1 (see Fig. 1.4.1). The changes are supposedly slow; the kinetic energy will be neglected; all configurations considered are states of equilibrium. The cycle begins by adding an amount dQ_1 of heat at constant temperature T_1, moving the system from point A to point B. To prevent the temperature from rising, the volume is allowed to increase as the pressure decreases. Next, the temperature is reduced by reducing the pressure, ending at point C. No heat is added or withdrawn during this stage; the temperature is reduced to T_2, say. Next an amount dQ_2 of heat is withdrawn at constant temperature, which implies a reduction of the volume and an increase in pressure. We are at point D. Finally, by a process that does not involve a transfer of heat, the system is returned to the initial state by compression. The fact that this final state is the same as the initial one implies that the two changes in entropy must cancel,[9] thus

$$dS_1 - dS_2 = \frac{dQ_1}{T_1} - \frac{dQ_2}{T_2} = 0.$$

Since $T_1 > T_2$, it follows that $dQ_1 > dQ_2$, more heat is added to the system than is withdrawn. Energy conservation implies that the difference has been converted to mechanical energy. This description of the **Carnot cycle** is taken from Holman (1974).

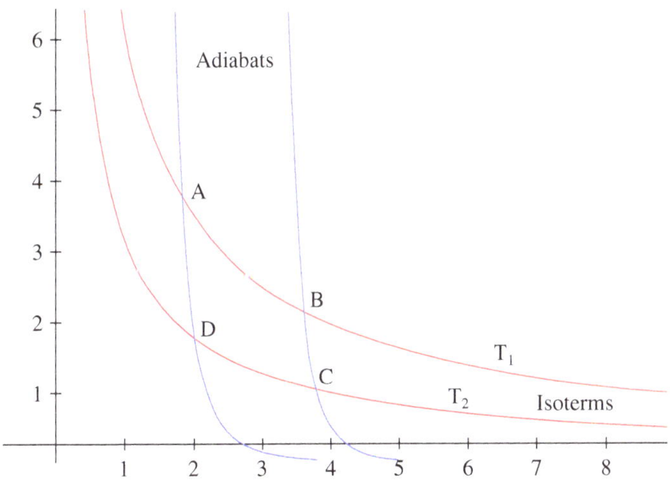

Fig. 1.4.1. The Carnot cycle. See the previous paragraphs for a description of the cycle $ABCD$. Heat engines reach maximum efficiency when operating in a Carnot Cycle. The abscissa is the volume and the ordinate is the pressure.

[9]We can express this fact in another way. Because S is a state variable, while Q is not, the integrated change $\int dS$ of S around a closed loop is zero while that of Q is not.

This very abbreviated account of one of the most dramatic developments in the history of science does not do it justice. But since I have nothing original to add to that story, I prefer to refer the reader to one of the excellent textbooks (see, e.g., Callen, 1960).

It is easy to understand the need to introduce an additional variable. We must recognize that, even in the simplest case, it is not enough to discover one equilibrium state for each elementary system. A system at equilibrium is capable of changing, for example, by varying the temperature. Heating will lead to another, distinct set of configurations and consequently there must be (at least) a one-parameter family of equilibria. They are distinguished by the value of the **state variable** S, the entropy. Cyclical heat engines can convert heat to mechanical energy because dQ is not a total differential.

Entropy is the most difficult subject in thermodynamics. The concept makes good sense for equilibrium configurations and, as we shall show, it easily extends to adiabatic changes and adiabatic configurations.[10] But a definition of entropy that applies to irreversible processes is still being debated.[11] In this book, it will not be neglected, far from it. We shall omit some material about entropy that may be found in standard textbooks and concentrate instead on our narrow topic: thermodynamics as a Eulerian field theory, and on the role that is played by entropy in that context.

The first law and free energy

The first law of thermodynamics is the statement that the function $U(V, S)$ exists. It involves the very difficult concept that is entropy.

The birth pangs of the **first law** of thermodynamics have been discussed in many places, including Truesdell (1980). I shall add some comments that may be new.

The difficulties that were finally resolved by the invention of entropy arise in the discussion of the piston experiment because an element is missing. We know that the system has additional variables, and yet we failed to specify which variables are being kept fixed as we changed the volume. Instead, it was assumed that the system is **thermally isolated**, that no heat enters or leaves the system. This is hazardous, for the concept of thermal isolation

[10] An adiabatic configuration is a solution of the Euler–Lagrange equations associated to a Lagrangian.

[11] See, for example, the discussion after a conference contribution by Meixner (1970), as well as references to Prigogine given below.

is as complicated as the concept of "heat flow". To introduce the first law of thermodynamics, it would perhaps be better to discuss another experiment.

Here is a proposal for another way to develop the foundations of thermodynamics. It avoids talking about heat as if it were a substance and it avoids the very difficult concept of an isolated system.

Consider an experiment in which a gas, at rest, is contained in a rigid vessel and the volume is changed with the help of a piston. But we shall not assume that the system is "isolated". Instead, we assume that the system is described by precisely two variables. There may be other ways to choose them, but consider this:

Assume that volume and temperature form a complete set of independent variables that suffices to fix the state of the system.

This is another way to approach the first law of thermodynamics. It implies that there is something that deserves to be called "energy", which can be expressed as a function of V and T. Call this "**free energy**" F, and assume that it has a differential

$$dF = -PdV - SdT.$$

This is the same as stating that

$$-P = \frac{\partial F}{\partial V}, \quad -S = \frac{\partial F}{\partial T}.$$

At this stage, "S" is simply a partial derivative of $-F$; nothing else is implied. In our experiment, the volume is changed just as before; the difference is that we fix the other variable: T. We put our vessel in thermal contact with a large heat bath with temperature T to ensure that T cannot change. By defining the experiment properly, we avoid all the problems.

In this first experiment, only V is changed, T is kept fixed, and the measured pressure is given as the coefficient $-P$ in

$$dF = -PdV, \quad -P := \left.\frac{\partial F}{\partial V}\right|_T.$$

This is a different experiment; there is no guarantee that the new "energy" F is the same as U. And it is not.

The last formula has a simple and direct intuitive content. It defines the **pressure** as the response of the system to a change in its volume, at constant temperature, given in the units of energy per degree.

Turning from this particular experiment, to one in which both the volume and the temperature are changed, we express the total variation of F as

$$dF = -PdV - SdT.$$

Here, naturally, appears a second coefficient,

$$-S := \frac{\partial F}{\partial T}\bigg|_V. \tag{1.4.5}$$

Just as P is a measure of the response of the system to a change in V (T fixed), S is a measure of the response of the system to a change in T (V fixed). The first coefficient is easily measured; gradually, methods will emerge that allow us to determine the other coefficient as well.

The function F is called **the free energy**; it is related to U by a **`Legendre transformation`**,

$$U(V,S) = \big(F(V,T) + ST\big)|_{S:=-\partial F/\partial T}.$$

The point of this digression is that, by simply focusing on another experiment, we have introduced the entropy S, now clearly a state variable, without tripping over an imperfect differential. It is defined by Eq. (1.4.5).

The first law of thermodynamics is the statement that the function F exists; it is of course equivalent to the existence of the function U.

The basic postulate that defines this development is the statement that the state of a system is completely defined by T and V. If these variables are fixed, and the expression for the free energy $F(T,V)$ is known, then S and P can be determined, but in general, when the expression for F has not been specified, then P and S are independent variables. It is natural to define an "**isolated system**" as one that is characterized by fixed values of S and P as additional independent variables.

The remaining sections of this chapter collect some additional remarks about hydrodynamics.

1.5. The `Kinetic Potential`, and a Warning

Equation (1.2.4),

$$\rho\dot{\vec{v}} + \rho\vec{\nabla}v^2/2 = -\vec{\nabla}\,p, \quad p := \rho\frac{\partial W}{\partial\rho} - W,$$

like Newton's equation, relates the time derivative of the momentum to various forces. There is the negative pressure gradient $-\vec{\nabla}p$ as well as the term $-\rho\vec{\nabla}(\vec{v}'2/2)$. This makes the expression $-\vec{\nabla}(\vec{v}^2/2)$ appear much like a conserved force derived from a potential

$$K := \vec{v}^2/2$$

and this latter expression has been called the "kinetic potential"; it is analogous to the centrifugal potential of particle mechanics.

One reason for wanting to let go of the restriction to irrotational motion is the hope of eventually developing a theory of **vorticity**. This would be beyond our scope at this point. Nevertheless, it is important to know that an irrotational velocity field is not an absolute requirement for setting up a Lagrangian formulation of **adiabatic thermodynamics**.

Assume that the density is defined over the complement M of a discrete set of lines in $\mathbb{R}^3$, that the field Φ is a section of a line bundle over M and that $\vec{v} = -\vec{\bigtriangledown}\Phi$ is well defined as a tangent vector field on M. For example, let M be the complement of the z-axis, and let $\Phi(x, y, z) = \arctan(y/x)$. The flow is horizontal and circular,

$$\vec{v} = \left(\frac{-y}{r^2}, \frac{x}{r^2}, 0\right), \quad r = \sqrt{x^2 + y^2},$$

and the curl is concentrated on the z-axis. Variation of Φ and ρ then gives the correct Hamiltonian equations of motion. The "vorticity",

$$\int_{\text{loop}} \rho\vec{v} \cdot \vec{dx},$$

is a constant of the motion; that is, the structure class of the section Φ is a constant of the motion. It follows that the singular line moves with the fluid. The centrifugal force,

$$K = \vec{v}^2/2 = r^{-2}/2, \quad -\vec{\bigtriangledown}K = \vec{r}/r^4, \tag{1.5.1}$$

is directed outwards, as it must.

Already, in a remark at the end of Section 1.2, it was pointed out that the requirement that $\vec{v}$ be a gradient can be relaxed. But there is one popular procedure that cannot be recommended. The two Euler–Lagrange equations of the Fetter–Walecka–Lagrange action are the equation of continuity (1.1.1) and the Bernoulli equation (1.1.3). A "theory" could be proclaimed, based on these two equations, without the restriction to potential flow. To fully appreciate the futility of constructing new theories based merely on a set of equations, it is sufficient to examine the consequences in this simplest possible case. Suppose that the fluid moves in the manner of a solid body, rotating around a fixed axis, with the vector field

$$\vec{v} \wedge (-y, x, 0).$$

Then K, defined as in (1.5.1) for the case of potential flow, would be

$$K := v^2/2 = r^2/2.$$

The force would be

$$-\vec{\bigtriangledown}K = -\vec{r},$$

directed towards the center, against all experience. Actually, for this kind of flow, the acceleration is the gradient of $\vec{v}^2/2$ and equal to $\vec{r}$. The kinetic potential is $-\vec{v}^2/2$. Clearly, it is dangerous to transplant equations from their natural domain to a wider context. We shall see that this warning should be heeded not only with respect to equations but also in the applications of concepts as well.

Something of the first importance has been learned here. With the study of action principles, with their high efficiency and strong aesthetical appeal, we become acquainted with certain concepts and with certain types of equations that eventually become part of our intuition. In the case at hand that includes the concept of kinetic potential and the idea that the negative gradient $-\vec{\bigtriangledown}K$ of the kinetic potential (multiplied by the density) constitutes a force, we are tempted to apply this idea to a wider field, looking at an equation of the type $\vec{a} = -\vec{\bigtriangledown}K$ as having general validity and using it when there is no action principle to justify it. We have seen that this easily leads to serious error (see also Section 3.9 in Chapter 3, Chapter 10 and the paper "Stability Analysis..." (Fronsdal, 2018a)).

Let us not use terms like "kinetic potential" or "energy" unless we are working within an action principle.

1.6. A Lagrangian in the Case of Dissipation

The importance of a Lagrangian formulation can be judged from the amount of effort that has been expended over the years to achieve an action principle for dissipative phenomena. That this is not an absurdity can be seen from the elementary example of a particle moving in one dimension in a force field derived from a potential and a frictional force of the type $-\mu\dot{x}$. The Lagrangian is

$$\mathcal{L} = \int dte^{-\mu q}(\dot{q}^2 - V(q)),$$

and the equation of motion is given by

$$\frac{d}{dt}\frac{\partial\mathcal{L}}{\partial\dot{q}} - \frac{\partial\mathcal{L}}{\partial q} = e^{-\mu t}\left(\ddot{q} + \mu\dot{q} + \frac{\partial V}{\partial q}\right) = 0.$$

Helmholz was the first to attempt a Lagrangian formulation of dissipative thermodynamics. His work was reviewed by Poincaré (1908)[12] who found that it was of inadequate generality. But the subject was still alive in 1970 (Prigogine, 1970). See also Prigogine (1965) and Kaufman (1984).

1.7. Hydrodynamics of Potential Flow

Hydrodynamics can be divided into three parts or stages. The first part deals with fluids at rest; about this pure hydrodynamics, does not have much to say; it is the classical domain of thermodynamics.[13] The second part is concerned exclusively with potential flow, a surprisingly rich subject. An important example is that the vortices of superfluid Helium seem to be of this type, except for a tiny neighborhood of the center, a vortex string. This is a subject that will come up in Chapters 8 and 10. Here, we shall discuss another application. The third part is the topic of Chapter 10.

It turns out that a sustained flow, a stationary but not static system, can exist only if the Cartesian components of the flow velocity vector field are harmonic,

$$\Delta v_i(x, y, z) = 0,$$

where Δ is the three-dimensional Laplacian. (The justification for this statement will come later, from the **`Navier-Stokes equation`**.) This is a case that can be dealt with very efficiently in two dimensions, when the velocity is independent of the third dimension for reasons of symmetry or when the system is a horizontal film with negligible extension in the vertical direction. In two dimensions, a harmonic function is the real or the imaginary part of a complex analytic function. We can take the velocity potential to be the imaginary part of any complex analytic function of $z = x + \mathrm{i}y$, such as

$$\ln z = \ln r + \mathrm{i}\theta,$$

where r, θ are polar variables. Then

$$\vec{v} = -\vec{\nabla}\theta = \frac{1}{r^2}(-y, x, 0)$$

describes a unique circular flow with angular speed $1/r^2$. This represents a whorl, such as is seen in the wake of ships.

---

[12] I have not found the references.

[13] An important exception: the theory of the shape of drops, and **`capillary action`**.

In order to keep the discussion as simple as possible, let us suppose that the fluid is **incompressible**. What is meant or implied by this is that, though the pressure is determined by the need to prevent motion in the radial direction, the effect that the pressure has on the density is negligible. No theory of the origin of the pressure is needed; the pressure is just assumed to be what it takes for the fluid to execute the motion considered. In the case of circular flow, the pressure gradient must be radial, originating from a cylindrical boundary wall.

By fixing the value of one of the canonical variables, we would have lost the Hamiltonian structure.

Perhaps, there is no need for the vessel to be circular. To illustrate, take

$$\Phi = \Re(\ln z + z^4). \tag{1.7.1}$$

This does not have circular symmetry, its symmetry consists of rotations by one, two, three or four right angles, as is reflected in the flow diagram (red lines) shown in Fig. 1.7.1. Counting outwards from the smallest circle, six

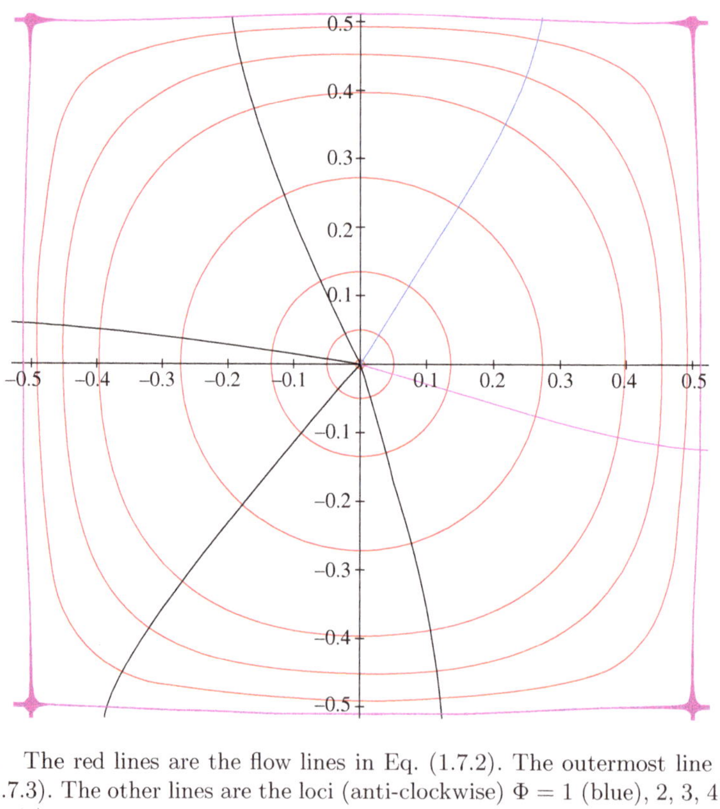

Fig. 1.7.1. The red lines are the flow lines in Eq. (1.7.2). The outermost line shows the locus of (1.7.3). The other lines are the loci (anti-clockwise) $\Phi = 1$ (blue), 2, 3, 4, 5 (black) and 6 (purple).

orbits are flows determined by fixing

$$\begin{aligned}\Phi &= \ln r + x^4 - 6x^2y^2 - 6x^2y^2 + y^4 \\ &= -5, -3, -2, -1.3, -0.9, -0.75, -0.65,\end{aligned} \tag{1.7.2}$$

respectively. The outermost (purple line) flow line is the locus of

$$\Phi = \Re(\ln z + z^4 + 0.24z^8) = -0.597; \tag{1.7.3}$$

it is indistinguishable from a square. A perfect fit would perhaps be obtained with an elliptic function (?). Possibly, an elliptic function can represent an exact, doubly periodic stack of compact flows of this type.

A more general ansatz for the velocity potential is

$$\Phi = a\theta + \Im A(x + iy) = \Im(\ln z + A(z)),$$

with A complex analytic.

The red lines are the flow lines, $\ln r + x^4 - 6x^2y^2 - 6x^2y^2 + y^4 = -5, -3, -2, -1.3, -0.9, -0.75, -0.65$ and -0.597 (this last one is purple). Since div $\vec{v}$ is zero, they are loci of constant density ($\vec{v} \cdot \vec{\nabla}\rho = 0$). The other

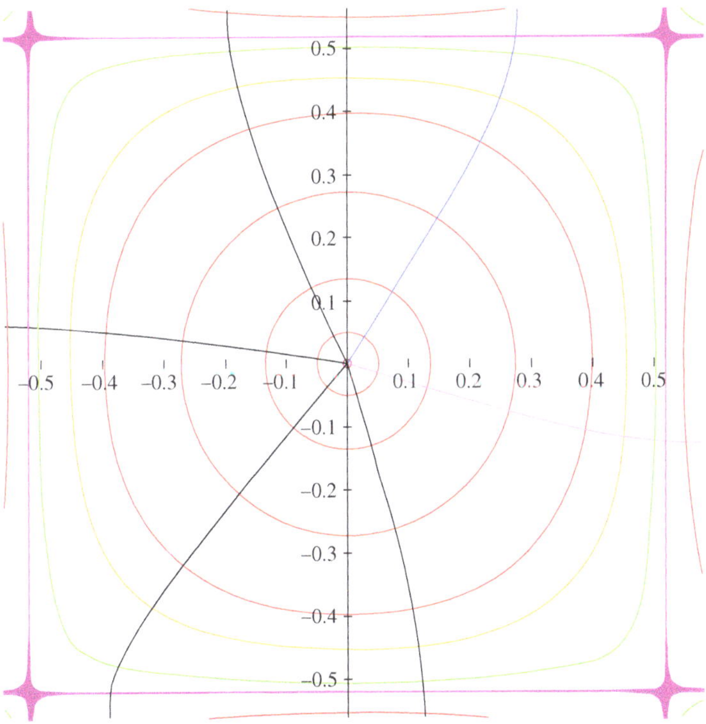

Fig. 1.7.2. An improvement of the previous figure, with the term $0.24\,z^8$ added to the function A in order to straighten the sides of the square. This one is nearly perfect.

lines are the loci (anti clockwise) $\Phi = 1$ (blue), 2, 3, 4, 5 (black) and 6 (purple). They show a modest deviation from the relation $v^2 \propto 1/r^2$.

For a range of values of the constant, these lines very nearly fill a square with corners at $(\pm 0.5, \pm 0.5)$. (The vertical side crosses the axis at $x = 0.514$.) This can be improved by adding higher powers. In Fig. 1.7.2, we have used $A(z) = \mathrm{Re}(\ln z + z^4 + 0.24z^8)$. The variation of the horizontal coordinate along the vertical side is now reduced to 0.004.

Figure 1.7.3 shows our best approximation to an equilateral triangle.

When, as in the examples, the velocity is divergenceless (since Φ is harmonic), the equation of continuity demands that, if the viscosity is neglected, the density can be expressed in terms of the harmonic conjugate function χ. In the case shown in the first figure, the centrifugal potential is one half of

$$\vec{v}^2 = \frac{a^2}{r^2} - \frac{2a}{r^2}(x^4 + y^4 - 6x^2y^2) + r^6.$$

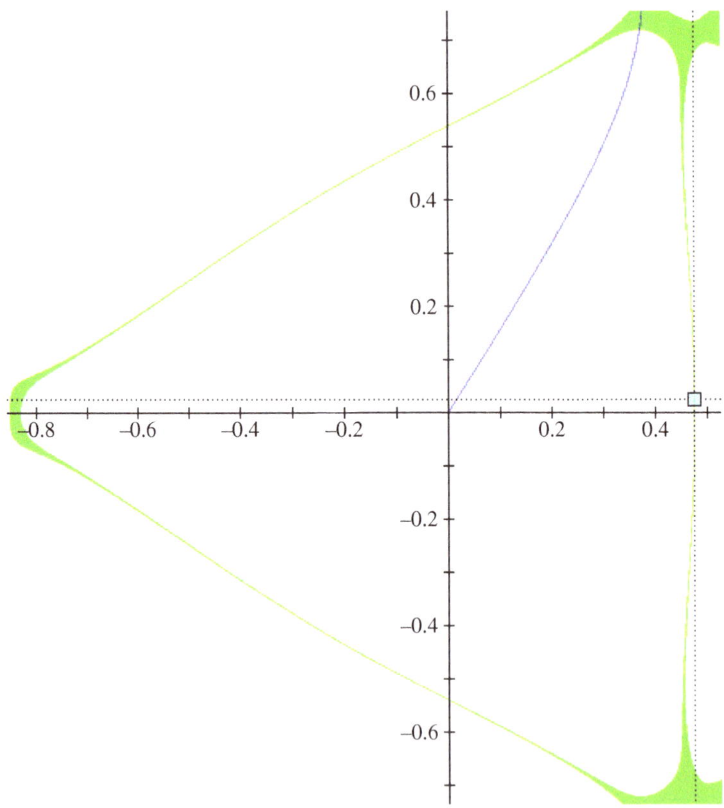

Fig. 1.7.3. Here, $A = \theta + \Im(z^3 + 3.96z^6)$.

The equations of motion, if the equation of state is uniform and there are no external forces, require that the density be expressed algebraically in terms of $\vec{v}^2$, while the equation of continuity is solved most simply by having it expressed in terms of the conformal conjugate field χ. It is perhaps encouraging that the first two terms of the expansion of the function $a^2 e^{-2\chi/a}$ agree with those of $\vec{v}^2$ (see Fig. 11.5.2).

1.8. The "Lagrangian" Form of Hydrodynamics

Up to now, we have been explaining what is called the Eulerian version of hydrodynamics, restricted for the most part to the special case of potential flows, $\vec{v} = -\vec{\nabla}\Phi$. It is a "field theory" in which the dynamical variables are the scalar field ρ and the vector field $\vec{v}$. There was no need to mention particles.

Another version of hydrodynamics is more directly related to the particle concept; it is known as the "**Lagrangian version**" and this name suggests that it is equivalent to the "**Eulerian version**". But, in fact, the two theories are anything but equivalent; they are complementary.

Like the theory of potential flows, the "Lagrangian" theory represents the velocity field as a derivative, this time a time-derivative,

$$\vec{v} = \dot{\vec{x}}.$$

We can think of the vector field $\vec{x}$ as a kind of potential, but that is not the standard interpretation. Instead, one evokes the motion of a particle that finds itself at the position $\vec{x}(0)$, $\dot{\vec{x}}(0) = \vec{a}, \vec{b}$, say (a point in phase space) at time $t = 0$, and moves along a curve

$$C(\vec{a}, \vec{b}) : t \mapsto \vec{x}(t).$$

This interpretation only affects the way that one thinks about the problem; it is permissible to take the other point of view, according to which $\vec{x}$ is a dynamical field variable. Taking that interpretation, we change the notation from $\vec{x}$ to $\vec{X}$.

The representation of the velocity field as a time derivative has the same effect as the restriction to potential flow: it makes it possible to develop an action principle. The simplest Lagrangian density is

$$\mathcal{L} = \rho\dot{\vec{X}}^2/2 - W[\rho]. \tag{1.8.1}$$

Variation of the density gives

$$\dot{\vec{X}}^2/2 - \frac{dW}{d\rho} = 0$$

and variation of $\vec{X}$,

$$\frac{d}{dt}(\rho\dot{\vec{X}}) = 0. \tag{1.8.2}$$

This theory differs from the Eulerian theory in several respects and it has its own limitations. First, it does not include an equation of continuity; instead, it incorporates a concept of integrability. With a chosen initial value $\vec{X}(0) = \vec{a}$, $\dot{\vec{X}}(0) = \vec{b}$, one can calculate a curve $t \mapsto \vec{X}(t)$. Under stationary conditions, one may expect that the particles move on closed orbits and such orbits give occasion to integrate, as in the formula for the "loop vorticity"

$$\int_C \vec{\nabla} \wedge \vec{v} \cdot d\vec{x}.$$

The integral makes sense for any closed curve C, but here one selects a curve that happens to be a particle orbit.

It will be recalled that the Lagrangian density of potential flows has a term similar to the first term in (1.8.1), but with the opposite sign. In the expression for the Hamiltonian, the term $\rho\vec{v}^2/2$ appears with the same, positive sign in both theories. If both types of velocities are present, we shall have the total kinetic energy

$$\rho\vec{\nabla}\Phi^2/2 + \rho\dot{\vec{X}}^2/2.$$

But there is no cross term, the sum $-\vec{\nabla}\Phi + \dot{\vec{X}}$ does not appear. In the equation of motion, we have seen that $\vec{\nabla}\Phi^2/2$ plays the role of a kinetic potential, but in this other theory the kinetic potential is $-\dot{\vec{X}}^2/2$, as the reader is encouraged to verify.

This very curious circumstance becomes a scandal when we recall that both types of velocity field play prominent roles in elementary applications. One example of the role played by field that is not a gradient is this. A glass partially filled with water is placed on the center of a turntable and made to rotate about its axis. The surface of the water assumes a parabolic shape. Elementary textbooks explain this by asserting that the water rotates with the glass as if it were a solid body,

$$\vec{v} \propto \vec{r}.$$

This statement is only approximately true.

What I find very hard to understand is that the profound difference between this type of flow on the one hand, and irrotational flow on the other, seems never to have been noticed. It is stated everywhere that the kinetic energy density is $\rho\vec{v}^2/2$ and that the kinetic potential is $\vec{v}^2/2$ in both cases.

The limitations of potential hydrodynamics

Among the most important, elementary applications of hydrodynamics are the flow of water, oil and **gas in pipes**, and **aerodynamic lift**. One would expect that, though our subject is thermodynamics, some space should be devoted to those applications. The reason that this is not the case is that neither can be treated adequately in terms of irrotational flow. The needed theory, Conservative Hydrodynamics, will be presented in Chapter 10 and some applications are in the last chapters.

Aerodynamic lift is an important, classical subject, treated in many books and perhaps most revealingly in a little treatise by Birkhoff (1950). There the subject is treated as a collection of paradoxes of the following type.

In a **wind tunnel**, insert a wing, or more precisely an obstruction that may have the shape of a sphere, a cylinder or, indeed, an airplane wing. The **lifting force** to which this object is subject, as a result of the horizontal air stream, is easily measured, and the result used to design wing shapes with the greatest lift.

To develop a theory that would explain the experiments, and thus make them unnecessary and save much money, recourse was taken to the theory of potential flow. This theory, as was seen, is formulated as an action principle, which makes the calculations relatively simple and implies a high degree of predictability. It implies, in fact, that there is no lift: Paradox! Paradox? No, the only conclusion that can be made is that the flow around an object is not irrotational.

Again, according to Birkhoff, a generalization is needed and what is proposed is that the three-dimensional flow is only piecewise irrotational. We are presented with examples where the flow domain is divided into several parts with individual, irrotational flows, separated by two-dimensional surfaces of discontinuity. Vorticity is thus concentrated on these surfaces. This new assumption turns out to be fairly successful, but it is *a posteriori* to the experiments; it is not predicted.

Nor does it help to refer to the **Navier–Stokes equation**, for that equation, even if supplemented by an equation of continuity, is not a complete theory. Other assumptions are needed and most frequently, an expression for an "energy" density is invoked, an expression that is necessarily *ad hoc* since no action principle is available; see Chapter 10.

A simpler phenomenon (because of symmetry) is low-speed, cylindrical Couette flow. Here, the Navier–Stokes equation has an initial success, but when it comes to understanding the limits of **stability** of the simplest flows, it has little to offer (see page 10).

When it comes to understanding turbulence, the book by G.K. Batchelor (1953) offers an interesting glimpse into the state of affairs. In the preface he says: "... we have got down to the bedrock difficulty of solving nonlinear partial differential equations". Unfortunately, we are not there yet — the equations are almost certainly not known.

Chapter 2

Action Principle for Global Thermodynamics

By a **global thermodynamic system**, we mean one that is described by a finite number (usually 4) of variables, such as the equilibrium configurations of a fluid contained in a rigid vessel.

The role of an action functional as a depository of all information about a system is ubiquitous in theoretical physics, including Eulerian thermodynamics. In statistical physics, the partition function plays a similar role. It has been maintained that a full expression for the internal energy as a function of the natural variables V and S provides a complete definition of a global thermodynamic system (Massieu, 1876).

The internal energy does not reign without competition; a number of Legendre transformations lead to other possibilities that, according to Gibbs (1878), and Callen (1960), contain precisely the same information.

2.1. The Internal Energy and the Free Energy

Fundamental relations: Legendre transformations

Because of the very abstruse nature of the entropy, we shall begin the discussion of the potentials with the free energy F, a function that, like the **internal energy** U, contains all intrinsic information about the system. The natural variables of F are V and T, much easier to grasp than V and S.

The relation between F and U is a **Legendre transformation**[14]

$$U(S,V) = (F(V,T) + ST)|_{S+\partial F/\partial T=0}. \tag{2.1.1}$$

The inverse transformation is

$$F(V,T) = (U(S,V) - ST)|_{T=\partial U/\partial S}.$$

From the expression for the differential dU, Eq. (1.4.2),

$$T = \frac{\partial U}{\partial S}\bigg|_V, \quad p = -\frac{\partial U}{\partial V}\bigg|_S, \tag{2.1.2a}$$

we obtain

$$p = -\frac{\partial F}{\partial V}\bigg|_T, \quad S = -\frac{\partial F}{\partial T}\bigg|_V. \tag{2.1.2b}$$

The advantage of basing the first law on F was pointed out in the last part of Section 1.4.

The **free energy** was introduced by Helmholtz (1883).[15] The two other potentials that make up the complete collection, first discussed comprehensively by Massieu and by Gibbs, are **the Gibbs function** $G(P,T)$ and **the enthalpy** $H(P,S)$. Two potentials with one common natural variable are related by a Legendre transformation.

The relation (2.1.1) is our second example of a Legendre transformation. Its significance lies in the fact that it correctly indentifies the natural variables of the new function:

$$dU(S,V) = \frac{\partial U}{\partial S}dS + \frac{\partial U}{\partial V}dV = dF + SdT + TdS.$$

The two dS terms cancel, showing that

$$dF = \frac{\partial U}{\partial V}dV - SdT = \frac{\partial F}{\partial V}\bigg|_T dV + \frac{\partial F}{\partial T}\bigg|_V dT.$$

That is, provided the function U is well defined, the variation of F (Section 1.4) can be expressed in terms of the variables V, T without having to solve equations that may turn out to be multivalued and hence ambiguous.

---

[14] As indicated, the variable V can be eliminated by using the relation $S = -\partial F/\partial T$. It is thus assumed that this equation implies a unique value for T, when V and S are given. It is useful to keep this in mind; see Section 4.3.

[15] See Section 4.3.

Normalization of the energy

Energy, strictly speaking, is not a well-defined physical quantity. We can add a constant term to our formula for the energy and the physics remains unaffected, only changes are important. But in thermodynamics there is an additional ambiguity; if we add a term cT (c constant) to the free energy, it will apparently have no effect on U, nor on P. The only effect is to renormalize the specific entropy; that is, shift this variable by the constant c. Until S has been normalized; that is, until the zero point has been fixed, such a shift has no physical meaning. It is a basic tenet of thermodynamics that all physical information about the nature of a substance is contained in the expression given for either U or F, but the said term cannot contain any physical information. This is a type of "gauge invariance" that can be used to standardize our formulas (see Sections 3.10 and 4.4).

2.2. The Ideal Gas

The **ideal gas** is the favorite paradigm of all textbooks on thermodynamics. It is a gas that satisfies two conditions: (1) the pressure is obtained from the formula $p = \mathcal{R}\rho T$ and (2) the internal energy is $U = n\mathcal{R}T$, with n constant. These properties are satisfied when the free energy is given by the following expression:

$$F(V,T) = -\mathcal{R}T\ln(VT^n), \quad \mathcal{R} = \frac{1}{m}.8314 \times 10^8, \tag{2.2.1}$$

where m is the molecular weight of the gas. The exponent n is related to the "**adiabatic index**" γ of the gas by

$$\gamma = 1 + \frac{1}{n}.$$

According to the statistical interpretation, $2n$ is the number of degrees of freedom of the molecules that make up the gas; in our present framework, it is a number: $n = 3/2$ for a monatomic gas, $n = 5/2$ for a diatomic gas.[16,17]

In some texts, the ideal gas is defined by the gas law alone, while the formula for the internal energy is derived. But this "derivation" appeals to concepts that are outside our context.[18] To remain within our subject, we

[16]Only translational and rotational degrees of freedom are counted in this connection. Vibrational and other degrees of freedom and other types of excited states are less relevant. See Clausius (1864), Landau and Lifshitz (1959) and Brush (1976).

[17]The ideal gas is a product of the kinetic theory of gases, first proposed by Herapath in a paper that was rejected by the Royal Society in 1820 (Brush, 1976).

[18]An attempt at a derivation within the context fails to convince (Reif, 1965).

shall define *the ideal gas* in terms of the formula for the free energy, from which both the gas law (the thermal equation of state) and the formula for the internal energy (the caloric equation of state) are derived, as we shall see.

Additional properties are sometimes imputed to the ideal gas, properties that are not implied by the definition that we have given. We intend to stick to our definition: An ideal gas is one that has a free energy function of the form (2.2.1); that is all.

One should have thought that this simple formula (2.2.1) for the free energy of an ideal gas would be presented early and prominently in any text dealing with ideal gases, along with the statement that many properties of ideal gases follow from it. Instead, the ideal gas is often defined by its properties, while the miraculously all-encompassing formula for F is not mentioned.

From Eqs. (2.1.2a) and (2.1.2b), and the above expression for F, there follows, respectively, that

$$p = \mathcal{R}T/V$$

and

$$S/\mathcal{R} = \ln(VT^n) + n. \tag{2.2.2}$$

The first is what is called *the ideal gas law.* The expression for the entropy implies that the **polytropic relation**, $VT^n =$ constant, is valid for any fixed value of the entropy and this is confirmed experimentally for some gases. When this expression is substituted into the definition $U = F + ST$, one gets the statement that $U = n\mathcal{R}T$.[19] This result is often accompanied by the confusing comment that, for an ideal gas, the internal energy is independent of the volume. It must be remembered that the natural variables for U are V and S and not V and T.[20] In terms of V and S, the internal energy of an ideal gas is

$$U(V,S) = n\mathcal{R}V^{-1/n}\exp\left(\frac{S}{n\mathcal{R}} - 1\right) \tag{2.2.3}$$

and not independent of the volume.

[19]More simply but less clear,

$$U = (1 - T/\partial T)(F + ST) = (1 - T/\partial T)F = n\mathcal{R}T.$$

[20]See, for example, Rowlinson (1959, p. 13) for a collection of statements that are difficult to interpret.

In this book, we shall follow the practice of defining a fluid by its free energy and derive all required properties from that. An ideal gas is a gas that has the free energy given by Eq. (2.2.1).

Some real gases behave in a way that is similar to the ideal gas, especially at high temperature and low density. At lower temperature and higher density, the modified ideal gas known as the van der Waals gas is a better approximation to real gases; it is sometimes used for liquids as well. The problem of finding the free energy and an equation of state for liquids is often avoided by using rough approximations (White, 1999, pp. 21–22).

2.3. The Two Timescales of Thermodynamics

The traditional point of view (and ours) envisages two quite different types of processes, with two different timescales. For the first type of process, the timescale is relatively short and certain phenomena, as radiation and dissipation, are often neglected. The propagation of sound in an ideal gas is an example of this type of adiabatic process. It is characterized by the fact that the entropy remains constant, thus in the case of an ideal gas, VT^n is constant in time. When treating such phenomena, one usually assumes that there is no interaction with the environment or, rather, that any such interaction is weak enough to be neglected. Processes of the second kind are usually characterized by much longer timescales; they can be ignored over short time intervals, but over longer time periods their effects are often cumulative. For simplicity, by choice, we limit our consideration of dissipative processes to the case when the system can be considered to be in equilibrium with respect to the faster adiabatic processes; the system passes through a sequence of equilibrium states. A process that is considered to be slow in this sense does not necessarily involve interactions with the environment. Consider a fluid in some kind of regular motion, as in flowing through a circular tube. If the energy and the entropy are both preserved then the adiabatic equations of motion apply (see below). But experience shows that the motion will eventually slow down and finally cease altogether. What happens is that the kinetic energy gradually converts to heat by a process that we shall refer to as internal dissipation. This too is a comparatively slow process, neglected in the context of processes of the first type. See Prigogine (1970), where the existence of two timescales is related to the short reaction time and the much longer relaxation time in collision theory. For a rare example of a text that discusses the same point, see Palatnik and Landau (1964, p. 17). Most textbooks take much less care to explain the meaning of the ubiquitous "variations" and this is one of the reasons why the study of thermodynamics can be painful.

Perhaps, it is better to explain this dichotomy in another way. The better to navigate in a difficult terrain, we choose to address situations in which one of the two simplifying factors is operating.

Adiabatic process

An **adiabatic process** is one in which the effects of dissipation are unimportant; this may happen because dissipation produces negligible changes over the period of time covered by the experiment. Ideally, such processes are characterized by **conservation laws**. An important example is the propagation of sound at high frequencies; the rapid oscillations of pressure are but little affected by dissipation. In this case, the same can be said about conduction; the half period during which the temperature is higher at point a than at point b is much too short for a significant amount of heat to flow from point a to point b by conduction. Then the temperature gradient changes sign and what little heat transfer took place is erased.

Dissipative process

A **dissipative process** is one in which dissipation (or conduction or other slow processes) is the only change that is taking place. In practice, one prepares a state that is as near to equilibrium as one can; ideally, this means that the configuration is at equilibrium with respect to all adiabatic changes. Then dissipation can be viewed as progressing sedately through a sequence of adiabatic equilibria. A slow process is not necessarily dissipative; when a piston is removed slowly, no flow of consequence results and the system effectively moves along a continuous family of adiabatic equilibria indexed by the pressure.

Figure 2.3.1 is an attempt to create an intuitive image of this situation. Let three-dimensional Euclidean space stand for all possible virtual states or configurations of the system. The actual physical configurations form a subset that is represented as the union of blue parabolas. States on any parabola form an adiabatic system; a system with fixed entropy. For each value of the entropy, there is a unique equilibrium state, a state of lowest energy; this state is the lowest point on the corresponding parabola. When we venture to describe non-adiabatic processes, we limit our attention, by prudence, to these equilibrium configurations, to changes that progress along the red line.

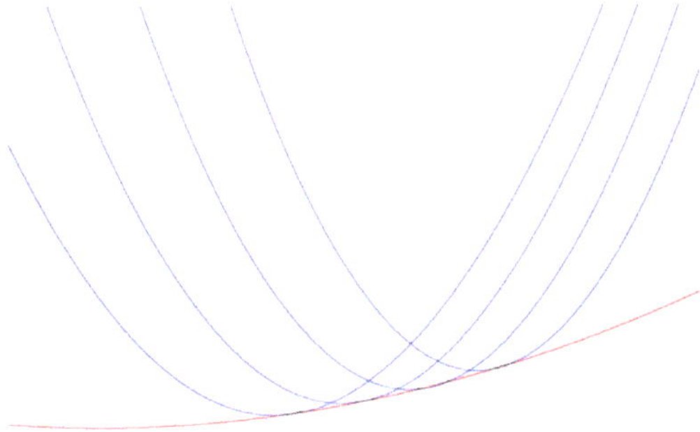

Fig. 2.3.1. In this sketch, a three-dimensional "space of configurations" is associated with a fixed expression for the Lagrangian. Each parabola represents states with fixed Lagrangian, entropy and pressure. The red line represents the set of equilibrium states.

It may be remarked that the preceding lacks clarity and precision. We shall return to the matter in a context where the definitions can be made more precise (Section 3.3.3). Without further work, we have to be content with this: that there are two kinds of processes that are sufficiently simple to be dealt with and that are usually dealt with independently.

"A process that is slow enough to be considered as passing through a continuous series of equilibrium states is called a quasi-stationary process" (Palatnik and Landau, 1964). What we call a (slow) dissipative process is quasi-stationary but not adiabatic.

2.4. Towards an Action for Thermodynamics

The thermodynamics of a simple system is defined by two **Fundamental Relations**, namely,

$$S = -\left.\frac{\partial F}{\partial T}\right|_V, \tag{2.4.1}$$

$$P = -\left.\frac{\partial F}{\partial V}\right|_T. \tag{2.4.2}$$

For any chosen set of values of the pair (S, P), they constrain the possible equilibrium values of the pair (V, T). Taken together, the set of allowed values of (S, P, V, T) defines a two-dimensional surface in the total four-dimensional Euclidean space with coordinates V, T, S, P. According to Maxwell, one of the most inspiring ideas in Gibbs' thermodynamics is to call attention to these two spaces: the total, four-dimensional space and the two-dimensional surface of physical states defined by the Fundamental Relations (see Rukeyser, 1942).

Gibbs' axiom of minimal energy (or the equivalent axiom of maximal entropy) aims to identify the actual equilibrium states on a two-dimensional surface.

The famous Action Principles of Physics are supplemented by a principle of lowest energy for the identification of equilibrium configurations in the manifold of solutions of the Euler–Lagrange equations. To turn Gibbs' thermodynamics into a principle of least action, we must find an action $A(V,T,P,S)$ with Euler–Lagrange equations that coincide with the Fundamental Relations.

As the title of Gibbs' paper, *On the Equilibrium of Heterogeneous Substances,* makes clear, it is not about simple systems. After stating the axioms, he seems to take the application to simple systems for granted, and, without further clarification, he applies them in a context that is a collection of simple systems, each with its own independent variables

$$(V,T,P,S),(V',T',P',S'),(V'',T'',P'',S''),\dots.$$

This leaves a large gap for the reader to fill and the result has been that the potential impact of the paper has not been realized. The most important void to fill is the associated Action Principle.

The remainder of this chapter attempts to fill the gaps that were left by Gibbs' paper; we shall see that its full implications have been to a large extent overlooked.

An action functional

The following action will serve us well:

$$A(V,T,P,S) = F(V,T) + ST + VP. \tag{2.4.3}$$

This expression applies to a fixed amount of material; that is, to a system with prescribed mass, or to a fixed number of particles, but variable volume and pressure. The walls of the vessel are not fixed, but the mass of the physical system in it is. What variations are we to carry out? If we vary T and V with S and P held fixed, we obtain Eqs. (2.4.1) and (2.4.2) that are basic for any fluid with a chosen expression for the free energy:

$$\left.\frac{\partial A}{\partial T}\right|_{V,P,S} = \left.\frac{\partial F}{\partial T}\right|_{V} + S = 0, \tag{2.4.4}$$

$$\left.\frac{\partial A}{\partial V}\right|_{T,P,S} = \vec{\nabla}\, p\left.\frac{\partial F}{\partial V}\right|_{T} + P = 0. \tag{2.4.5}$$

This action principle showed up in an attempt to introduce the temperature into the theory of Fetter and Walecka (Fronsdal, 2014) and a connection to the work of Gibbs was immediately recognized. It bears a clear imprint of Gibbs' ideas, but it is not his principle of minimum energy. Instead, it is a restatement of the laws of thermodynamics. Gibbs' axioms refer to an energy functional: we propose to identify his energy functional with the Hamiltonian of this action principle. After appropriate generalizations, it will lead us to Gibbs' principle of minimum energy as well as his principle of maximal entropy, for **it is a setting in which the two principles are equivalent.**

At this point, we shall try to arrive at an understanding of the action principle in its present form without taking into account its potential connection to the energy principle.

Interpretation of the action principle

Variation of S would lead to $T = 0$ and variation of P to $V = 0$; clearly therefore, the principle makes sense only if S and P are treated as fixed parameters while only V and T are varied. Under these conditions, the action principle neatly summarizes the laws that govern the properties of the fluid. This provides the fundamental insight into the role of the **entropy**.

> **The variational equations are not primarily to be regarded as formulas that allow to calculate entropy or pressure, but as constraints on the variables V and T for assigned values of P and S.**

Exactly the same point of view can be discerned in Gibbs' work on the principle of minimum energy. We must specify the value of entropy and pressure before we shall have a well-defined energy functional to vary. Both S and P are parameters that can, in principle, be chosen freely and that must be fixed to give a meaning to the variational equations. The central, difficult problem is not to discover the formula that allows the entropy to be calculated when the pressure and the density are known, but to *assign values to the entropy* so as to obtain the correct constraint on those variables. So far, S and P play similar roles.

The states contemplated at this point are uniform, equilibrium distributions; there is expected to be exactly one such for each pair of values of the two parameters S and P. The interpretation of P is the pressure on the walls, assumed the same everywhere. The second equation constrains the variables V and T so as to maintain this pressure. The first equation shows that for each value of P there is not one unique equilibrium state, but a one-parameter family of such states, labeled by the value assigned to the parameter S.

This insight, that the variables P and S are fixed parameters of the variational principle, is very important. It is easily forgotten or obfuscated by subsequent interpretations of the entropy.

As important is the fact that the action principle, by itself, gives no information whatever about the value of the entropy, or the value of the pressure. Nor is there any *a priori* guarantee that the variational equations have solutions for all values of these parameters.

Gibbs postulates that, at equilibrium, among states with fixed entropy and pressure, the energy must be at a minimum. That is, it is at a (local) minimum with respect to certain virtual changes. The proper interpretation of this statement is a difficult question. It is the central issue in equilibrium thermodynamics and it occupies an important place in adiabatic thermodynamics as well.

Gibbs' geometrical formulation of thermodynamics was enthusiastically promoted by Maxwell (see Rukeyser, 1942). Its major role is to introduce the raw and mutually independent variables. The action functional is, to use a term from quantum mechanics and quantum field theory, **off shell** and subject to no constraints. The laws of adiabatic thermodynamics are the equations of motion of the variational principle and it is of no interest to use these laws to transform or evaluate the action functional. To approach a specific thermodynamical problem, we must determine the off-shell action functional, choose the expression for the free energy and the values of entropy and pressure; here, experience and physical insight are most important. From then on, the only equations that can and should be invoked are the equations of motion. There is no danger that a crucial relation has been overlooked and there is less need for ingenuity or additional intuitive insight.[21] What has just been said applies to adiabatic processes with fixed entropy and pressure. For non-adiabatic processes, the situation is much more difficult.

Important. A Legendre transformation relates the free energy to the internal energy,

$$U(S,V) = (F(T,V) + ST)|_{S=-\partial F/\partial T|_V}, \tag{2.4.6}$$

[21] This is an overstatement but it shall stand, for it is a succinct summary of the attitude that is adopted in this book. But see Section 3.5.

but note that this defines the internal energy on shell.[22] Similarly, since A is stationary with respect to variations of V and T, the values of A at stationary points,

$$A(V, T, P, S)|_{\text{on shell}},$$

define a function of P and S, the enthalpy. But that is not to say that our action functional can be identified with the enthalpy. The variational principle is concerned with virtual variations, emphatically not restricted to the physical states of solutions. The thermodynamic potentials, with the exception of the free energy, are defined on shell via Legendre transformations.

Free energy and Gibbs function

The function F plays a central and distinguished role in this formulation of adiabatic thermodynamics. Other formulations are possible, that give this prominent place to one of the other potentials. In the limited context that is being envisaged at this stage, all four choices give rise to exactly the same variational equations, *modulo* Legendre transformations, and all four versions are absolutely equivalent, according to Gibbs (1878) and to Callen (1960). With localization, the functions $F(T, V)$ and $U(S, V)$ become densities $f(T, \rho)$ and $u(s, \rho)$; both have the density ρ as one of their two natural variables and that is clearly advantageous since the density is a major variable of hydrodynamics. Rowlinson (1959, p. 79) works with F and gives some hints as to why this choice is preferred. In Landau and Lifshitz (1959), "energy" usually means free energy; the practice of referring to "energy" without precision is widespread. Callen and others emphasize the equivalence of formulations based on either of the four potentials, but his context is global thermodynamics. The integration with hydrodynamics that follows in Chapter 3 favors the present formulation, based on F.

Experiments are usually performed with either the volume or the pressure kept constant. The first situation is realized when the fluid is confined within rigid walls; the other when the volume is determined by the atmospheric pressure. In that case, the Gibbs function takes center stage. We shall discuss both possibilities later.

In global thermodynamics, the entropy is a parameter; in the local theory to be examined next, it is a field. This is one of the reasons why localization is not a straightforward matter.

[22] See above.

Historical remark. The most interesting attempt to base thermodynamics on an action principle was made by Helmholtz, an outstanding scientist of the 19th century. Poincaré devotes a chapter to evaluate this attempt and finds that it falls short. Helmholtz's ideas were not very different from those presented here, but he looked for an action that would depend only on two variables, in the first place, one of the fundamental potentials. That is why, the work of Gibbs is essential.

2.5. Lowest Energy and Maximal Entropy

Gibbs axioms of **minimum energy** and **maximal entropy** are stated in the opening of his monumental paper (Gibbs, 1878):

> *The criterion of equilibrium for a material system which is isolated from all external influences may be expressed in either of the following entirely equivalent forms:*
>
> (I) *For the equilibrium of any isolated system it is necessary and sufficient that in all possible variations of the state of the system which do not alter its entropy, the variation of its entropy shall either vanish or be negative. If ϵ denote the energy, and η the entropy of the system, and we use a subscript letter after a variation to indicate a quantity of which the value is not to be varied, the condition of equilibrium may be written*
>
> $$(1) \qquad (\delta\eta)_\epsilon \leq 0.$$
>
> (II) *For the equilibrium of any isolated system it is necessary and sufficient that in all possible variations of the state of the system which do not alter its energy, the variation of its energy shall either vanish or be positive. This condition may be written*
>
> $$(2) \qquad (\delta\epsilon)_\eta \geq 0.$$

The nature of the proof is of interest:

> *That these two theorems are equivalent will appear from the consideration that it is always possible to increase both the energy and the entropy of the system, or to decrease both together, viz., by imparting heat ...*

This is very abstract, but the last sentence gives a hint of a quantity of the nature of "energy plus entropy" that is invariant with respect to the joint variation of energy and entropy. We shall see that this quantity is the action. How shall we interpret Gibbs' "energy"? In modern thermodynamics of isolated systems, the functional that is minimal at equilibrium is usually the free energy F or the Gibbs energy G.[23] Thus, at equilibrium, F can only increase. When we take this to be valid, it follows that $A - ST$ can only decrease and a variation for which the action is stationary can only increase the entropy, or at least it can only increase the value of ST.

Much has yet to be done to complete the formulation of the action principle. The problem of achieving a thorough understanding of the principle of minimum energy will be taken up again in Section 3.9.

[23]It is sometimes assumed that experiments are carried out over a long time, long enough that the system can be seen as being in thermal equilibrium with the environment. On this basis, a case has been made for assuming that it is the enthalpy that is minimized. For simplicity, we shall restrict ourselves to deal with isolated systems and assign the role of energy to F.

Chapter 3

Localization

The term *thermodynamics* is often limited to the study of systems in equilibrium. The present goal is to unite thermodynamics with hydrodynamics in a framework that we shall refer to as local thermodynamics. The result is a vastly enlarged dynamical theory that is no longer limited to the study of equilibria. The most usual approach starts with the **main assumption** (Callen 1960), valid when no external fields are present,

> **Relations between the thermodynamical variables, at each point in the fluid, are the same as at equilibrium of the corresponding uniform system.**

In Gibbs' paper (1878), this assumption is used throughout though it is not clearly stated as an axiom. For example, the entropy and any other thermodynamical variable may vary from point to point, but the fundamental relations do not contain any reference to position. At equilibrium, Eqs. (2.1.2a) and (2.1.2b) hold and for them to make sense locally, it is enough to interpret V as the inverse density; in what follows, V is replaced by $1/\rho$. The question of the range of its validity is discussed, briefly, by Fitts (1962) who refers to the work by Prigogine (1949). In this paper, a fundamental expression for the entropy is shown to be valid in certain situations away from equilibrium.

We have shown that, in the case of a simple system, with a specific choice of the action functional, the variational principle is equivalent to the fundamental relations of thermodynamics. Here, as we apply the ideas to heterogeneous systems, we shall continue to base the theory on an explicitly defined action functional.

3.1. Hamiltonian Density and Adiabatic Equations of Motion

The total energy is obtained by integrating an energy density over the extent of the fluid, and the same must be true of the free energy and the other potentials. The densities related to F, S and U will be denoted as f, s and u. What we have so far is this

$$A[\rho, T, s, P] = \int_\Sigma d^3x h + PV, \quad h = f(\rho, T) + sT.$$

For the time being, we shall assume that the system is confined within rigid walls; then the term PV can be ignored, its role being taken over by the boundary conditions; it will be discussed in Section 3.5.

In the case that the fields are uniform over the volume,

$$f(\rho, T) = \rho F(V, T), \quad V = 1/\rho.$$

Here, $f(\rho, T)$ is the all important **free energy density**. The density ρ and the temperature T are scalar field variables. The pressure P is the pressure on the boundary, assumed uniform for the time being. The extensive nature of the entropy[24] tells us to interpret $\int d^3xs$ as **the total entropy** of the system.

With regard to the entropy density s, it is hoped that the **specific entropy density** $S = s/\rho$ is independent of ρ (and of T). For we shall see, when this is true, the term sT makes no contribution to the pressure and in this case only does the expression that we shall obtain for the thermodynamic pressure agree with the "main assumption". The correct approach to entropy is a major preoccupation of this book; it will be undertaken in specific circumstances, beginning with the simplest. At this point, we shall only assume that s can be expressed as a function of ρ and S, with ρ, T and S as basic variables, but we shall be led to conclude that a situation in which the specific entropy is not uniform will be encountered only in special circumstances, for example, when a chemical interaction is taking place (combustion) and in the presence of external forces.

Hence, S, and not s, is a field variable independent of the density. The role that is played by entropy in the local variational principle is the same as in the case of the global variational principle, but instead of being a pre-assigned number that is held fixed under the variations of T and ρ, it is a

[24] "The total entropy is, of course, equal to the sum of the entropies of all subsystems of the isolated composite". Quoted from Modell and Reid (1974). Let us try to avoid such vague generalizations (see Section 3.4).

pre-assigned function of ρ. This circumstance can give rise to any number of misunderstandings; it must be understood that s is a definite function of the density and that, for each choice of this function, the adiabatic condition (see below) gives a well-defined constraint on the pair T, ρ. We insist on this point because it will help us confront difficult issues later on.

Variation with respect to T gives the "**adiabatic condition**"

$$\frac{\partial f}{\partial T}\Big|_\rho + s = 0, \tag{3.1.1}$$

in agreement with the main assumption. Variation with respect to the density, with fixed boundary conditions, and fixed mass, leads to

$$\rho\frac{\partial f}{\partial \rho} - f = p,$$

where p is the local version of the pressure. Equivalently,

$$\vec{\nabla}\left(\mu + T\frac{\partial s}{\partial \rho}\right) = 0, \tag{3.1.2}$$

where μ is known as the **chemical potential** and is defined in either of two equivalent ways,

$$\mu := \frac{\partial f}{\partial \rho}\Big|_T \tag{3.1.3}$$

or

$$\mu := \frac{\partial u}{\partial \rho}\Big|_s = 0.$$

Here, u is the **internal energy density**; it will be discussed in Section 3.7. The first definition is more straightforward; in the second form appears the **adiabatic derivative**, where T is to be eliminated in favor of ρ and S using (3.1.1).

The hydrostatic condition

A simple equation that is convenient and popular relates Eq. (3.1.2) to the gradient of the **thermodynamic pressure**, defined by

$$p = \rho\frac{\partial f}{\partial \rho} - f. \tag{3.1.4}$$

Equation (3.1.2) is, under certain assumptions, the same as

$$\vec{\nabla}p = 0.$$

We shall use it often, but we need to know its limitations and implications.

From the definition of the thermodynamic pressure, we get

$$\begin{aligned}\vec{\nabla} p &= \vec{\nabla}(\rho\mu - f)\\ &= \rho\vec{\nabla}\mu + \mu\vec{\nabla}\rho - \left(\frac{\partial f}{\partial \rho}\vec{\nabla}\rho + \frac{\partial f}{\partial T}\vec{\nabla}T\right)\\ &= \rho\vec{\nabla}\mu + s\vec{\nabla}T. \qquad (3.1.5)\end{aligned}$$

We used the adiabatic condition $\partial f/\partial T = -s$. Hence,

$$\rho\vec{\nabla}\mu = \vec{\nabla}p - s\vec{\nabla}T$$

and (3.1.2) becomes

$$\rho\vec{\nabla}\left(\mu + T\frac{\partial s}{\partial \rho}\right) = \vec{\nabla}p + \left(\rho\frac{\partial s}{\partial \rho} - s\right)\vec{\nabla}T + \rho T\vec{\nabla}\frac{\partial s}{\partial \rho} = 0, \qquad (3.1.6)$$

which is the general form of the **hydrostatic condition** in the present context, in the absence of external forces and so far without taking into account any effect of flow or kinetic energy.

(1) Unless s is a linear function of the density, $s = \rho S$, there is a force proportional to the gradient of the temperature.
(2) Unless $\partial s/\partial \rho$ is uniform, the gradient of the specific entropy S gives rise to an **entropic force** that must be included in the balance of forces.

We conclude that to agree with the universally accepted thesis that the pressure, at equilibrium and in the absence of external forces, is uniform, we need to postulate

1. that the entropy density s is a linear function of the mass density, and
2. that the specific entropy density S is uniform.

Nevertheless, neither statement will be given the stature of an axiom, for special conditions may exist where the linearity of the entropy will be violated. There are circumstances where the specific entropy fails to be uniform, as in the case of atmospheric combustion. But both statements will be assumed to hold, as working assumptions, so long as they are not contradicted by experiment.

The justification for referring to expression (3.1.4) as the thermodynamic pressure will be discussed in Section 3.5. The familiar and expected result, the hydrostatic condition (in the absence of external forces), in the naive form $\vec{\nabla}p = 0$, is verified when (a) $s = \rho S$, with S independent of ρ (and of T) and (b) S is uniform. This result is not in disagreement with hydrodynamics, but an extension of it. Conventionally, hydrodynamics is developed under the implicit assumption that the entropy density $s = \rho S$, with S uniform, that is, the assumption that there are no entropic forces. It is not an axiom, but an observation based on experiments. We remain open to the possibility that we may encounter situations in which the specific entropy is not uniform and that entropic forces become a reality.

The term with $\vec{\nabla} S$ is an entropic force present whenever the specific entropy is not uniform. The simple way to look at it is this: if the specific entropy is not uniform, then a force is present that tries to make it so; just as is the case with the pressure.[25] It is a surprise to find that the unquestioned use of the relation grad $p = 0$, in the absence of external forces, implies these strong assumptions about the entropy. In the context of atmospheres or stellar structure, the relation is replaced by $\vec{\nabla} p + \vec{\nabla} \phi = 0$, where φ is the gravitational potential. This relation is widely used in astrophysics; the fact that it has implications for the entropy is very rarely pointed out, but recent publications show an increased awareness of this important point. It implies that gravity does not ordinarily give rise to entropic forces.

Once this restriction on the entropy density is accepted, a remarkable integration of thermodynamics with hydrodynamics ensues. The Gibbsean variation of the energy with respect to the density gives the hydrostatic condition of hydrodynamics; in the absence of other forces, it is the statement that the pressure is uniform. This key feature of hydrodynamics has now merged with one of the two central relations of thermodynamics. Variation with respect to the temperature gives the second principal relation of thermodynamics; it is the adiabatic relation that lies behind the polytropic condition of hydrodynamics.

It is also significant that if S is uniform, and only in that case, then the total entropy $\int d^3x\, s = \int d^3x\, \rho S$ is a constant of the motion as a direct consequence of the conservation of the total mass (the equation of continuity). In the interpretation of the fluid as a collection of particles, this assumption means that the entropy is assigned to each particle unrelated to the variation of density and temperature from one point to another, but subject to change through the addition of heat to the system. An additional argument in favor of a uniform specific entropy density will be discussed in Section 3.9.

[25] Special situations may call for an additional contribution to the entropy density, independent of the mass density. Thus, $s = \rho S + s_{\rm ex}$, the second term is treated as an external field. An idea adapted from Landau and Lifshitz (1959).

Thermodynamic pressure and the pressure on the walls

So far, the **pressure** term VP or $\int d^3P$ has not been adequately explained. If we assume that the system is confined to rigid walls, the volume is fixed and the pressure term is replaced by **boundary conditions**. To understand the pressure, as it appears in the form of the thermodynamic pressure defined in Eq. (3.1.5), it is necessary to allow for variations of the total volume. We have considered **local variations** of the density. A complimentary, uniform **variation** of the density, accompanied by a **variation of the volume**, gives the additional result that $p = P$ on the boundary; the thermodynamic pressure is a field that extends the pressure on the walls into the bulk (see Section 3.5). It is this circumstance that justifies referring to p as "the pressure". But if we choose to fix the spatial extension of the system, then the term VP in the energy functional becomes redundant. A detailed discussion of the pressure is in Section 3.5.

3.2. The Kinetic Energy

Here, we attempt to complete the synthesis of hydrodynamics and thermodynamics. The hydrodynamical picture of a fluid, as a collection of particles, described by density and flow velocity, suggests that the total energy must include the **kinetic energy**,[26]

$$A[\rho, T, S, P, \vec{v}] = \int_\Sigma \delta^3 x\, h + VP,$$

$$h = \rho\lambda + \rho\vec{v}^{\,2}/2 + f(\rho, T) + sT, \quad s = \rho S. \tag{3.2.1}$$

This is the Fetter–Walecka Hamiltonian of hydrodynamics, with the potential W now replaced by $f + sT$. It is an action principle for thermodynamics limited to the case of irrotational velocity fields. Let us examine the variational equations.

Note. We insist, throughout the following discussion, and in most of this volume, that we are treating the case of potential flows exclusively. This limitation will be lifted in Chapter 10.

---

[26]Gibbs' *vis viva*. The kinetic energy will always appear as an adjunct to the free energy, but it shall be added explicitly. Later, when we generalize the theory to include rotational velocity, the structure of the kinetic term will change, see Chapter 10.

Variation with respect to T, treating T, S and ρ as independent variables off shell, leads as before to

$$\left.\frac{\partial f}{\partial T}\right|_\rho + s = 0, \tag{3.2.2}$$

in accordance with the main assumption. If we know the expression for f and the value of S, we can solve this equation for T, eliminate T from the Hamiltonian, and revert to hydrodynamics. All that remains of the thermodynamic description is a "**hydrodynamical equation of state**", a relation between density and pressure. The most commonly used is the polytropic relation $p/\rho^\gamma = a =$ constant. One interpretation of this formula is that the fluid is an ideal gas and that the entropy is uniform, but information has been lost as a result of eliminating the temperature (a partial on-shell projection) and other interpretations are possible. The expression obtained for the Hamiltonian density will depend on the parameter a and hence on the entropy.

The other hydrodynamical equations are the Hamiltonian equations of motion, namely,

$$\dot{\rho} = \{\rho, H\} = \partial_i \frac{\partial H}{\partial v_i} = -\vec{\nabla} \cdot (\rho \vec{v}) \tag{3.2.3}$$

and

$$\dot{\vec{v}} = \{\vec{v}, H\} = -\vec{\nabla} \frac{\partial H}{\partial \rho} = -\vec{\nabla} \left(\frac{\vec{v}^2}{2} + \frac{\partial f}{\partial \rho} + ST \right). \tag{3.2.4}$$

The second equation reduces, in the case of equilibrium ($\vec{v} = 0$), to Gibbs' variational equation. With a little work (Section 3.1), it turns into the **Bernoulli equation**,

$$\rho \frac{D\vec{v}}{Dt} = -\vec{\nabla} p, \quad p = \rho \frac{\partial f}{\partial \rho} - f, \tag{3.2.5}$$

the dynamic generalization of the hydrostatic condition. This result depends on the assumption that s is linear in the density, $s = \rho S$ with S uniform, see the preceding section.

The Poisson bracket implied by Eqs. (3.2.3) and (3.2.4) is defined for functions ρ, Φ of the density and the velocity potential,

$$\{f, g\} = \int d^3x \left(\frac{\partial f}{\partial \rho} \frac{\partial g}{\partial \Phi} - f, g \right).$$

Now that the context explicitly includes motion, the meaning of the term VP in (3.2.1) is unclear. This has been discussed earlier, in Section 3.2, and it will be taken up again in Section 3.5.

We have just examined the simplest example of a hydro-thermodynamic system that combines the Gibbs variational principle with hydrodynamics. In the absence of **external forces**, it follows directly from the Euler–Lagrange equations that at equilibrium the pressure is uniform. If we admit from the start that the specific entropy is uniform as well, then it follows that all the fields, including the temperature, are uniform at equilibrium. There is no need for another postulate to reach this conclusion. In the presence of an external force, the temperature may not be uniform (see Chapter 9).

In the thermodynamics of equilibria, great importance is attached to the profound, **statistical interpretation** developed especially by Boltzmann. This valuable insight does not include an accepted definition (interpretation) of entropy for systems that are out of equilibrium. But in the theory of adiabatic fluids, an interpretation of this sort is not needed. We define an adiabatic system as one governed by a Lagrangian, anticipating that viable Lagrangians will be found that describe real, physical systems.

3.3. Lagrangian and Adiabatic Dynamics

The notion of minimal energy applies to equilibria; to deal with systems that are out of equilibrium, the classical procedure is Hamilton's Principle. The idea originated with Maupertui (1698–1759), who was the first to apply a variational principle to a dynamical system away from equilibrium, one of the most important suggestions ever made within theoretical physics. Here, we apply the idea to hydro-thermodynamics. In our context, it is a simple generalization of the variational principle of hydrodynamics.

The Lagrangian formulation requires the introduction of the **velocity potential** Φ,

$$\mathcal{A} = \int d^3x \, (\mathcal{L} - P), \quad \mathcal{L} = \rho(\dot{\Phi} - \vec{v}^{\,2}/2 + \lambda) - f(\rho, T) - sT, \qquad (3.3.1)$$

as in hydrodynamics (see Section 1.2). The equations of motion derived from this action include the equations of Hamiltonian dynamics derived above. It is perhaps necessary to repeat that the variational principle defined by (3.3.1) is not a statement about minimum energy, for it describes a complete Lagrangian system that affects much more than the states of equilibrium; that is, the configurations in which all the fields are time independent.

The constant λ is a Lagrange multiplier, needed to fix the total mass $\int d^3x\rho$. It can be absorbed into $\dot{\Phi}$ and it will not appear in later expressions for the Lagrangian. It shall be seen that this association between the Lagrange multiplier and the velocity potential is natural (see Section 4.3).

We are now in a position to give a precise meaning to the term "adiabatic". By **adiabatic dynamics**, we mean dynamics that is defined by a Lagrangian, with a definite choice of entropy and external pressure (or spatial extension). Adiabatic dynamics is expected to be a good approximation for dealing with phenomena that are characterized by a short timescale and by conservation laws. It is contrasted with dissipative processes, normally evolving slowly, during which the system takes on a sequence of adiabatic equilibria with different values of the entropy (Section 2.3). Note that the entropy profile, not just the total entropy, is an **adiabatic invariant**.

Some processes that include changes in the pressure on the walls are usually termed adiabatic. In our terminology, they are neither adiabatic nor dissipative, they require a special treatment.

Remark. The on-shell value of the Lagrangian density is

$$\mathcal{L}|_{\text{on shell}} = \left(1 - \rho\frac{\partial}{\partial\rho}\right)\left(\rho\left(\dot\Phi - \frac{\vec v^2}{2} - \lambda\right) - f(\rho,T) - sT\right)\Big|_{\text{on shell}} \tag{3.3.2}$$

since the derivative term vanishes on shell.[27] Consequently, the on-shell value of the Lagrangian density is equal to p, the thermodynamic pressure. A proposal to identify the Lagrangian density with the pressure, in the context of General Relativity, was put forward by Taub (1954). It seems as if the identification of pressure with Lagrangian density is very natural for reasons that we do not yet fully understand (see Section 3.8); this adds support for our contention that the entropy is usually linear in the density.

Similarly,

$$\mathcal{L}|_{\text{on shell}} = \left(1 - T\frac{\partial}{\partial T}\right)\left(\rho\left(\dot\Phi - \lambda - \frac{\vec v^2}{2}\right) - f(\rho,T) - sT\right)\Big|_{\text{on shell}}$$

$$= -u + \rho\left(\frac{\dot\Phi - \vec v^{\,2}}{2 - \lambda}\right). \tag{3.3.3}$$

In particular, in the case that the velocity field is zero, the value of $(u+p)/\rho$ is a constant. ($u + p$ is the enthalpy density.) If the interaction with the gravitational field is added, we get instead

$$gz + \frac{u+p}{\rho} = \text{constant}.$$

Combining Eqs. (3.3.2) and (3.3.3),

$$\rho(\dot\Phi - \lambda - \vec v^2/2 - gz) = -p - u, \tag{3.3.4}$$

[27]The argument depends on the fact that our Lagrangian includes a Lagrange multiplier; there are no constraints on the variation of the density. Otherwise, we could conclude only that the relation holds up to an additive constant.

and for an ideal gas

$$\dot{\Phi} - \lambda - \vec{v}^2/2 - gz = (n+1)\mathcal{R}T. \tag{3.3.5}$$

Alternatively, these useful formulas can be derived by integrating the Bernoulli equation under the conditions of validity of the latter. Equation (3.3.4) is universal, so long as the entropy density is linear in the density.

A Lagrange multiplier comes automatically with the velocity potential. It makes it possible to fix the extent of the system. The equations obtained by variation of the density will always contain this constant or, equivalently, take the form of Eq. (3.1.2).

Since the time derivatives appear linearly in the Lagrangian density, the term $\rho\dot{\Phi}$ in the **Legendre transformation** that defines the Hamiltonian density is slightly unfamiliar:

$$h = \dot{\Phi}\frac{\partial \mathcal{L}}{\partial \dot{\Phi}} - \mathcal{L} = \rho\dot{\Phi} - \mathcal{L}.$$

The variable that is canonically conjugate to Φ is ρ. The inverse transformation is

$$\mathcal{L} = \rho\frac{\partial h}{\partial \rho} - h = \rho\dot{\Phi} - h.$$

3.4. The Second Law

The **second law** of thermodynamics represents the highest achievement of the physics of the 19th century. A simple statement of the law can only serve as an introduction to the subject, for the real meaning is extremely subtle.

> **The second law: No change within an isolated system results in a decrease of the entropy.**

It is a property that has a strong appeal in the context of statistical mechanics, but that is beyond the scope of this book.

An example will serve to highlight the difficulties and perhaps clarify the meaning of this statement.

Consider a body that consists of two parts, each an adiabatic system and each at adiabatic equilibrium. In the simplest case, it consists of two systems with fixed temperatures T_1 and T_2. Suppose a small amount of heat is transferred from system 1 to system 2, with no appreciable change in either temperature. The total system is isolated so there is no net transfer of heat to it, but an amount dQ is added to the first system and the same amount taken away from the second system. Such a change could be adiabatic, caused by transfer of liquid from one vessel to another, but we now assume that no transfer of mass takes place. This leaves a transfer by **conduction or radiation**, but note that no attempt is made to describe the mechanism of transfer. It is assumed that each system makes a transition from one adiabatic equilibrium state to another. Then the following rule holds:

$$dS = dS_1 + dS_2 = \frac{-dQ}{T_1} + \frac{dQ}{T_2}.$$

That is, it is assumed that entropy is an additive property. This is a logical extension of the fact that entropy is an extensive property.

Additivity assumption. The entropy of a system that consists of several parts, that do not influence each other in any way, is defined to be the sum of the entropies of the parts.

The second law says that, if the transfer is spontaneous, then $dS \geq 0$; here it implies that $T_1 \geq T_2$, that is, heat flows out of the warmer and into the colder of the two subsystems. It is an example of "internal dissipation" of the total system. Here, we must reject an attempt to prove that $T_1 = T_2$ at equilibrium, as that requires an additional assumption. The additional assumption that is needed to reach that conclusion is that, whenever the temperatures are unequal, then heat will necessarily flow.

The internal energy is conceived as a sum of heat energy that increases with the addition of heat and a mechanical energy that decreases when the system performs work. A system is "degraded" when its capacity for doing work is reduced; if this happens when the internal energy is maintained constant, then the entropy must increase. A variation of this example is provided by the free expansion of an ideal gas (see below). Free contraction, which is not known to occur, would result in a decrease in the entropy.[28]

We have already, in Section 2.5, expressed our doubts about extending the variational principles to include exchange of entropy between parts of the system. Here, we are taking up the question once more in the local context. The original variational principle, and the Lagrangian variational principle, considers variations that hold the entropy distribution fixed. It leads to equations of motion that fix, to a very large extent, the physical configurations. Any suggestion to consider a wider class of variations must verify that it does not lead to internal contradictions. Since the principle of increasing entropy is of the first importance, we propose to allow variations

[28]This concept of "degradation" was introduced by Lewis and Randall (1936). It is the main idea of their explanation of the second law.

of the entropy (and other variations that may be proposed) but to avoid contradictions we should add a proviso as follows:

An isolated system that exists in a certain configuration will move to one of greater entropy, *if one is available.*

In the case of an ideal gas that is described in terms of the Euler–Lagrange equations, no equilibrium configuration with constant temperature is available when a gravitational field is present.

Classical, **statistical mechanics** steps out of its domain of competence when it is argued that the temperatures, when equilibrium between the two parts of the system is reached, must be equal. As strongly emphasized by Prigogine, we cannot compensate entropy production in one part of the system with entropy reduction in another; entropy is not a conserved quantity. We do not have a general proof that two bodies in thermal contact must have the same temperature. If you allow variations of the entropy in the formulation of the variational principle, then you are excluding the case of an ideal gas in the presence of a gravitational field. The tendency for entropy to become maximal is limited by the equations of motion. Again, a shift to a state of higher entropy is favored, provided that one is available.

The fixation on the isothermal equilibrium is strange. General axioms cannot be formulated as conditions that affect only equilibrium states. In statistical mechanics, the energy is a property of the system as a whole; it does not address problems in which the temperature is a field that varies from point to point. The existence of systems with variable temperature is acknowledged but removed from consideration as being, by definition, out of equilibrium.

3.5. Work and Pressure

We have seen that **entropy and pressure** play similar roles in the variational principle. But while entropy is a many faceted and subtle concept that we shall continue to study throughout this work, pressure is comparatively easy to understand. Or is it?

In the case of the global system that we started out with, in Section 2.4, P was just a number. In the localized system, some refinement is needed. In certain situations, the pressure on the walls may be the same everywhere, but that is not generally true, even for equilibrium configurations. If the walls are fixed, then the term P in the action is irrelevant, its roles taken over by the boundary conditions, so it remains to consider situations where the volume can change.

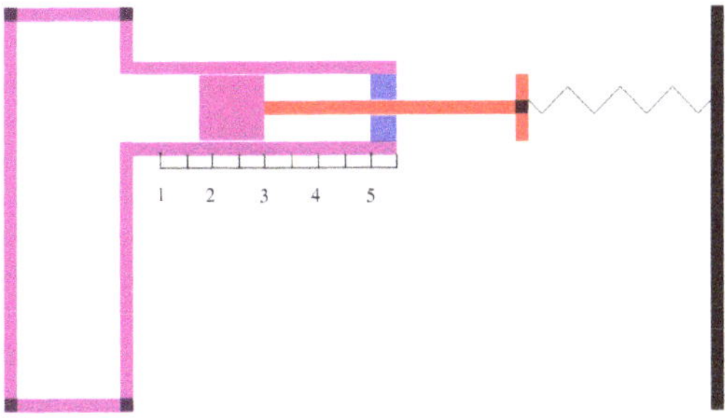

Fig. 3.5.1. Vessel with an additional degree of freedom, including piston with mass m and a spring with spring constant k. See Eq. (3.5.1).

The simplest situation is that of a vessel furnished with a piston, moving in a cylinder and having a face that is perpendicular to the direction of motion, as in Fig. 3.5.1. At equilibrium, the pressure of the gas against the face of the piston has to be balanced by an externally applied force. When this external force is varied slightly, the piston will be set in motion, a motion that can be controlled so as to be slow enough that the flow of the gas is minimal and the kinetic energy is negligible. When the piston is withdrawn, the volume increases and, *if the entropy does not change*, the internal energy of the gas decreases; the gas "does the work" and $\Delta U = -p\Delta V$. Since $U = n\mathcal{R}T$, this brings about a calculable change in T; this prediction was confirmed by a crucial set of experiments by Joule and Kelvin. An essential part of our understanding is that we can interpret the change as a progression through a sequence of adiabatic equilibria on a timescale that is very slow compared to what is characteristic of adiabatic dynamics. Perhaps, the most important lesson is the realization that this is in close analogy with the phenomenon of slow dissipation. It is assumed, and verified, that during the slow motion of the piston, no heat transfer takes place and that, consequently, the total entropy does not change. The context is **quasistatic changes** of a system with a fixed Lagrangian, as in Section 2.8. In the of case of a homogeneous system with entropy density $s = \rho S$, with uniform mass density ρ, it follows that S does not change. But it should be understood that, with a slightly different setup, this simple statement may not always apply. That is, under different conditions, both pressure and entropy may undergo changes.

Here, we are in danger of overlooking some of the most interesting and subtle aspects of gas dynamics. It is indeed true that, under the circumstance just described, when the piston was moved slowly, the entropy did not change. But this is not an inevitable property of every ideal gas; that is, it does not follow from the expression for the free energy density. It is a fact deduced from experiments with certain real gases, approximately ideal, where it has been found that the energy gained from allowing the gas to expand (slowly) is accounted for by the term $-PdV$ in the expression for the change of the internal energy, which implies that the entropy does not change. It tells us something very important about the gases used in the experiment: that the specific entropy density is an adiabatic invariant.

Henceforth, we add a working assumption in all our work: that the specific entropy of a homogeneous fluid is an adiabatic invariant.[29]

In this book an adiabatic system (with one component) is characterized by a fixed specific entropy density.

This leads to the assertion that the process under discussion is "reversible". In fact, it is found that by pushing the piston back in and reverting to the original volume, the gas seems to revert to the original configuration. But "reversibility" does not always imply that $dS = 0$; it is a term that cannot be rigorously defined and we shall not use it. In short, a process is reversible if it can be reversed; but what means are we allowed to employ?

Consider the opposite extreme, the rapid removal of a wall that results in a finite change of volume. In this case, "no work is done by the gas", which means that no useful work is done to profit the environment, in contrast with the work done on the piston in the case of slow motion.[30] What happens is that the gas is set in motion. This motion can be accounted for, in principle, over the short term by means of the equation of motion of the adiabatic Lagrangian. Energy conservation means that the value of the total Hamiltonian, that includes the kinetic energy, is unchanged. (This is an idealization; we have to agree to neglect the loss of energy due to friction against the walls.) If the system is isolated there can be no heat exchange with the environment and no change in energy. The flow can change but it cannot cease by virtue of the adiabatic dynamics. Yet, one observes that the motion eventually comes to rest by some mechanism that is not understood within the adiabatic framework. To give it a name, we can call it "**`internal dissipation`**" or "conversion of the kinetic energy to heat" (Callen, 1960).

---

[29] According to Gibbs himself, only the total entropy is an adiabatic invariant.

[30] It is also assumed, tacitly, that no work is done by removing the piston. This is patently false if the piston is withdrawn along its axis, but it may be true if it is replaced by a wall that opens by moving it in its own plane. Alternatively, a weightless piston may be used, released rather than withdrawn. Kelvin and Joule used a stationary, permeable foam.

Again, this is a development that cannot be predicted on the basis of the definition of the ideal gas; it is an observation to be added to our fund of knowledge of the properties of certain real gases.

During the turbulent phase that sets in when the piston is withdrawn, some of the energy is kinetic, contained in the term $\rho\vec{v}^2/2$. But at the end, when the macroscopic motion has ceased, it is observed that the temperature has not changed. Consequently, neither has U, since $U = n\mathcal{R}T$; the term $-PdV$ in the expression for dU must have been compensated by an increase in the entropy. What (apparently) takes place during dissipation is a change of entropy; the net result being conservation of the value of the internal energy,

$$dU = T\Delta S - p\Delta V dt = 0,$$

or more precisely

$$\int (TdS - pdV) = 0.$$

If now we try to recover the original configuration of the expanded gas, we can only do so by decreasing the entropy.

It is observed that there takes place a spontaneous conversion of kinetic energy into heat.[31] To justify the neglect of this process in the context of adiabatic dynamics, we have to assume that its effect is very slow, that dissipation is a slow process.

In the case of an ideal gas, the internal energy density is $n\mathcal{R}T$ so the temperature does not change in this type of process. Also, in the case of an ideal gas, $p = \mathcal{R}T/V$, so that

$$\Delta S = \int \frac{\mathcal{R}}{V} dV = \mathcal{R}\Delta \ln V.$$

In our experience, a fluid given over to itself tries to expand, never will it contract. This exemplifies the most important property of entropy: in a spontaneous process in which there is no interference from the outside, it can only increase.

Pressure and Lagrangian density

Returning to the system illustrated in Fig. 3.5.1, we wish to consider some generalizations. The coordinate z gives the position of the piston. The volume can change only by virtue of the motion of the piston, therefore the term

[31] See Callen (1960) for a very authoritative discussion of this point.

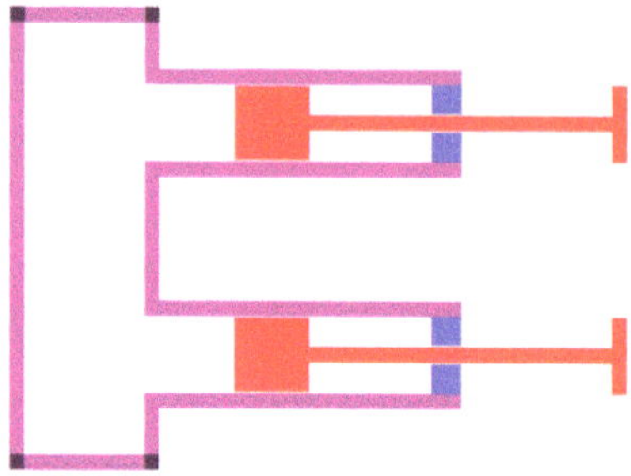

Fig. 3.5.2. A vessel with two pistons.

$\int d^3xP$ can be replaced by APz. Now, let us review the derivation of the equations of motion, more precisely the variation of the density. So far, except briefly in Section 3.3, we have only considered variations that vanish at the boundary. Here, we shall repeat the discussion in Section 3.3 in more detail. A complimentary variation[32] of the volume leads to

$$\delta\int d^3x(\mathcal{L}-P)=\delta\int d^3x\,(p-P)=\int d\sigma\, p\delta z-AP\delta z=0, \qquad (3.5.1)$$

where the two-dimensional integral is over the face of the piston and A is the area of the piston. By replacing $\mathcal{L}$ by p, we have assumed that the entropy density is linear in the density. It follows that P is equal to the average thermodynamic pressure on the piston. If this were not the case, then the identification of P with the thermodynamic pressure would not be valid. It is also worthwhile to note that the identification of the Lagrangian density (on shell) with the pressure on the walls is of great utility and significance.

Quite generally, it follows from the variational principle that the pressure on the wall of a homogeneous system is equal to the value of the Lagrangian density, provided only that the entropy is a linear function of the density.

In this example, the expression VP in the Hamiltonian was replaced by AzP, because Az is, up to a constant that plays no role, the volume, and P is the pressure against the face of the piston. The pressure against the wall elsewhere in the vessel is not recorded since the energy does not depend on it, and indeed it is not defined. The expression VP in the Hamiltonian is a stand-in for a more accurate expression for a part of the energy.

Returning to the system in Fig. 3.5.1, let us load the piston with an inertial mass m and add the term

$$m\dot{z}^2/2-kz^2/2$$

[32] In Gibbs' paper, where all results obtained by local variations are supposed to be expressed by the fundamental relations, these global variations are the only one discussed.

to the Lagrangian. The last term represents the energy of a spring that provides the force that holds the piston against the pressure. Now, z is a dynamic variable and the variation of the total action with respect to z gives

$$Ap - kz - m\ddot{z} = 0.$$

The pressure on the piston is no longer an assigned parameter but a variable that is determined by the dynamics. A small initial value of $Ap - kz$ will start a slow change in the value of z, and this motion is adiabatic. A large value of m will mimic the slow motion of the piston and a small value of m will imitate the sudden removal of the piston by an external force.

This motion too is (approximately) adiabatic, but the observed, slow relaxation of the system is not. This approach is appropriate for dealing with balloons and drops held together with surface tension and with self-gravitating heavenly bodies.

By now, it must be clear that the term VP in the Hamiltonian is symbolic. The pressure on the wall may be constant in certain cases of equilibrium, but certainly not so in the general case.

Consider the case that there are two pistons, as in Fig. 3.5.2. It may happen that the thermodynamic pressures are different at the two locations. Most notably, this will happen if gravitation is taken into account. Then, to ensure the existence of a solution, we must replace

$$\int d^3x\, P \to A_1 P_1 z_1 + A_2 P_2 z_2,$$

with a choice of the values P_1, P_2 that is not arbitrary. A wrong choice will have the consequence that there will be no equilibrium solutions.

There are cases when the pressure on the wall is not the same everywhere, even at equilibrium, as in the problem of a vertical column of gas under the influence of gravity. If the walls are rigid, the simple solution is to impose boundary conditions. This happens in the last example in the limit when the mass m of the piston tends to infinity.

The general lesson that should be learned from this example is that boundary conditions have to be compatible with the equations of motion, and that this needs to be verified in each case. The most important case is the one that one encounters when considering a system with non-uniform boundary temperature specified (see Section 9.2).

3.6. Important Lesson

The expansions of an ideal gas, the slow expansion and the free expansion, have been covered in the manner of the textbooks for the most part.

But the main lesson that should be learned from these processes deserves more emphasis.

There is an interesting system called **entropic rubber**. It may be assumed that a rubber band has a free energy and an entropy, the former is a function of the length and the temperature, the latter depends only on the length.[33] There is a mechanical model of the rubber band that compares it with a conventional spring; stretching the rubber band increases the tension and the free energy, entropy is not involved. Entropic rubber is another model. We must imagine a system of long polymers of twisting and interlaced molecules, to which a very large set of states is available. Stretching the band results in straightening the strings, and now a reduced number of states is available; in a word, stretching results in a reduction of the entropy. Writing $F = U - TS$, the result of stretching is attributed, in the first model, to a change in the free energy and, in the second model, to an increase of the entropy.

To decide which model is right, or at least closest to reality, there is only one sure way: experiment. Returning to the expansion of the ideal gas, we now realize the following central truth: That slow expansion preserves the entropy, and that free expansion does not, and that these facts are known to be true only because they have been established by experiments for certain gases.[34] To say that the gas does no work in the case of free expansion is not an argument that can be advanced as part of the theoretical treatment of the process; it is a statement that summarizes the result of the experiment. Some of the textbooks make this point quite clear but, in my opinion, with insufficient emphasis. The analogy between entropic rubber and the free expansion of an ideal gas was pointed out by Baez (2012).[35]

Remark. Having thus come to know of the existence of entropic forces, we are in a better position to appreciate the last term in Eq. (3.1.6). The gradient of the entropy is a force that tends to reduce the entropy gradient, just as a gradient of pressure is a force that tends to reduce the pressure gradient.

Some questions

Let us reconsider the phenomenon of free expansion. It is an unequivocal result of the Joule–Kelvin experiment that the net result is an increase of the entropy. But when does this increase occur?

[33] I am not sure how this statement is justified.
[34] Historically the most important by Joule and Kelvin.
[35] https://johncarlosbaez.wordpress.com/2012/02/01/.

For $t < 0$, the gas is confined in a small volume. The **entropy**, according to the statistical interpretation, is the logarithm of the number of states with the given energy. It can be calculated for the two equilibrium configurations; suppose it is S_0 for the gas at rest in the small volume at negative times and S_2 for the gas at rest in the larger volume at a much later time. In the meantime, two events have taken place.

1. We open the gate; this is a severe interference with the system. The volume is suddenly larger and the entropy should therefore have changed. Remember that the entropy is determined by the number of available states, occupied or not. The difficulty is that, with the gate open, the gas is not at equilibrium and, according to statistical mechanics, the entropy is not defined. Suppose that the entropy at this time, if it can be defined, is S_1.
2. What happens after that is that severe **convection** is taking place. Gradually, the fluctuations take on a smaller scale until finally becoming identified only as heat (Callen, 1960). Again, since the system was not at equilibrium at the beginning of this process, the initial entropy was not defined, so we cannot ask how much it has changed. Thus, though both S_0 and S_2 can be determined, S_1 cannot be determined; perhaps, we are free to decide what value to assign to it.

And what, after all, prevents us from asserting that the expansion is adiabatic?[36] Adiabatic configurations include equilibrium states, and also states with densities that vary from point to point, including the extremely unlikely possibility of a large part of the vessel containing no particles?

Do I dare to suggest that the entire change of entropy happens during the first phase by the act of removing the gate?[37] It is certain that, by the mere act of defining the hitherto undefined value of the entropy S_1 of the system immediately afterwards, the expansion can be brought into the province of adiabatic dynamics. The issue cannot be reduced to a question of reversibility; the process can be reversed or not depending on what methods are allowed.

Conventional wisdom says that $S_1 = S_0$ so that the entire change from S_0 to S_2 is attributed to the expansion. But this makes no sense when it comes from a tradition where S_1 is not defined! See Section 9.8.

[36]I now think that the flow may be adiabatic, but not irrotational, and that it involves strong internal stress. Note added in proofs.
[37]For an answer to this question, see the last paragraph of this section.

To Joule and Kelvin it is the attributed measurements that show that the internal energy U_2 at the end of the experiment is the same as the internal energy U_0 at the beginning. (The Navier–Stokes equation suggests a loss of energy due to viscosity, but the meaning of "energy" in that context is unclear.)

The overall transition, from the initial equilibrium in the smaller vessel to final equilibrium in the larger one, includes a stage where the statistical entropy is not defined. If no dynamical theory accounts for this transition, how are we to normalize the entropies? The experiments tell us that both systems, the initial and the final, have the same physical properties, with the same expression for the entropy in terms of density and temperature up to normalization! The answer is that there are other paths, quasi-static and "reversible", that connect the two configurations: slow expansion and slow expansion in a heat bath.

Finally, what is the distribution of entropy in the system immediately after the gate was opened? With the extra volume available, the particles find new states available to them and a purely statistical argument shows that their entropy is greater (cf. Baez, 2012, p. 65).

In the light of these questions, it would be useful to know what accuracy has been achieved in the experiments of Joule and Kelvin or by others with more modern equipment.

For an answer to the question. No, I do not go so far as to suggest that the increase of entropy takes place, in its entirety and instantly, as a result of the opening of the gate. There is strong intuitive support for the idea that the entropy increases as the temporary convective motion quiets down. But I do think that the question should be asked and discussed. It was asked, and widely debated, during the second half of the 19th century, but the question remains poorly understood.

3.7. Local Thermodynamic Potentials

Localization, the passage from global thermodynamics to continuum thermodynamics, puts a new face on the theory of thermodynamic potentials. What is discussed here will become important later on, when the set of dynamical variables is expanded to include several densities.

Recall that the global theory deals with the following quartet

$$\begin{array}{ccc} U(V,S) & ST & F(V,T) \\ VP & & VP \\ H(P,S) & ST & G(P,T) \end{array}$$

of **potentials** and associated **Legendre transformations**. The letters ST between the potentials U and F remind us of the substitution $S \mapsto T$ in the Legendre transformation. The potentials that have appeared so far refer to a fixed quantity of fluid, a fixed total mass as one gram or one mole. To convert them to **densities**, we multiply each by the mass density, $\rho U = u, \rho F = f$ and so on. In addition, we change variables, setting $V = 1/\rho$ and $S = s/\rho$. Then we find that

$$s = -\frac{\partial f}{\partial T}\bigg|_\rho = -\frac{\partial g}{\partial T}\bigg|_p \tag{3.7.1}$$

and

$$g(p,T) = f + p, \quad h(p,s) = u + p,$$
$$p = \rho\frac{\partial f}{\partial \rho} - f, \quad \hat{p} = u - \rho\frac{\partial u}{\partial \rho}. \tag{3.7.2}$$

The first line brings no surprise but the second line informs us that the pressure has joined the family of potentials. Instead of the quartet of global potential energies, we now have a quartet of potential densities,

$$\begin{matrix} \hat{p}(s,\mu) & sT & p(T,\mu) \\ \rho\mu & & \rho\mu \\ u(s,\rho) & sT & f(T,\rho) \end{matrix}$$

where

$$\mu := \frac{\partial f}{\partial \rho}\bigg|_T = \frac{\partial u}{\partial \rho}\bigg|_s. \tag{3.7.3}$$

For historical reasons, this function μ is called the **chemical potential**. It will play an important role in the theory. The natural variables for p are T and μ. The densities related to G and H are

$$h(s,p) = u(s,\rho) + p, \quad g(p,T) = f(T,\rho) + p.$$

Among the partial derivatives,

$$\frac{\partial p}{\partial T}\bigg|_\mu = s, \quad \frac{\partial p}{\partial \mu} = \rho. \tag{3.7.4}$$

The first one is related to the **Clapeyron equation**. See Section 5.1.

In this book, we shall rely principally on the density potentials u, s and f, functions of the densities, the entropy density and the temperature. The total number of particles will not be employed as a variable and there will be no distinction between **open and closed systems**.

In terms of densities, the **main relations** of thermodynamics take the form

$$\frac{\partial}{\partial T}(f+sT)\Big|_{\rho,S}=0,\quad \dot\Phi-\vec v^2/2-\frac{\partial}{\partial\rho}(f+sT)\Big|_{T,S}=\text{constant}\qquad(3.7.5)$$

or

$$\frac{\partial f}{\partial T}\Big|_{\rho}+s=0,\quad \dot\Phi-\vec v^2/2-\mu-ST=\text{constant}.\qquad(3.7.6)$$

3.8. Conservation Laws and the Stress Tensor

What more than any other property characterize action principles are conservation laws. The most familiar examples are in analytical mechanics and some of them are examples of **Noether's theorem**: If the Lagrange function is invariant under a variation $\delta q_1,\dots,\delta q_n$, then the function

$$\sum_i \delta q_i\frac{d\mathcal{L}}{d\dot q_i}$$

is a constant of the motion. But to prove the most important conservation law, the time independence of the Hamiltonian, one needs a generalization: If the variation of $\mathcal{L}$ is a time derivative, $\delta\mathcal{L}=\dot K$, then the function

$$\sum_i \delta q_i\frac{d\mathcal{L}}{d\dot q_i}-K$$

is a constant of the motion.

This fundamental statement has applications in field theoretic action principles as well.

To apply the technique to the Fetter–Walecka action (1980), let $\delta q_i=\frac{d}{dt}q_i$, $i=1,2$, and replace q_1, q_2 by Φ, ρ. With $\vec v=-\vec\nabla\Phi$,

$$\delta A=\int_{t_0}^{t_1}dt\int_\Sigma d^3x\,\frac{d}{dt}\mathcal{L},\quad \mathcal{L}=\rho(\dot\Phi-\lambda-\vec v^2/2)-W[\rho].$$

The calculation of δA, on shell, proceeds exactly as in the derivation of the **Euler-Lagrange equations**, with the partial integrations, but in this case only the "integrated parts", or surface integrals, that are ignored in that context, remain, since the Euler–Lagrange equations are satisfied on shell. The result is

$$\int_{t_0}^{t_1}dt\int_\Sigma d^3x\,\frac{d}{dt}\mathcal{L}=\int_{t_0}^{t_1}dt\int_\Sigma d^3x\left(\frac{d}{dt}(\rho\,\dot\Phi)+\vec\nabla\cdot(\rho\vec v\dot\Phi)\right),$$

that is,

$$\frac{d}{dt}\int_{\Sigma} d^3x(\rho\dot{\Phi} - \mathcal{L}) = \int_{\partial\Sigma}(\rho\dot{\Phi})\vec{v}\cdot\vec{d\sigma} = \int_{\partial\Sigma}(h+p)\vec{v}\cdot\vec{d\sigma}. \tag{3.8.1}$$

This identifies the Hamiltonian density $h = \rho\dot{\Phi} - \mathcal{L}$. The last equality comes from the fact that the value of $\mathcal{L}$, on shell, is the pressure, $\mathcal{L} = p$. Note that what we have called "Hamiltonian density" is the internal energy density, while the scalar factor in the boundary term, the "energy flux density" is actually the enthalpy density, which on reflection is found to be intuitively sensible. Normal boundary conditions make $\vec{v}\cdot\vec{d\sigma} = 0$ and then $d/dt\int d^3x h = 0$.

The structure of (3.8.1) mirrors that of the global formulation of the equation of continuity,

$$\frac{d}{dt}\int_{\Sigma} d^3x\,\rho = \int_{\Sigma} d^3x\,\dot{\rho} = -\int_{\partial\Sigma} \rho\,\vec{v}\cdot d\vec{\sigma}. \tag{3.8.2}$$

The integral on the right of this equation is the total mass flow out of the volume Σ. The local version of this is the equation of continuity,

$$\dot{\rho} + \vec{\nabla}\cdot(\rho\vec{v}) = 0.$$

The corresponding local version of (3.8.1) is

$$\frac{d}{dt}(\rho\dot{\Phi} - \mathcal{L}) + \vec{\nabla}\cdot((h+p)\vec{v}) = 0,$$

where the "energy flow" is $\vec{\nabla}(h+p)\vec{v}$.

Next, consider the case that the variation is generated by a translation, $\xi = \vec{\xi}\cdot\vec{\nabla}$ or the infinitesimal rotation $\xi = \vec{\omega}\wedge\vec{x}\cdot\vec{\nabla}$ with $\vec{\xi}$ and $\vec{\omega}$ constant. The variations of a scalar field are

$$\delta\phi = \xi\phi, \quad \delta\dot{\Phi} = \xi\dot{\Phi}, \quad \delta\partial_i\phi = \partial_i\xi\phi.$$

The effect on the action of (3.8.1) is

$$\delta A = \int dt \int_{\Sigma} d^3x\;\xi\mathcal{L}.$$

With the help of the equations of motion, one obtains

$$\int d^3x\,\xi\mathcal{L} = \int d^3x\left(\frac{d}{dt}(\rho\xi\Phi) + \vec{\nabla}\cdot(\rho\xi\Phi)\vec{v}\right)$$

or

$$\frac{d}{dt}\int_{\Sigma} d^3x(\rho\vec{\xi}\cdot\vec{v}) + \int_{\partial\Sigma}\left(\rho(\vec{\xi}\cdot\vec{v})\vec{v} + \mathcal{L}\vec{\xi}\right)\cdot\vec{d\sigma} = 0. \tag{3.8.3}$$

Note that the term on the left-hand side of the first equation now is a space boundary term, it appears as the last term in Eq. (3.8.3).

Using the equation of continuity, we convert (3.8.3) to

$$\rho\frac{d}{dt}(\vec{\xi}\cdot\vec{v})+\rho(\vec{v}\cdot\vec{\nabla})\vec{\xi}\cdot\vec{v}+\vec{\xi}\cdot\vec{\nabla}p=0.$$

When $\xi=\vec{\xi}\cdot\vec{\nabla}$, $\vec{\xi}$ constant, this is the **Navier-Stokes equation**,

$$\rho\frac{d}{dt}\vec{v}+\rho(\vec{v}\cdot\vec{\nabla})\vec{v}+\vec{\nabla}p=0. \tag{3.8.4}$$

The conclusion is that the use of "angular momentum", and more so "angular momentum conservation" in the construction of intuitive arguments about fluids requires careful circumspection.

The local form of the result, Eq. (3.8.3), is an example of a **conservation law** that involves the **stress tensor**. As a phenomenological concept, it goes back to Cauchy, but its connection to action principles is the work that was done by Emmy Noether when she was asked by Einstein and Hilbert to clarify the conservation laws of General Relativity. After Noether's contribution, conservation laws is a subject that is inseparable from action principles. To illustrate, we shall develop an alternative analysis of Fetter–Walecka theory using concepts and notation from Noether.

Consider first a field theory Lagrangian, a function of a scalar field ϕ and its first-order derivatives. Without invoking relativity, we shall use a convenient notation, writing $\partial_\mu\phi=\phi_\mu$, $\mu=0,1,2,3$ for the "4-vector" $(\dot{\phi},\vec{\nabla})$. What follows can easily be generalized to the case that there is any number of fields.

The relativistic stress tensor, or **energy-momentum tensor**, is defined by

$$T_\mu^\nu=\phi_\nu\frac{\partial\mathcal{L}}{\partial\phi_\mu}-\delta_\nu^\mu\mathcal{L}.$$

Repeating the calculation that led to (3.8.3) in this more general case, we get

$$\int d^3x\xi\mathcal{L}=\sum\int d^3x\left(\frac{\delta\mathcal{L}}{\delta\phi}\delta\phi+\frac{\delta\mathcal{L}}{\delta\phi_\mu}\partial_\mu\xi\phi\right).$$

The sum is over all the scalar fields in $\mathcal{L}$. A partial integration of the second term, and using the equations of motion, leaves the integrated part

$$\int d^3x\xi\mathcal{L}=\sum\int\delta^3x\;\partial_\mu\left(\frac{\delta\mathcal{L}}{\delta\phi_\mu}\xi\phi\right).$$

If the variation $\xi\phi$ is an infinitesimal translation (in the time or space coordinate) $\xi = \xi^\nu \partial_\nu$ with ξ^ν constant, then this result takes the form

$$\partial_\nu T^\nu_\mu = 0.$$

The simplest example is

$$\mathcal{L} = \frac{1}{2} g^{\mu\lambda}\phi_\mu\phi_\lambda,$$

with the Euler–Lagrange equation $\Delta\phi = 0$. Then

$$T^\nu_\mu = g^{\nu\lambda}\phi_\lambda\phi_\mu - \delta^\nu_\mu \mathcal{L}.$$

A completely straightforward calculation gives

$$\partial_\nu T^\nu_\mu = -\phi_\mu \Box\phi = 0, \quad \text{on shell.}$$

The operator $\Box$ is the d'Alembertian. It can easily be verified that this "conservation law" is valid for any Lagrangian, relativistic or not, that depends exclusively on the fields.

Another example is the Fetter–Walecka theory. The indices run over $0, 1, 2, 3$. The energy–momentum tensor has components

$$\begin{aligned} &T^0_0 = \rho\dot{\Phi} - \mathcal{L}, \quad T^i_0 = \dot{\Phi}\rho v_i, \quad T^0_i = \Phi_i\rho, \\ &T^j_i = -\Phi_i\rho\Phi_j - \delta^j_i\mathcal{L}. \end{aligned} \tag{3.8.5}$$

We can verify the conservation law $\partial_\nu T^\nu_\mu = 0$. It splits into these two cases:

$$\partial_0 T^0_0 + \partial_j T^j_0 = 0$$

is easy to verify, it reduces to

$$\frac{d}{dt}h + \vec{\nabla}\cdot(\rho\dot{\Phi}\vec{v}) = 0, \tag{3.8.6}$$

which is the statement of **energy conservation**, Eq. (3.8.1). The other one, with $\mu \to i = 1, 2, 3$, becomes

$$\frac{d}{dt}(\rho\Phi_i) - \partial_j(\Phi_i\rho\Phi_j) - \partial_i\mathcal{L} = 0, \quad i = 1, 2, 3,$$

which is the statement of **momentum conservation**. Making use of the equations of motion, we simplify this last equation to recover the **Bernoulli equation**.

$$\frac{d}{dt}\vec{v} + (\vec{v}\cdot\vec{\nabla})\vec{v} = -\frac{1}{\rho}\vec{\nabla}\mathcal{L}. \tag{3.8.7}$$

So, we have verified (and re-verified) that the 4-divergence of T is zero, and that the statement relates to the conservation of energy and momentum.

We have also learned that the conservation laws of the energy momentum tensor are related to equations of motion of the action principle, but that it contains less information. It has become usual to regard the Navier–Stokes equation as the core of the subject, but it does not give an expression for the energy and it does not relate to the equation of continuity.

In contrast, the Fetter–Walecka formulation of hydrodynamics is a complete, dynamical field theory (but limited to irrotational velocities).

Quite generally, Noether's discovery is a relation between invariance under a group of transformations and a conservation law. Invariance of what?...of the Action. And what is conserved: the "momentum" that is canonically conjugate to the coordinate that is being varied under the action of the group. The whole concept assumes that there is an action and without it there is no sense in talking about energy or any other kinematical conservation law.

The equalization of pressure

The subject of equalization of pressure was left in an unsatisfactory state in Section 2.5, we were unable to show that two systems separated by a movable wall will have the same pressure at equilibrium. This was due to the fact that the volume appears twice in the total energy. Now, we are in a better position to reach a reasonable conclusion.

By localization, the variable "volume" has been replaced by two new variables: the density ρ and the total volume V. If there are no external forces, then the free energy potential is a function of density, independent of total volume. For a divided system, the free energy depends on the two densities only; volumes V_1 and V_2 appear only in the term $V_1P_1 + V_2P_2$. And now minimizing the energy for fixed total volume evidently leads to $P_1 = P_2$ (see Section 3.5). When external forces such as gravity are present, the pressure is not constant but only continuous.

3.9. Minimal Energy and Maximal Entropy

We are ready to resume the study of the general structure of thermodynamics, including a central and most difficult issue: maximalization of the entropy.

We have an action, a Lagrangian density, a Hamiltonian density defined by the usual **Legendre transformation**,

$$h = \dot{\Phi}\frac{\partial}{\partial\dot{\Phi}}\mathcal{L} - \mathcal{L},$$

and Euler–Lagrange equations that summarize both hydrodynamics and thermodynamics. If we use the adiabatic equation to eliminate the temperature, we are left with a non-degenerate symplectic structure with one pair of canonical variables; the Hamiltonian is a first integral of the motion and the canonical generator of time translations. What we do not have so far is a **principle of minimum energy**.

The classical Lagrangian paradigm

So far, the action principle provides all the laws of traditional thermodynamics, including some that go beyond the description of the states of equilibrium. What is missing is part of the interpretation. To complete the picture in this respect, we can do no better than to seek inspiration from tradition, for equilibrium is the unique concern of a classical and highly successful theory.

We need to understand how it comes about that the Hamiltonian of an action principle tends to have the lowest possible value at equilibrium. The paradigm of an action principle is one that involves dynamical variables $q_1, \ldots, q_n$ and

$$\mathcal{L} = \sum \dot{q}_i^2/2 - V(q_1, \ldots, q_n) = \sum \dot{q}_i p_i - H(q_1, \ldots, q_n, p_1, \ldots, p_n). \tag{3.9.1}$$

This system is non-degenerate, each variable has an independent conjugate momentum,

$$p_i = \frac{\partial \mathcal{L}}{\partial \dot{q}_i}.$$

The equations of motion in Hamiltonian form are

$$\dot{q}_i = \frac{\partial H}{\partial p_i}, \quad \dot{p}_i = -\frac{\partial H}{\partial q_i}, \quad i = 1, \ldots, n. \tag{3.9.2}$$

Here, we see what we are looking for: If "**`equilibrium`**" is defined as a solution in which nothing varies with the time, $p_1, \ldots, p_n$, $q_1, \ldots, q_n$ being time independent, then Eq. (3.9.2) shows that the Hamiltonian must be stationary at all equilibrium solutions. In principle, the equilibrium may be stable or not, for the stationary point may be a maximum or a minimum.

Turning now to our action principle for hydrodynamics, with its one-dimensional canonical structure, we have

$$\int d^3x\mathcal{L} = \int d^3x\ \rho\dot{\Phi} - H[\rho,\Phi] \tag{3.9.3}$$

and the canonical equations of motion (3.9.2):

$$\dot{\rho} = \{\rho, H\}, \quad \dot{\Phi} = -\{\Phi, H\}. \tag{3.9.4}$$

This is a close, field-theoretic analogue and the conclusion is the same: At equilibrium, the Hamiltonian H is extremal (assuming the appropriate definition of the Lagrangian multiplier is included in Φ).

Note. Adding a boundary term to (3.9.3), we obtain instead the "symmetrized" expression

$$\int d^3x\mathcal{L} = \int d^3x\left(\frac{\rho\dot{\Phi} - \dot{\rho}\Phi}{2} - H[\rho,\Phi]\right).$$

The equations of motion are the same as before. In fact, nothing has changed.

In thermodynamics, we must recognize that it does not have a non-degenerate canonical structure, for the temperature does not have a conjugate momentum. Variation of the action with respect to T is technically not an equation of motion but a constraint, which is fine for most purposes, but it is, perhaps, the source of the following difficulty.

It has become established that the "energy" functional that takes a minimal value in many applications of thermodynamics to an isolated system (with fixed total volume) is the free energy

$$F = \int d^3x f(\rho, T).$$

The "energy" is the integral of the free energy density f. The Hamiltonian density h, on the other hand, is defined by

$$\mathcal{L} = \rho\dot{\Phi} - h,$$

that is, see Eq. (3.3.1), $h = f + sT$. The problem is that the "energy" of classical, equilibrium thermodynamics is *not* the Hamiltonian of the Action Principle. A solution is to interpret the entropy as a time derivative.

This should tell us something about the physical interpretation of entropy. The entropy is a record of past interactions between our system and the environment, an interaction that is no longer active. That is why neither temperature nor entropy appears in the Lagrangian as a time derivative.

We shall have to generalize this situation, and we must try to do it without altering the essential properties of the theory.

Consider replacing, in the Lagrangian density, the term sT:

$$sT \to T\rho\dot{\sigma} + j(\sigma). \tag{3.9.5}$$

We have introduced a new field variable σ and a source density j. Variation of T gives

$$\frac{\partial f}{\partial T} + \rho\dot{\sigma} = 0.$$

This confirms that $\dot{\sigma}$ is the **specific entropy density** under a new name. Variation of the mass density gives the familiar equation with S replaced by $\dot{\sigma}$. We have one new equation, from variation of σ,[38]

$$\frac{d}{dt}(\rho T) = \frac{\delta j}{\delta \sigma}.$$

The source $j(\sigma)$ is for us to choose and we shall take advantage of that by demanding that the systems decouple. That is, we interpret the last equation as a condition that affects the left side, with no implications for the right side. The external part of the enlarged system simply allows the original system to obey its own equations of motion. This is reasonable, for the only thing that we know and want to know about the extension of the system by the new degree of freedom represented by σ is that it should maintain and justify the assumptions that were made about the "isolated" system. This is exactly the same attitude that validates the way that the pressure is treated in the case of an incompressible gas. The relation of this equation with the Fourier heat equation is obvious but we have not yet explored that direction.

After the replacement (3.9.5), we have two canonical pairs of dynamical variables and a non-degenerate symplectic structure. The Hamiltonian density is now defined by,

$$\mathcal{L} = \rho\dot{\Phi} - \rho Y\dot{\sigma} - h, \quad h = \rho\dot{\vec{v}}^2/2 + fj(\sigma).$$

The Hamiltonian is extremal for allowed variations of the equilibrium state, variations that do not affect the field $S = \dot{\sigma}$. Finally, the functional $H = \int d^3x(\rho\vec{v}^2/2 + f)$ is extremal with respect to variations that do not affect the external source $j(\sigma)$. The term $j(\sigma)$ is inconsequential and will be dropped from now on.

[38]Here, there is an opportunity to make contact with Fourier's heat equation, but we leave that for another time.

In the new formulation, our Lagrangian and Hamiltonian densities are related by

$$\mathcal{L} = \rho\dot{\Phi} - \rho T\dot{\sigma} - h, \quad h = \rho\vec{v}^2/2 + f. \tag{3.9.6}$$

Finally, just as choosing the value of the total mass leads to fixing the value of $\dot{\Phi}$, the fact that the system is isolated implies fixing the value of $\dot{\sigma}$.

Summary

Prompted by the experimental evidence, we have moved the term sT out of the Hamiltonian. The field σ is canonically conjugate to ρT, the canonical structure is non-degenerate.

We have developed the interpretation of the entropy in a direction that helps us get a better grasp on its nature. The parameter S acts like a Lagrangian multiplier in that the actual value is determined by a one-sided influence of the environment. The system approaches an equilibrium state by giving up energy to the environment while it remains constrained by the principle of least action; this necessarily leads to adopting the highest value of the total entropy.[39] The environment has infinite entropy and is not affected in a measurable way; just as the reflection of a small mass from a very large one has no measurable effect on the latter because the momentum change is measurable only through observation of the velocity $\Delta v = M^{-1}\Delta p$.

These conclusions are not the result of speculation but are based on the experimental fact that it is the free energy, rather than the internal energy, that is minimal for equilibrium configurations of an isolated system.

3.10. The Third Law and the Normalization of Entropy

So far, nothing that has been done defines a zero point of entropy; only differences have a direct physical meaning in terms of heat transfer. But perhaps, to reach the absolute zero of temperature, we must take all the heat out of it.

The **third law** is an invention of Walther Hermann Nernst (1969), who dedicated a monograph to it. It postulates that the entropy is zero at zero absolute temperature, $T = 0$. Combined with the second law, it has interesting consequences. Consider a system of particles, where each particle has

[39] Actually, the highest value of $\int sT$.

precisely two atomic levels, with energy difference ϵ. Each level has a well-defined entropy per particle. A state of equilibrium will have a number of particles in each level. This number can change; typically the population of the higher level will increase as the temperature is raised. It follows that each particle is subject to migrate between the two levels. Then at equilibrium, the energy must be stationary with respect to this level switching,

$$T(S_1 - S_2) + \epsilon = 0.$$

The binding energy ϵ is very nearly independent of the temperature; therefore, this gives us a very rare example where the assignment of the total entropy of a mixture, between the two components of a mixture, can be predicted.

3.11. Galilei Invariance

The behavior of a thermodynamical system under Galilei transformations has played an important role in the development of thermodynamics, for there are situations where the question of invariance imposes itself.

The Galilei group is a group of coordinate transformations. We need to understand how this group acts on fields. The geometry of a simple thermodynamic system is three-dimensional, the fields are defined over $\mathcal{R}^3$ or a part thereof. Galilei invariance is possible when the system extends over all of space.

The most important fields are **scalar fields** and the prototype is the temperature T. Under any transformation of coordinates,

$$x^i \to x'^i = x'^i(\vec{x}), \tag{3.11.1}$$

a scalar field transforms as follows:

$$T \to T',$$

where the field T' is defined by

$$T'(\vec{x}') = T(\vec{x}).$$

Nothing could be simpler than this fundamental transformation property.

Our theory also employs vector fields, most of them are gradients and transform like gradients of scalar fields. Under the transformation (3.1.1) we have, using the summation convention,

$$\frac{\partial T'}{\partial x'^i} = \frac{\partial x^j}{\partial x'^i}\frac{\partial T}{\partial x^j}, \quad i = 1, 2, 3.$$

The velocity field transforms similarly,

$$v'^i = \frac{\partial x'^i}{\partial x^j} v^j, \quad i = 1, 2, 3.$$

Examples of fields that transform as scalar fields are the density ρ, the temperature T, and the kinetic energy density.[40]

A transformation of the Galilei group relates the coordinates of two observers that move relatively to each other with a fixed velocity $\vec{u}$,

$$x_i \to x'_i = x'_i(\vec{x}') = x_i - tu_i, \tag{3.11.2}$$

with $\vec{u}$ a constant vector. This is a coordinate transformation in $\vec{x}$ space, but a complication arises from the dependence of the transformation on the parameter t.

For reasons that are related to the appearance of the time t in the transformation (3.11.2), the action of the Galilei group on our fields is not what one could expect. Instead,

$$\Phi'(x', t) = \Phi(x, t), \quad \partial'_i \Phi(x', t) = \partial_i \Phi(x, t) + u_i, \tag{3.11.3}$$

and

$$\dot{\Phi}'(x', t) = \dot{\Phi}(x, t) + \vec{u} \cdot \vec{\nabla}\Phi + \vec{u}^2/2. \tag{3.11.4}$$

At first sight, this seems incomprehensible and an explanation within this present context can never remove all doubts. The easiest way to demonstrate that this is the transformation law that we want is to obtain the result by reduction of a Lagrangian that is Lorentz invariant. The reader may not wish to follow the derivation; the alternative is to accept Eqs. (3.11.3) and (3.11.4) on faith. What really matters is to verify that our Lagrangian is invariant. We shall return to show that that is the case below.

The Lorentz group is the group of coordinate transformations of the form

$$x'^\mu = x^\nu \Lambda^\mu_\nu.$$

Here, the Greek indices run over 1, 2, 3, 0 and Λ is a four-dimensional matrix. The group includes rotations of the spatial coordinates, but the "proper"

[40]Our use of the term "**density**" is slightly abusive. Ultimately, our context always involves a metric. If we use non-Cartesian coordinates, we shall make it conventional to include a factor $\sqrt{g}$ in the differential volume dv. Then the "density" ρ in $\int dv\rho$ is a scalar field.

Lorentz transformations involve one space coordinate and the time. The parameter is a vector $\vec{u}$; when it points in the x-direction the transformation is

$$x' = \gamma(x + ut), \quad t' = \gamma(t + ux/c^2), \quad y' = y, \quad z' = z,$$
$$\gamma := 1/\sqrt{1 - (u/c)^2},$$

where c is the velocity of light. Any *four-dimensional, scalar* field ψ transforms as follows:

$$\psi'(\vec{x}', t') = \psi(x, t) \tag{3.11.5}$$

and the gradient transforms just as expected for four-dimensional coordinate transformations,

$$\begin{aligned}
\frac{\partial}{\partial x'}\psi'(\vec{x}', t') &= \gamma\left(\frac{\partial}{\partial x}\psi(x, t) - u\dot{\psi}(\vec{x}, t)/c^2\right),\\
\dot{\psi}(x'.t') &= \gamma\left(\dot{\psi}(x, t) - u\frac{\partial}{\partial x}\psi(x, t)\right),\\
\frac{\partial}{dy'}\psi'(\vec{x}') &= \frac{\partial}{\partial y}\psi(\vec{x}, t),\\
\frac{\partial}{dz'}\psi'(\vec{x}', t') &= \frac{\partial}{\partial z}\psi(\vec{x}, t).
\end{aligned} \tag{3.11.6}$$

The odd elements in (3.11.3) and (3.11.4) come from a simple change of notation. We replace the field ψ by a new field Φ according to the definition,

$$\psi(x, t) = c^2 t + \Phi(x, t).$$

With this change of notation, (3.11.5) takes the form

$$\Phi(x', t') + c^2 t' = \Phi(x, t) + c^2 t$$

from which

$$\Phi(\vec{x}, t) = \Phi'(x', t') + c^2(t' - t)$$
$$\frac{\partial}{\partial x}\Phi(\vec{x}, t) = \gamma\frac{\partial}{\partial x}\Phi'(\vec{x}', t') + u$$

and

$$\dot{\Phi}(x, t) = \gamma\frac{\partial}{\partial t'}\Phi(x', t') + \gamma u_i\frac{\partial}{\partial x^i}\Phi(x', t') + c^2(\gamma - 1).$$

In the non-relativistic theory, one ignores terms of order $1/c^2$, hence $\gamma = 1 + \vec{u}^2/2c^2$ and this result is the same as (3.11.3) and (3.11.4).

Finally, combining (3.11.3) and (3.11.4), we find that $\vec{v}(x) = \vec{v}'(x') + \vec{u}$ and, to first order in $\vec{u}$,

$$\dot{\Phi}(x,t) - \vec{v}^2(x)/2 = \dot{\Phi}'(x',t) - \vec{v}'^2(x')/2.$$

So, we see that the Fetter–Walecka Lagrangian is invariant under Galilei transformations; there is no preferred system in conservative hydrodynamics.

Other formulations of hydrodynamics are not Galilei invariant. Attempts to define an energy have not discovered one that is Galilei invariant. Indeed, the *Hamiltonian* of Fetter–Walecka theory is not Galilei invariant, but the Lagrangian is! Some very well-known textbooks are less than clear about this. See, for example, Khalatnikov (1965), Landau and Lifshitz (1959, p. 264), Putterman (1974).

A relativistic version of the Fetter–Walecka model will be presented in Chapter 10. It is the only demonstration, in a field theory, that the non-relativistic limit of General Relativity is the Newtonian theory of gravity of fluids.

3.12. The Space of Variations

There is a fundamental point to be understood and remembered. In any variational system, every allowed variation of the dynamical variables implies a condition that must be satisfied by the system; every enlargement of the **space of variations** implies an additional equation of motion, or an additional constraint on the solutions. **Of course, there is a risk that admitting more variations may lead to a system of Euler–Lagrange equations that do not allow solutions!**

Therefore, let us be aware.

According to **Gibbs**, the "variations" that are to be considered in applications of his energy axiom are "all possible variations of the system". Elaborating through several pages of text, he insists that constraints of various sorts limit the variations to those that are physically realizable. Now, it is possible to imagine a variation of S_1 and S_2 that keeps the sum fixed, but it is not physically realizable unless the temperatures are equal. What we are assured with great emphasis in the textbooks is that entropy is not conserved. When heat is transferred from a system with temperature T_1, then this system suffers a change $dQ_1 = T_1 dS_1$ of heat. If this heat is received by another body at temperature T_2, then the change of heat of that system is $dQ_2 = -dQ_1 = T_2 dS_2$. The total change of heat of the system as a whole is $0 = dQ_1 + dQ_2 = T_1 dS_1 + T_2 dS_2$. Unless the temperatures are equal, the sum $dS_1 + dS_2 \neq 0$. Consequently, when the temperatures are unequal, there are no changes in the individual entropies that keep the total entropy constant. The space of variations remains the same. **Prigogine** (1949) already reached the same conclusion. He writes, on pp. 17–18:[41]

> "We can therefore say that "absorption" of entropy in one part, compensated by a sufficient "production" in another part of the system is prohibited. This formulation implies that in every macroscopic region of the system the entropy production due to irreversible processes is positive. (...) Interference of irreversible processes is only possible when they occur in the same region of the system".

The importance that Prigogine attaches to this clarification can be judged from the sequel:

> "Such a formulation may be called a "local" formulation of the second law in contrast to the "global" formulation of classical thermodynamics. It's value lies in the fact that it permits a much closer analysis of irreversible processes and, as such, it will constitute the central postulate on which this book is based".

The central postulate! Apparently, **Prigogine** (Nobel Prize 1977) goes to some pain to avoid saying that **Gibbs** erred at this point.

> **We do not agree that the principle of minimum energy demands equalization of the temperature. Consequently, we must keep both S_1 and S_2 fixed for the class of variations considered in the principle of minimum energy.**

Let us turn to the axiom of maximum entropy. The total entropy is

$$S = S_1 + S_2 = \frac{1}{T_1}(E_1 - F_1 - V_1 P_1) + \frac{1}{T_2}(E_2 - F_2 - V_2 P_2),$$

and the variations of E_1 and E_2, with $dE_1 + dE_2 = 0$ make a contribution to the variation of the total entropy,

$$dS = \frac{dE_1}{T_1} + \frac{dE_2}{T_2} = dE_1 \left(\frac{1}{T_1} - \frac{1}{T_2} \right) + \cdots .$$

And here, it is clear that if it is possible to vary the partial energies without changing the values of the other variables, then this contribution to the change in the total entropy will vanish (making the total entropy extremal)

[41]See also pp. 20–21, where a clear distinction is made between entropy production and entropy convection by mass flow, and p. 39.

only if the temperatures are equal. Such changes are said to be physically possible even if not discussed in detail within the present context. Transfer of energy by radiation is often mentioned, but is there any reason to hold that radiation is not another example of heat transfer? In fact, what is required is a variation of the energies that does not affect the other terms in the last equation. This requires additional degrees of freedom, and when these are appropriately included in the expression for the total energy of the theory, we find that they must be kept fixed as well. So, there is no way to transfer energy without affecting the other terms, and no **general** conclusion is possible.

Taking stock

We have seen that the fundamental relations of thermodynamics, more precisely the thermodynamics of isolated systems, are the variational equations of an action principle. We have defined adiabatic thermodynamics as the mathematical structure defined by this action principle. To use it, we need an expression for the free energy and we must learn what values to assign to the specific entropy density. There is a strong indication that, for many homogeneous systems, this density takes the form $s = \rho S$, S constant.

But thermodynamics is more than adiabatic thermodynamics, and more than a theory of homogeneous systems. Although we shall postpone the study of dissipative processes, we shall very soon find out that it is essential to be able to determine the final result of entropy producing dissipation. Without formulating general axioms, we shall take up the study of important systems and use experimental input to try to deduce special results that we hope will merge into a general pattern that can be called "theory".

What does it mean to "find the entropy". What has already been emphasized deserves to be reiterated here, as we are on the threshold of the applications. There are several separate problems.

To most writers, "finding the entropy" means one of two things: to discover an expression, in terms of density and temperature, that gives the entropy of a configuration where the density and the temperature are known, or to evaluate this expression for a given configuration. There is a third aspect to "determination of the entropy", one that we consider to be more fundamental because it shall be confronting us at every turn: The more basic problem is to discover the correct value of **the assigned, specific entropy density**; that is, the off-shell expression for the entropy density in terms of the mass densities that appears in the variational principle as the term sT. The first instance of this problem is the case of a pure substance, where in most applications $s = \rho S$ with S usually uniform. For mixtures, we shall propose that the **assigned entropy** density s is linear in the densities.

This problem is not part of the traditional treatment; it is a principal innovation presented in this treatise. Lacking was a sufficiently precise specification of the class of allowed variations, as I have tried to explain above. It is a part of the program to determine the Lagrangians that encode all the properties (or as many as we can) of a given system.

The reader may be surprised to find no discussion of "interactions" in the sense of Carathéodory (1909). The first version of this book did develop this subject, but without a sufficiently precise specification of the class of allowed variations, the discussion was inconclusive and the subject was dropped for lack of interest.

Chapter 4

Applications to Simple Systems

Recall that a **simple system** is one for which the dynamical variables are a density, the temperature, the pressure and the entropy, all fields over a subset of $\mathbb{R}^3$, possibly with movable boundaries. The paradigm of a simple system is an ideal gas; in the first application, we consider the problem of propagation of sound in an ideal gas.

This process is not within the province of thermodynamics as it is usually understood, but it is an important part of a theory of the dynamics of fluids, and of adiabatic thermodynamics. The theory was developed by Laplace (1825). (See also Biot (1802).)

4.1. Propagation of Sound in an Ideal Gas

This is the first of several sections devoted to the **propagation of sound**. Here we deal with the simplest case, the propagation of a plane **pressure wave** in an ideal gas.

The difficult part was to know what theory to apply to this phenomenon. One starts with a gas at equilibrium, density, pressure and temperature uniform and time independent, and considers a first-order **perturbation** of this system

$$\rho = \rho_0 + \epsilon\rho_1, \quad T = T_0 + \epsilon T_1,$$

with **harmonic perturbations**, e.g., $\rho_0 \propto \exp i(kx + \omega t)$, where k and ω are constants.

We may be sure that there will be a fluctuation of density and pressure but it is possible to entertain some doubts concerning the temperature. Newton's isothermal theory of sound propagation (Proposition 49 of Book II of the

Principia) postulates that the temperature is fixed and that the equation of continuity as well as the Bernoulli equation apply. He also needs a relation between density and pressure and for this he uses the ideal gas law.

Here is the detailed paradigm for the use of the **`Euler-Lagrange equations`** in perturbation theory:

Write down the expression for the Lagrangian with the appropriate formula for f:

$$\mathcal{L} = \rho(\dot{\Phi} - \vec{v}^2/2) - \mathcal{R}\rho T \ln \frac{\rho}{T^n} - \rho S T, \tag{4.1.1}$$

then use the Euler–Lagrange equations:

(1) Variation of Φ gives the equation of continuity; to first order in ε,

$$\dot{\rho}_1 = -(\rho_0 v_1), \quad \ddot{\rho}_1 = -\rho_0 \dot{v}_1.$$

(2) Variation with respect to the temperature shows that ρ/T^n is uniform, see Eq. (4.1.3).

(3) Variation of the density gives

$$\dot{\Phi} = \mathcal{R}T\left(\ln\frac{\rho}{T^n} + 1\right) + ST. \tag{4.1.2}$$

Thus,

$$\dot{\vec{v}} = -\vec{\nabla}\mathcal{R}T\left(\ln\frac{\rho}{T^n} + 1\right).$$

It follows that, at equilibrium, when $\vec{v} = 0$, the density and the temperature are both uniform; there is no need to vary the entropy distribution to assure us.

In Newton's theory of sound propagation, T is held fixed and the calculation goes as follows, using the last two equations. If a prime denotes the spatial derivative in the direction of propagation, then

$$\frac{\ddot{\rho}_1}{\rho_0} = -\dot{v}_1' = \dot{\Phi}_1'' = \mathcal{R}T\frac{\rho_1''}{\rho_0},$$

and thus

$$c^2 = \omega^2/k^2 = \mathcal{R}T.$$

Using $T = 293$ and $\mathcal{R}_{\text{air}} = 0.8314 \times 10^8/28.96$ gives $c = 290\,\text{m/sec}$. The actual value of c is $343\,\text{m/sec}$.

A better result is obtained with the assumption, first proposed by Laplace (1806), that sound propagation is "**`adiabatic`**". **That is to say that it is governed by the Euler–Lagrange equations of our Lagrangian**, with a fixed value of S. In Laplace's theory, ρ and T are related by the

adiabatic condition; it is the Euler–Lagrange equations, or more precisely the **constraint**, obtained by variation of $\mathcal{L}$ with respect to T,

$$\mathcal{R}\rho\left(\ln\frac{\rho}{T^n} - n\right) + \rho S = 0. \tag{4.1.3}$$

The ratio ρ/T^n is thus fixed and therefore $dT/T_0 = (1/n)d\rho/\rho_0$, so the same calculation gives

$$\frac{\ddot{\rho}_1}{\rho_0} = -\dot{v}_1' = \dot{\Phi}_1'' = \left(1 + \frac{1}{n}\right)\mathcal{R}T\frac{\rho_1''}{\rho_0} = \gamma\mathcal{R}T\frac{\rho_1''}{\rho_0}$$

and

$$c^2 = \gamma\mathcal{R}T.$$

We shall always assume that the **specific entropy** is fixed. For this reason, it is useful to repeat the last calculation, that of Laplace, in somewhat greater generality.

In this book, all calculations begin by writing down the expression for $\mathcal{L}$:

$$\int d^3x\mathcal{L} = \int d^3x(\rho(\dot{\Phi} - \vec{v}^{\,2}/2) - W(\rho, T)), \quad W = f + sT.$$

The three Euler–Lagrange equations are:

$$\dot{\rho} + \vec{\nabla}\cdot(\rho\vec{v}) = 0, \quad \frac{\partial W}{\partial T}\bigg|_\rho = 0, \quad \dot{\Phi} - \vec{v}^{\,2}/2 = \frac{\partial W}{\partial \rho}\bigg|_T.$$

As we have seen, for this to simplify, we need (a) the entropy density s to be linear in ρ, $s = \rho S$ and (b) the specific entropy S to be uniform. The first condition is satisfied by an ideal gas; it is not always pointed out that the second condition (the isentropic assumption) is used as well to get the formula (see Section 3.1)

$$\begin{aligned} \vec{\nabla}p &= \vec{\nabla}\left(\rho\frac{\partial f}{\partial \rho} - f\right) = \rho\vec{\nabla}\frac{\partial W}{\partial \rho}\bigg|_S, \\ \frac{\ddot{\rho}_1}{\rho_0} &= -\dot{v}_1' = \partial_x\frac{p'}{\rho}\bigg|_S = \frac{\rho_1''}{\rho}\frac{dp}{d\rho}\bigg|_{\text{adiab.}}. \end{aligned} \tag{4.1.4}$$

The speed of propagation is thus the square root of the **adiabatic derivative** of p with respect to ρ; the partial derivative of the function $p(\rho, S)$ that is obtained from $p(\rho, T)$ when T is expressed in terms of ρ with the use of the adiabatic relation $\partial W/\partial T|_\rho = 0$,

$$\frac{dp}{d\rho}\bigg|_{\text{adiab.}} = \frac{\partial p}{\partial \rho}\bigg|_T + \frac{\partial p}{\partial T}\bigg|_\rho \frac{\partial T}{\partial \rho}\bigg|_S. \tag{4.1.5}$$

For an ideal gas, the equations are

$$\dot\rho + \text{div}(\rho\vec v) = 0, \quad \mathcal{R}\left(\ln\frac{\rho}{T^n} - n\right) + S = 0,$$
$$\dot\Phi - \vec v^{\,2}/2 = \mu + ST = (n+1)\mathcal{R}T, \tag{4.1.6}$$

where μ is the chemical potential (Section 2.7) and

$$\frac{dp}{d\rho}\bigg|_{\text{adiab.}} = \gamma\mathcal{R}T. \tag{4.1.7}$$

For air, $\gamma = 1.4$, which gives a speed of sound in air as 343 m/sec. The result is an excellent prediction for air and other gases at atmospheric temperature and pressure. See the web article "Laplace and the Speed of Sound".[42] Laplace's calculation is found in an article by his protegé Biot (1802).

Corollary. *If the adiabatic derivative of p is negative, the system is* **`unstable`**; *the perturbation grows exponentially.*

Remark. It deserves to be emphasized that the adiabatic condition was needed, which would be something of a surprise if it were interpreted as no more than a formula that determines the entropy. We do not need the value of the entropy, but we used the fact that it is uniform (the polytropic condition) and that it is constant in time. In other words, the specific entropy density, not only the total entropy, is an adiabatic invariant. At least, the assumption that it is so, is a successful assumption as far as the propagation of pressure waves is concerned. It should also be emphasized, once again, that this work of Laplace, 200 years old, when translated to the modern idiom is an undeniably successful use of entropy for a system that is not in equilibrium.

Radial wave

A radial wave, originating at the center of a Cartesian coordinate system and propagating in the direction of the radius vector, or else converging on the center, requires a source or an absorber at either end. The propagation of the wave from source to absorber can be discussed without a specification

[42] www.nd.edu/~powers/ame.20231/finn1964.pdf.

of the emitter or the transmitter; all that is needed is the boundary conditions. We postulate spherical symmetry; the boundaries are spheres of radii r_0, r_1, with $0 \leq r_0 < r_1 < \infty$.

We consider a first-order perturbation of a static configuration in which all the field variables are uniform and the flow velocity is zero. The appropriate generalization of (4.1.4) is, for an ideal gas,

$$\partial_t^2 \delta\rho = \gamma \mathcal{R} T_0 \Delta \delta\rho.$$

Setting $\partial_t \delta\rho = i\omega\delta\rho$, we get the radial, three-dimensional Helmholtz equation

$$(k^2 + r^{-2}\partial_r r^2 \partial_r) d\rho = 0, \quad k^2 = \frac{\omega^2}{\gamma \mathcal{R} T_0}.$$

It has two independent solutions, both singular at the origin,

$$\delta\rho(t, r) = \frac{1}{r} e^{i\omega t \pm ikr}.$$

With the freedom to choose the sign of k, we can construct a wave traveling into or out of the origin. These solutions will be singular at the origin, evidence of a source or a sink. In the case of a converging wave, we can speculate about the mechanism by which the energy can be accounted for; **sono-luminescence** is one way.

4.2. The van der Waals Fluid

Van der Waals (1873) proposed a physically motivated modification of the ideal gas law. For 1 gram or one mole of gas, it takes the form

$$p = \frac{\mathcal{R}T}{V - b} - \frac{a}{V^2}. \tag{4.2.1}$$

In this section, V stands for $1/\rho$. For any amount of gas, it is

$$p = \frac{\mathcal{R}\rho T}{1 - b\rho} - a\rho^2. \tag{4.2.2}$$

The main interest of this formula is that it accounts for the condensation of the gas at low temperatures, but at first we shall examine the pure, gaseous phase only.

The parameter $b > 0$ is "**the excluded volume**", intended to take into account the fact that the volume available to a particle of the gas is diminished by the fraction of the total that is occupied by the other particles. This arises from the point of view that regards a gas as a collection of classical particles, which is not wholly justified since the particles obey quantum statistics. A dramatic effect of this change in the expression for the free energy density is that, when the gas is strongly compressed, the density increases as before, but now it is limited by the value $\rho = 1/b$. As this limit is approached, the pressure grows without bounds.

The second term is meant to take into account a direct interaction between the particles. This is probably the first example of a suggestion to model an interaction by the addition of a polynomial in the density to the free energy density. It was justified by a study of interparticle interactions, but attempts to predict the values of the parameters by particle dynamics have been unsuccessful, to the point of disfavoring, for a time, the work of van der Waals (Uhlenbeck, 1937). In modern times, the parameters a and b are treated as phenomenological, to be adjusted to a best fit of observed thermodynamical properties, independently for each type of gas. The parameter a is normally positive, which indicates an **`attractive force`**. To make the pressure positive, we may restrict the value of a,

$$\frac{a}{\mathcal{R}T} < 4b,$$

but negative pressures actually occur, as **`metastable`** configurations (Ramsey, 1956) and van der Waals' theory is successful in that domain as well.

The van der Waals expression has the great merit that it never leads to unreasonable predictions. There have been many proposals to improve the agreement with experiment but here we shall work with the original proposal, Eq. (4.2.2). Some generalizations will be considered later in this chapter.

Figure 4.2.1 shows the pressure as a function of the volume for several values of the temperature. At very high temperatures or large volumes, the

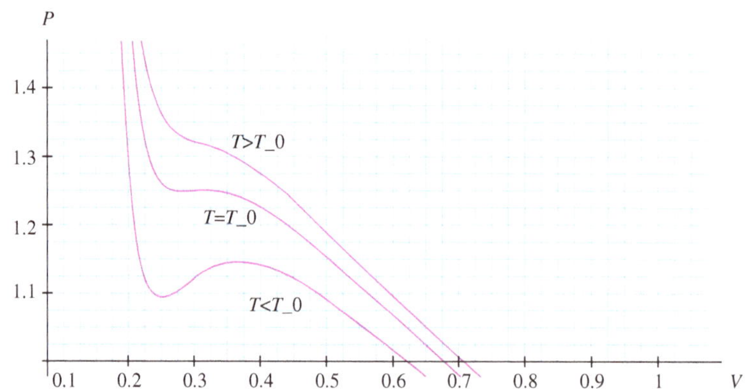

Fig. 4.2.1. Loci of constant T in the V, P plane according to van der Waals. The abscissa is the specific volume $V = 1/\rho$, the ordinate is the pressure.

pressure is given fairly accurately by the ideal gas law, and there is just one real value (and two complex values) of the volume for each value of (p, T). At lower temperatures, when

$$T < T_c,$$

there are three real values of the volume for each value of (p, T).[43] This is the domain within which condensation takes place and a liquid phase coexists with the gas. The "**critical temperature**" T_c is the highest temperature at which the gas and the liquid can coexist. At this temperature,

$$\left.\frac{\partial p}{\partial V}\right|_T = 0 \quad \text{and} \quad \left.\frac{\partial^2 p}{\partial V^2}\right|_T = 0. \tag{4.2.3}$$

The critical temperature T_c, and the associated critical pressure and density are, respectively, given by

$$\mathcal{R}T_c = \frac{8}{27}\frac{a}{b}, \quad p_c = \frac{1}{27}\frac{a}{b^2}, \quad \rho_c = \frac{1}{3b}.$$

For He_4, the experimental values in cgs units are

$$T_{\rm cr} = 5.19, \quad \rho_{\rm cr} = 0.0696, \quad p_{\rm cr} = 0.227 \times 10^7, \quad V_{\rm cr} = 14.35,$$

and in the same units, the parameter values are reported to be

$$a = 2162.5 \times 10^6, \quad b = 5.9.$$

This is not quite right, it improves if we change the critical density to 0.05625, then

$$p_{\rm cr} = \frac{0.8314 \times 10^8 \times 5.19 \times 0.05625}{4(2/3)} - 2162.5 \times 10^6 \times 0.05625^2$$
$$= 2.2596 \times 10^6.$$

In the more popular units,

$$\rho_{\rm cr} = 56.25/4 = 14.0625 \text{ mol/L}, \quad a = 2162.5 \times 16 = 34600 \text{ PaL}^2/\text{mol}^2,$$
$$b = (3 \times 14.0625)^{-1} = 0.02370 \text{ L/mol},$$
$$T_{\rm cr} = 5.19, \quad \rho_{\rm cr} = 14.0625 \text{ mol/L}, \quad p_{\rm cr} = 0.227 \times 10^7.$$

Now,

$$p_{\rm cr} = \frac{0.8314 \times 10^5 \times 5.19 \times 14.0625}{(2/3)} - 34600 \times 14.0625^2 = 2.2596 \times 10^6.$$

[43]The values of T_c for real gases range from 5K for helium "... to values well beyond measurement" (Rowlinson, 1959).

Finally, in bars, multiply all three terms by 10^{-5}. Then

$$p_{\rm cr}\frac{0.8314\times 5.19\times 14.0625}{(2/3)} - 0.3460\times 14.0625^2 = 22.596\,\text{bar}.$$

The principal application of van der Waals' formula is to study the **change of phase**. But a relation between pressure, volume and temperature is not sufficient to determine all the thermodynamical properties of the system; in particular, it is not enough information to determine the formula for the free energy, nor the internal energy, and it is not enough to predict the speed of propagation of sound.

Van der Waals' formula does not determine the free energy. The equation $p = -\partial F/\partial V$ has the general solution

$$F(V,T) = -\mathcal{R}T\ln(V-b) - \frac{a}{V} + \psi(T).$$

This can also be expressed as

$$F(V,T) = -\mathcal{R}T\ln\{(V-b)T^n\} - \frac{a}{V} + \tilde{\psi}(T).$$

This form is suggested by the expectation that the gas becomes an ideal gas in the limit when $a = b = \tilde{\psi} = 0$. If $\tilde{\psi}(T) = 0$ then

$$S = -\frac{\partial F}{\partial T} = \mathcal{R}\ln\{(V-b)T^n\} + n\mathcal{R}, \tag{4.2.4}$$

where V is the molar volume, and if S is uniform, this gives a modification of the polytropic relation obeyed by an ideal gas,

$$k := \frac{1}{(V-b)T^n} = \text{constant}.$$

The internal energy is simply

$$U = n\mathcal{R}T - \frac{a}{V}$$

or, in terms of the natural variables,

$$u(\rho,s) = n\mathcal{R}\rho\left(\frac{\rho}{1-b\rho}\right)^{1/n} k^{-1/n} - a\rho^2, \quad k = e^{S/\mathcal{R}-n}. \tag{4.2.5}$$

From now on, we shall take $\tilde{\psi} = 0$ as our definition of the van der Waals gas, thus

$$F_{\rm vdW}(V,T) = -\mathcal{R}T\ln(V-b)T^n - \frac{a}{V}. \tag{4.2.6}$$

Let us compare this approach with that of Rowlinson (1959). He does not use an expression for the free energy, or the internal energy, but he says

(p. 71) that $\partial U/\partial V|_T \propto 1/V^2$ and he reports that this is approximately verified for five liquids between the triple point and the critical point. The choice $\hat{\psi}(T) = 0$ is thus indicated for a gas that satisfies van der Waals' relation while preserving some properties of the ideal gas. We shall refer to this fluid as a van der Waals' fluid. Later, we shall see how Rowlinson gets some of his results by postulating the mere existence of an analytic expression for F. The above formula for the free energy is used by Lemmon *et al.* (2000).

See Fig. 4.2.1. When the temperature is fixed below the critical temperature, and the pressure is fixed in the interval between the two stationary points, then the relation (4.2.2) allows for three values of the density,

$$\rho_1 < \rho_0 < \rho_2,$$

as is illustrated in Fig. 4.2.1. The middle value is an unstable equilibrium[44] and will be ignored from now on. What happens to the gas as the temperature is lowered past the critical point is complicated. But, by action of surface tension and gravity, there results, if one waits long enough, a separation of the fluid into two phases, the liquid taking up the lower part of the vessel. We are not prepared to account for this transformation and limit ourselves to describe the result. The separation is observed, not predicted, but once it is admitted to occur, the theory has much to say about it.

4.3. Saturation

Consider a homogeneous body of gas with total mass 1, temperature $T < T_c$. As the gas is compressed, the temperature is maintained by contact with a reservoir, the process carried out slowly so that the system goes through a sequence of equilibria. The pressure increases until the point A in Fig. 4.3.1 is reached. One observes that, as the volume is reduced further, the gas gradually converts, at constant pressure, to the liquid phase at point B, until all the gas has been converted. In the case of water at normal atmospheric pressure, the density increases by a factor of 1700.[45]

[44] It is unstable in a mechanical sense (Rowlinson, 1959, p. 20) and also in the sense of the preceding section.

[45] We are ignoring the interesting phenomenon of metastable states. Some catalyst is needed for the process of condensation to begin. Apparently, it is possible to reach the maximum on the pressure curve by increasing the pressure, and the minimum by decreasing it. The pressure at the minimum is not necessarily positive. See Fronsdal (2019a), Rowlinson (1959, p. 21) and Ramsey (1956).

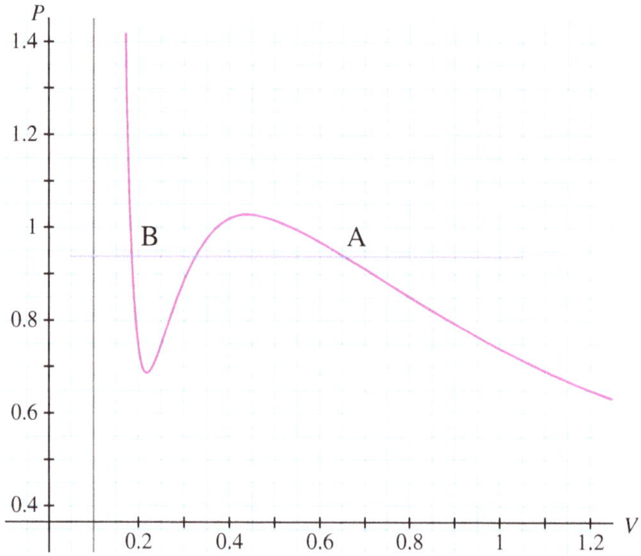

Fig. 4.3.1. **Maxwell's rule**. The saturation pressure (= vapor pressure) is the value at the horizontal line that makes the two enclosed areas equal. The coordinates are as in Fig. 4.2.1. B is the **bubble point**, and A is the **dew point**.

In what follows, we shall make frequent reference to "the temperature". We have found that the Euler–Lagrange equations, for a homogeneous system at rest, demand a uniform temperature. During condensation, the system separates into two parts. A discontinuity in the temperature would give rise to an entropic force. The experiments are unequivocal: at equilibrium, both phases have the same temperature.

The van der Waals gas at the point A is characterized by a uniform, specific entropy density S_A. The same is true for the van der Waals liquid at the point B, though with a different value, S_B, say. Formula (2.2.3) in Section 2.2 shows that $S_A > S_B$.

We are considering configurations of adiabatic equilibrium only, so the flow velocity vanishes. The effective Lagrangian equation is thus

$$\int d^3x(\rho_\ell \dot{\Phi}_\ell + \rho_{\rm g}\dot{\Phi}_g - f - sT), \tag{4.3.1}$$

where $\dot{\Phi}_\ell$ and $\dot{\phi}_g$ are real numbers. The integrals extend over the region occupied by the liquid and over the region occupied by the gas. We have here a first example of a rule that we expect to apply universally:

> **The Lagrangian of a compound system is, at least in a first approximation, the algebraic sum of the Lagrangians of the components.**

The field $\dot{\Phi}$ takes different values in the gas and in the liquid. The velocity potential is uniform for equilibrium configurations, except for a possible discontinuity at the meniscus. We consider the equations of motion, restricted to the stationary case, all fields time independent.

Local variation with respect to the density gives, with T and P fixed, in either phase,

$$\dot{\Phi} - ST - \frac{\partial}{\partial \rho} f = 0, \tag{4.3.2}$$

and variation of the temperature gives the adiabatic relation

$$\frac{\partial}{\partial T}(f + sT)\Big|_\rho = 0. \tag{4.3.3}$$

For a stationary state with $\vec{v} = 0$, this implies that all the fields are uniform within each phase and, except for density and entropy, continuous throughout.

This is a complete description of the stationary system for given values of temperature and pressure. Global variation of both densities leads to the conclusion that both parts have the same temperature and pressure. A complementary global variation of the density will help to specify the **vapor pressure**; i.e. the location of the point A in the figure, the pressure of coexistence. This point is determined by Maxwell's rule, that we shall discuss next, before completing the discussion of the equations of motion.

The above Lagrangian also applies to metastable configurations (Reisman, 1971).

Interpretation of the velocity potential

In a homogeneous fluid, at rest, the velocity potential fulfills the role of Lagrange multiplier; in a fluid that is separated into a liquid and a gaseous phase, it has a more interesting interpretation. The discontinuity in $\dot{\Phi}$, across a boundary of thickness x, is

$$\Delta\dot{\Phi} = \dot{\Phi}_g - \dot{\Phi}_\ell = \Delta\vec{x} \cdot \vec{a} = \vec{x} \cdot \vec{F}/\rho;$$

it is the work required to lift one gram of matter out of the liquid. It is equal to $\Delta\vec{v}^2/2$ and we thus have

$$\Delta\dot{\Phi} = \frac{1}{2}\vec{v}^2_{\text{esc}}, \tag{4.3.4}$$

where $\vec{v}_{\text{esc}}$ is the escape velocity required to overcome the surface tension. The interpretation of $\rho\Delta\dot{\Phi}$ as the energy of evaporation of one cm^3 of the liquid is thus natural. In a famous experiment, designed for education and employing the simplest means, the **surface tension** was measured and related to the **evaporation energy** (also measured). From these data, Avogadro's number could be calculated. The value is about 6×10^{23}. Here are some web posts.[46,47]

Van der Waals has developed a more detailed model for the transition region, the "**squared gradient model**", where the above picture is made more explicit (van der Waals, 1893).

With two independent densities and associated velocity potentials, we have two independent equations of continuity. This is because, in the bulk, whichever component is present is conserved. Conversion takes place only on the phase boundary. In the squared gradient model (van der Waals, 1893), the single field $\dot\Phi$ varies continuously from $\dot\Phi_g$ to $\dot\Phi_\ell$ in the finite boundary layer.

Because the two spatial divisions of the system each contain only one fluid component, either liquid or gas, the adiabatic condition holds separately in both,

$$S_B = -\mathcal{R}\left(\ln\frac{\rho_g}{(1-\rho_g)T^n} - n\right), \quad S_A = -\mathcal{R}\left(\ln\frac{\rho_\ell}{(1-\rho_\ell)T^n} - n\right). \tag{4.3.5}$$

In the pure gas domain, as in the pure liquid domain, there is only one velocity potential and, in particular, in the stationary case, only one independent parameter $\dot\Phi$. As the gas condenses, all properties of each component remain constant, except for the total mass of each. Consequently, $\dot\Phi_g$ and $\dot\Phi_\ell$ remain constant during isothermal compression. The two parameters are different because the fixed values of p and T allow for two stable values of the density.

4.4. Dew Point, Bubble Point, Maxwell's Rule

We turn to the question of predicting the values of the densities and the pressure at A and B in Fig. 4.3.1, the state of equilibrium or coexistence of gas and liquid.

Variation of the density in either region gives the information that the dynamical variables are constant in each; nothing more, since the term $\dot\Phi$ in (4.3.2) contains a Lagrange multiplier. Additional information is obtained by

---

[46] Ishita Patel, M. S., Surface tension, wsimson.tripod.com/OFFLINE/surf_tension.pdf.

[47] National Physical Laboratory, resource.npl.co.uk/mtdata/surfacetension.htm.

the supplemental, piecewise uniform variation with $\delta\rho_1 + \delta\rho_2 = 0$, namely,

$$\text{Disc } \dot{\Phi} = \text{Disc}(\mu + ST). \tag{4.4.1}$$

Here,

$$\mu = \frac{\partial f}{\partial \rho}\bigg|_T$$

is the chemical potential. This relation is standard thermodynamics. We shall assume that the entropy density is $s = \rho_g S_B + \rho_\ell S_A$, which is also standard since the two phases are separated. This implies that the specific entropy difference is $S_B - S_A$ and Boltzmann's theory then identifies $T(S_B - S_A)$ with ϵ at equilibrium (De Groot and Mazur, 1962).

It remains only to discover the state of lowest "energy". An unequivocal result of 200 years of study is that the "energy" that characterizes the ground state in this isolated system is the free energy $F = \int d^3 f$, with the *caveat* that the traditional Lagrange multiplier must be taken into account. The action principle suggests that "energy density" is synonymous with Hamiltonian density, in this case $f + sT$. To overcome this conflict, we have suggested replacing the entropy, so far treated as a mere constant, by $\dot{\sigma}$ (see Section 2.9).

Now, Eq. (4.4.1) should better be written as

$$\text{Disc}(\dot{\Phi} - \dot{\sigma}T) = \text{Disc } \mu. \tag{4.4.1$'$}$$

The principle of **maximum entropy** at equilibrium implies that for a fixed value of the temperature, the left side should vanish, making μ continuous. It is nevertheless instructive to work out the implications of the equivalent minimization of the energy. It too implies the continuity of the **chemical potential**. It remains only to consider the variation of the respective volumes of liquid and gas. The "energy" is the total Hamiltonian

$$H = V_1 f_1 + V_2 f_2 + \epsilon V_2. \tag{4.4.2}$$

The last term, with ϵ an arbitrary constant, allows for the normalization of the Hamiltonian. It would be needed to make this expression for H minimal with respect to variations of the partial volumes, but actually it is irrelevant since the values of the dynamical variables are already fixed by the Euler–Lagrange equations — the state of equilibrium is uniquely determined in terms of the fixed parameters S_A, S_B, $\dot{\Phi}_g$ and $\dot{\Phi}_\ell$. To alter the relative amounts of liquid and vapor, we must change the value of the Lagrange multipliers.

The state of equilibrium is thus characterized by

$$\text{Disc}\ \mu = \text{Disc}(\dot{\Phi} - ST) = 0. \tag{4.4.3}$$

Both statements are verified experimentally. Because we are not aware of any convincing theoretical argument that the free energy must be minimal, and because it can only be proved within the framework of an action principle, or by experience, we prefer to regard this result as an experimental fact. It has been supported by numerous other experiments, and for this reason we have tweaked our action principle so as to identify the free energy, rather than the internal energy, with the Hamiltonian (Section 3.9).

The conventional point of view is essentially the same. The on-shell value of μ is the specific Gibbs potential,

$$\mu = \frac{\partial f}{\partial \rho} = \frac{f + p}{\rho} = F + VP = G(p, T).$$

Because "this function is a function of T and p", it follows that the slope of the free energy density has the same value in either phase. But this statement is not trivial. The familiar argument is

$$d(F + VP) = \frac{\partial F}{\partial T} dT + \frac{\partial F}{\partial V} dV + VdP + PdV = -SdT + VdP.$$

This shows only that the value of the function $F + VP$ is defined *locally* by T and P. It does not imply that the value is the same at the points A and B.

Callen (1960), in his highly admired textbook, makes the same point. His strategy throughout the book is to invoke maximal entropy to arrive at an operative statement of the principle of minimum energy, with the interpretation of "energy" (of an isolated system) equal to free energy. In this case, the axiom of greatest entropy makes the left side of (4.4.1′) equal to zero and this implies that $\mu_1 = \mu_2$. The principle of maximal entropy is thus confirmed by experience.

From Callen's point of view, the statement $\text{Disc}(\dot{\Phi} - ST) = 0$ reflects the fact that the two phases are in a type of diathermal contact where the heat is exchanged by means of transmutation.

It is an experimental fact that $F + VP$ (F if the volume is fixed) does have the same value for both components of the saturated phase. It implies that the horizontal line in Fig. 4.4.1 divides the enclosed area in two equal parts and that the slope $\partial f/\partial \rho$ (the chemical potential) has the same value at the points A, B. This condition fixes a unique relation between p and T. An actual example, with numbers and figures, will be presented later.

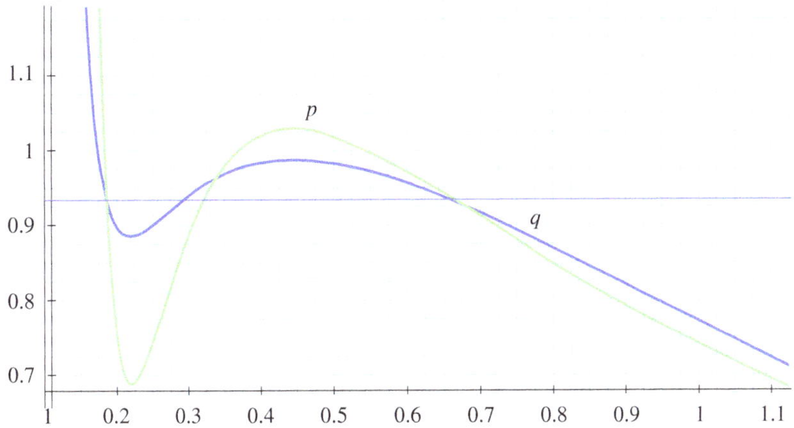

Fig. 4.4.1. Coexistence, schematic. The curves are the values of p and μ, plotted against the molar volume.

The difference $\epsilon = \dot\Phi_{\rm g} - \dot\Phi_\ell$ is the specific heat of evaporation, or latent heat; for

$$\epsilon = (S_{\rm g} - S_\ell)T = \Delta Q, \tag{4.4.4}$$

where ΔQ is the amount of heat needed, per gram, to convert the liquid to gas. This last equation is sometimes presented without justification. It means that the energy that is supplied to bring about the evaporation at constant temperature and pressure is entirely entropic. This is in accord with the principle that the entropy tends to assume the highest possible value. In the context of the action principle, it implies that the free energy is minimum.

The **adiabatic relation** is local, so it applies separately in both regions,

$$S_i = \mathcal{R}\left(\ln\frac{\rho_i}{(1-b\rho_1)T^n} - n\right), \quad i = 1(g), 2(\ell).$$

Consequently,

$$S_1 - S_2 = \mathcal{R}\ln\frac{\rho_1}{\rho_2}\frac{1-b\rho_2}{1-b\rho_1},$$

which determines the evaporation energy ϵ. See Saha's formula in Section 5.8.

When applied to the evaporation of water at 373 K, this gives a value for the heat of evaporation that is comparable to the actual value of 2260 J/g.[48]

[48] I used $\mathcal{R}_{\rm air} = 0.46$ J/g and $V_{\rm liq}/V_{\rm gas} = 1700$.

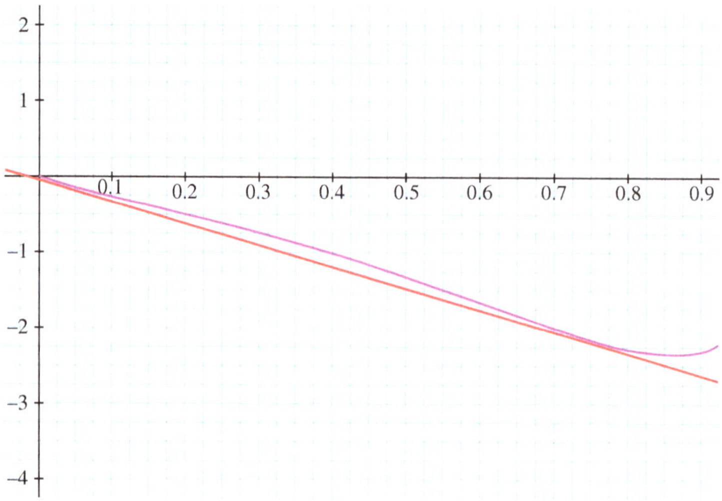

Fig. 4.4.2. The common tangent. This is a graph showing the free energy density against the density. The common tangent identifies the two coexistent densities. They become minima if a suitable, immaterial linear term is added to the definition of f.

Finally, since both $p = \rho\partial f/\partial\rho - f$ and $\partial f/\partial\rho$ are continuous, it follows that

$$f(\rho_\ell) - f(\rho_{\rm g}) = (\rho_\ell - \rho_{\rm g}) \left.\frac{\partial f}{\partial \rho}\right|_{\rho=\rho_{\rm g}}, \tag{4.4.5}$$

so that the two points on the curve $f(\rho)$ have a **common tangent**, as illustrated in Fig. 4.4.2.

Conversely, when a straight line is drawn so as to touch the f locus at two points, then the pressure has the same value at both points. This could have been predicted on the basis of the fact that the function $f+p$ has zeros at both points.

The function $f(T,\rho)$ thus plays exactly the role that Rowlinson attributes to the free energy while at the same time giving it a curious twist since he says that there is no analytic expression for the free energy. Indeed, the free energy is the function,

$$\int_{\rm liq} d^3x\ f(T,\rho_1) + \int_{\rm gas} d^3x\ f(T,\rho_{\rm g}),$$

of the three variables T, ρ_1, ρ_2. It depends on the function f but also on the amount of vapor that is present. The analytic *density* $f(T,\rho)$ is nevertheless the key.

The location of the line of condensation, connecting the points A and B in Fig. 4.4.1, cannot be deduced from the expression for the pressure, nor from the variation of the Gibbs function. The principle of maximum entropy or minimum (at least stationary) energy has to be invoked. This conclusion is not novel. What may be new is the perspective that allows to transfer the lesson to other contexts.

Normalizing the free energy density

We return to the concept of minimum energy. For most purposes, the free energy is defined up to a term linear in the density. The action contains only

$$f + \rho ST = f - \rho\delta ST + \rho(S + \delta S)T =: \hat{f} + \hat{s}T.$$

The value of δS is immaterial. But when we discuss the transformation from one phase to the other it is convenient to arrange it so that no change of free energy occurs. For this reason, we find it convenient to normalize the free energy density in such a way that the common tangent is horizontal. To accomplish this, given any expression for f, we must calculate the equilibrium value of μ and use it in defining the **`normalized free energy`** density and the normalized specific entropy by

$$\hat{f} = f + \rho(S - \mu)T, \quad \hat{S} = S + \mu. \tag{4.4.6}$$

In Section 4.5, we show how this can be helpful.

This has the effect of re-normalizing the entropy as well. As we said, during condensation and during evaporation at constant temperature the normalized free energy remains constant and only the entropy is affected.

Finally, to avoid any misunderstanding, it should be pointed out that there is absolutely no material difference between our treatment of saturation and that of Callen, for our use of the action principle reduces to the Euler–Lagrange equations and they are the fundamental relations of thermodynamics used by Callen. Our sole advantage, as we see it, is that we never need to have any doubt about what equations to use: the Euler–Lagrange equations and nothing else.

Calculating the `dew point` for Argon

At the point where the liquid is in equilibrium (coexistence) with the gas, the pressure depends only on the temperature; this pressure is the *vapor pressure*. To find the **`vapor pressure`** of Argon as a function of the temperature, we must determine the two densities at which both the pressure,

$$p = \frac{\mathcal{R}T\rho}{1 - b\rho} - a\rho^2,$$

and the chemical potential,

$$\mu = \frac{\partial f}{\partial \rho} = \mathcal{R}T\left(\ln \frac{\rho}{(1 - b\rho)T^n} + \frac{1}{1 - b\rho}\right) - 2a\rho,$$

have the same values. Data for Argon are as follows:

$$m = 40, n = 3/2, \; T_c = 150K, \; p_c = 4.9 \text{ MPa}.$$

Setting $a = 1.355\,\mathrm{L}^2\,\mathrm{bar/mol}^2$, $b = 0.03201\,\mathrm{L/mol}$ in the van der Waals' formula gives $T_c = 152.6$, $p_c = 4.96\,\mathrm{MPa} = 0.003912\;\mathcal{R}T_c$ and $\rho_c = 0.0104$.

The simplest way to determine the dew point, (i.e., the saturation pressure for a chosen value of T) is to plot, in the x, y plane, the loci $p(x) = p(y)$ and $\mu(x) = \mu(y)$, namely, with the densities x and y in kg/cm^3,

$$\ln \frac{x}{1 - 32x} + \frac{1}{1 - 32x} - \frac{16478}{T}(2x) = \ln \frac{y}{1 - 32y} + \frac{1}{1 - 32y} - \frac{16478}{T}(2y),$$

$$\frac{x}{1 - 32x} - \frac{16478}{T}x^2 = \frac{y}{1 - 32y} - \frac{16478}{T}y^2.$$

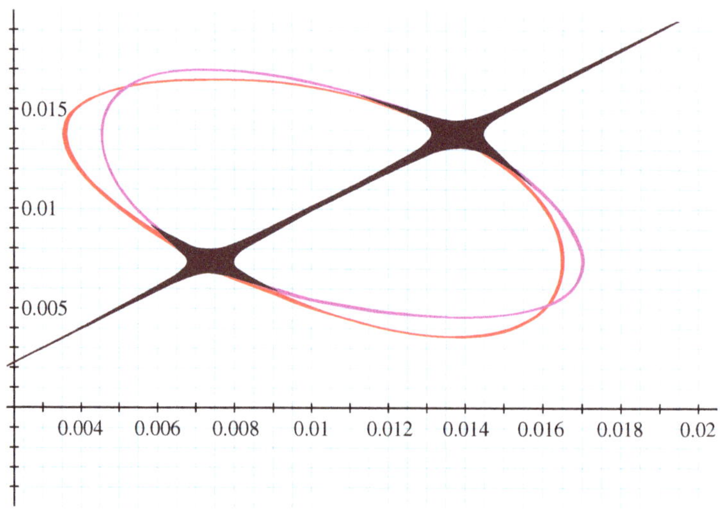

Fig. 4.4.3. *Argon*. The two curves are loci of points (x, y) where $p(x) = p(y)$ and $\mu(x) = \mu(y)$. The values of x and y at the intersection of the two curves are the liquid and gas densities at saturation, in grams per liter. Here $T = 140\,\mathrm{K}$.

The dew point is found by inspection of Fig. 4.4.3. This figure shows the loci of points (x, y) where $p(x) = p(y)$ and $\mu(x) = \mu(y)$ for the case that $T = 140$ (see Fig. 4.4.4). They intersect at the points where x, y are the densities of the dew point, in grams per liter,

$$\rho_\ell = 0.01662 \quad \text{and} \quad \rho_g = 0.00492.$$

Another way to find the dew point is to plot the free energy and find the two points with a common tangent (see Fig. 4.4.5). To read the figure, we normalize f and S, then expand the vertical coordinate, to convert Fig. 4.4.5 to Fig. 4.4.6. The basic formulas can be written

$$\frac{b}{RT}p = \frac{x}{1-x} - \frac{27}{8}\frac{T_{cr}}{T}x^2,$$
$$\frac{b}{RT}f = \frac{x}{1-x} - \frac{27}{8}\frac{T_{cr}}{T}x^2 + \Lambda(x).$$

Here $x = b\rho$ and $\Lambda(x)$ is a first order polynomial that is chosen so that the double tangent of F is horizontal.[49]

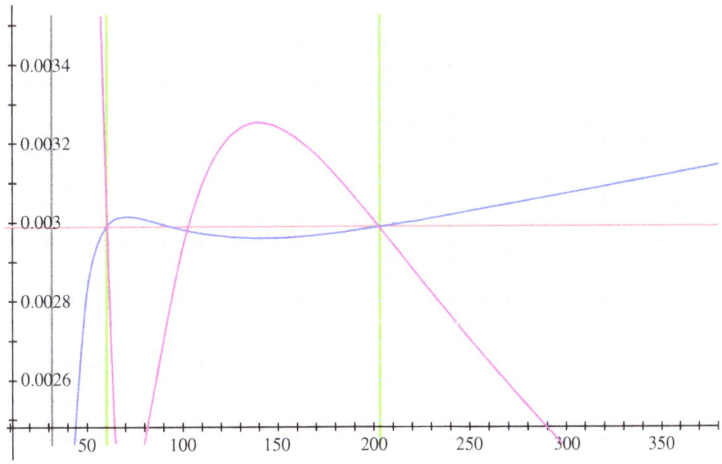

Fig. 4.4.4. The same information is shown. The abscissa is the density. The two curves are the pressure (purple) and the chemical potential (blue), the latter scaled to pass through the dew point densities. The horizontal line is determined by Maxwell's rule.

[49]In the illustration $\Lambda(x) = 0.150722x + 0.0767582$.

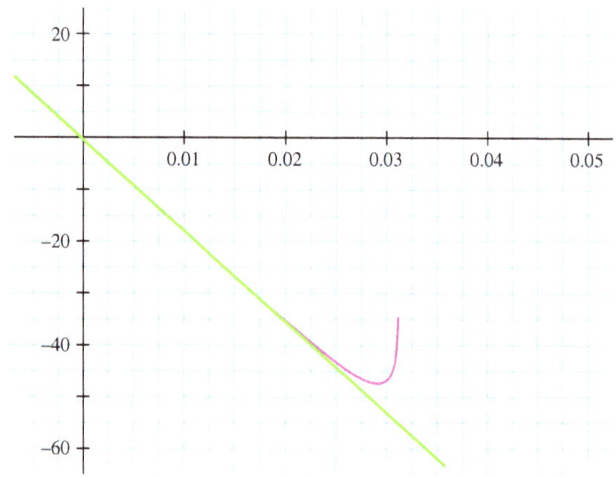

Fig. 4.4.5. *Argon.* The common tangent. A plot of the free energy density of Argon at $T = 140\,K$ against the density.

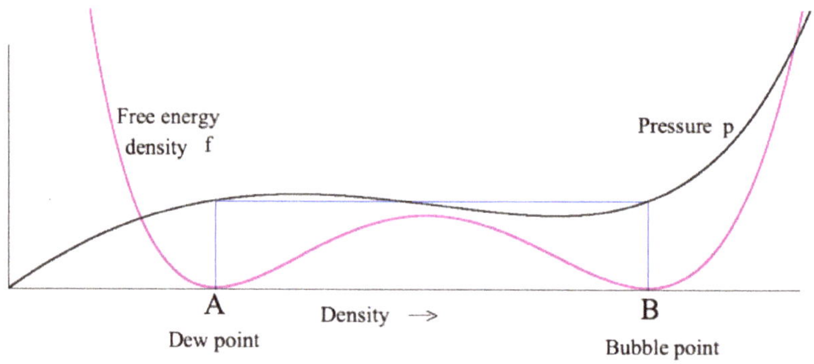

Fig. 4.4.6. To get an accurate reading of the points of contact add a linear function to make the tangent horizontal.

4.5. The Critical Point

The nature of the **critical point** is of great interest. It has been the object of very extensive investigations, most notably with the help of the renormalization group.

The critical point, a fixed point where ρ_1, ρ_2, T and p have well-defined values, is the state with the highest value of T at which coexistence occurs.

The van der Waals equation fails to give the correct relation between temperature and pressure in the saturation region. Nevertheless, we begin by studying this equation in the neighborhood of the critical point. We introduce "reduced" variables as follows:

$$\mathcal{R}T = \mathcal{R}T_c + t, \quad p = p_c + dp, \quad \rho = \rho_c + d\rho,$$

where the critical values are

$$\mathcal{R}T_c = \frac{8}{27}\frac{a}{b}, \quad p_c = \frac{1}{27}\frac{a}{b^2}, \quad \rho_c = \frac{1}{3b}. \tag{4.5.1}$$

In terms of these variables, the van der Waals equation takes the following form (exactly, the reduced variables are not differentials but finite displacements from the critical point)

$$\frac{1}{3b}(t - 2bdp) + (t + bdp)d\rho + ab\, d\rho^3 = 0. \tag{4.5.2}$$

For fixed temperature and pressure, this equation for $d\rho$ has either one or three real roots. In the region of one real root, this root is the density, and the relation is a true equation of state. In the domain of three real roots, the temperature and pressure are related and the (average) density is undetermined in the interval between the smallest and the largest roots.

Expanding the equation $q_A = q_B$, namely,

$$\mathcal{R}T\left(\ln\frac{\rho_1}{1-b\rho_1} + \frac{1}{1-b\rho_1}\right) - 2a\rho_1 = \mathcal{R}T\left(\ln\frac{\rho_2}{1-b\rho_2} + \frac{1}{1-b\rho_2}\right) - 2a\rho_2$$

in the neighborhood of the critical point, we find, to lowest order in $d\rho$, the simple statement that $d\rho_1^2 = d\rho_2^2$. If the three real solutions of (5.5.2) are $d\rho_1 < d\rho_0 < d\rho_2$, then we have $d\rho_1 + d\rho_2 = 0$. In addition,

$$d\rho_1 + d\rho_2 + d\rho_0 = 0,$$

$$d\rho_1 d\rho_2 d\rho_0 = \frac{-1}{3ab^2}(t - 2bdp), \quad (d\rho_1 + d\rho_2)d\rho_0 + d\rho_1 d\rho_2 = \frac{t + bdp}{ab}.$$

This implies that, to lowest order in $d\rho$, $t - 2bdp = 0$, so that only one direction in the t, dp plane leads from the critical point into the saturation region (Fig. 4.5.1).

We obtain direct confirmation of this conclusion by writing down the condition for the existence of three real roots of Eq. (4.3.2),

$$\frac{(t + bdp)^3}{3a} + \frac{(t - 2bdp)^2}{4b} < 0. \tag{4.5.3}$$

This defines a curve with two branches that meet at the origin with a common tangent, the line $t = -2bdp$.

In the saturation region, there is a fixed relation between temperature and pressure, so that there is only one path along which the critical point can

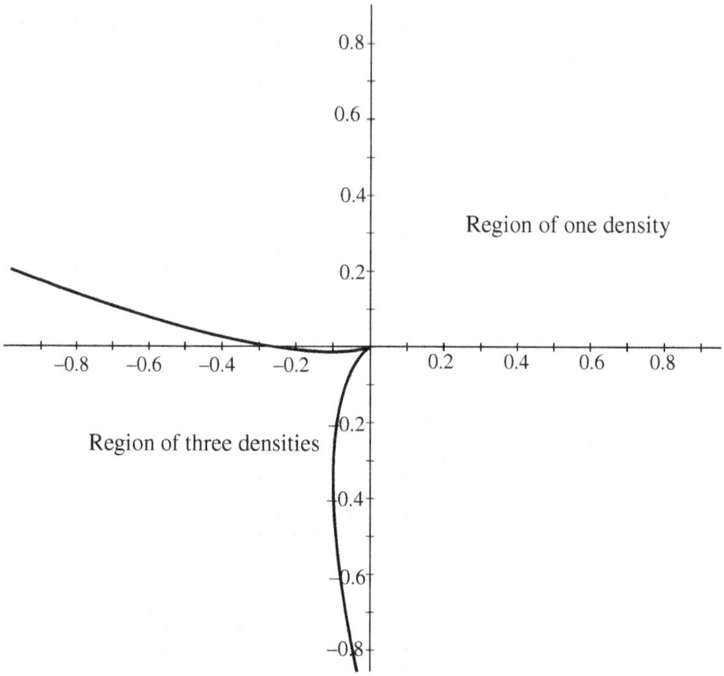

Fig. 4.5.1. The two curves mark the border of the region (above and right) of the pure phase in the reduced variables t, dp.

be approached. On this path, the average density is related to the entropy and varies between ρ_1 and ρ_3. Any path that leaves the critical point in another direction begins in the region of a single real root, in the region of a pure phase. There is no condensation in a (sufficiently small) neighborhood of the critical point. The linearized theory and the description of small perturbations around the critical point are therefore unexpectedly simple and regular, to be regarded as problems of a single phase governed by the Lagrangian equation (4.3.1).

4.6. Sound Propagation at the Critical Point

For a first application, let us calculate the speed of propagation of sound in Helium according to van der Waals. Figure 4.6.1 shows the speed of sound as measured.

Figure 4.6.2 shows values obtained from van der Waals' formula with critical parameters

$$\text{Helium}: \ a = 0.0346\,\mathrm{L}^2\mathrm{bar/mol}^2, \quad b = 017.4\,\mathrm{L/mol}.$$

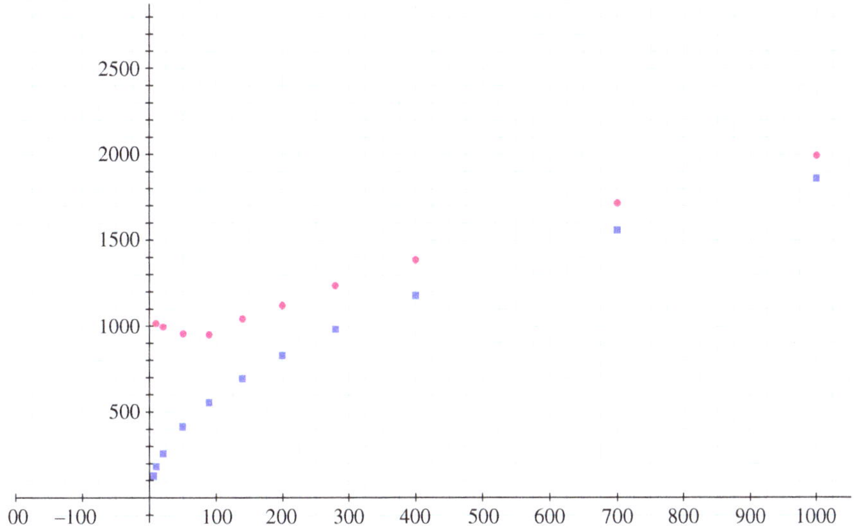

Fig. 4.6.1. Helium sound against temperature for fixed pressure, $p = 0.01\,\text{atm}$ (red) and $p = 50\,\text{atm}$ (blue), against temperature, from $T = 5.19$ to $T = 1000$.

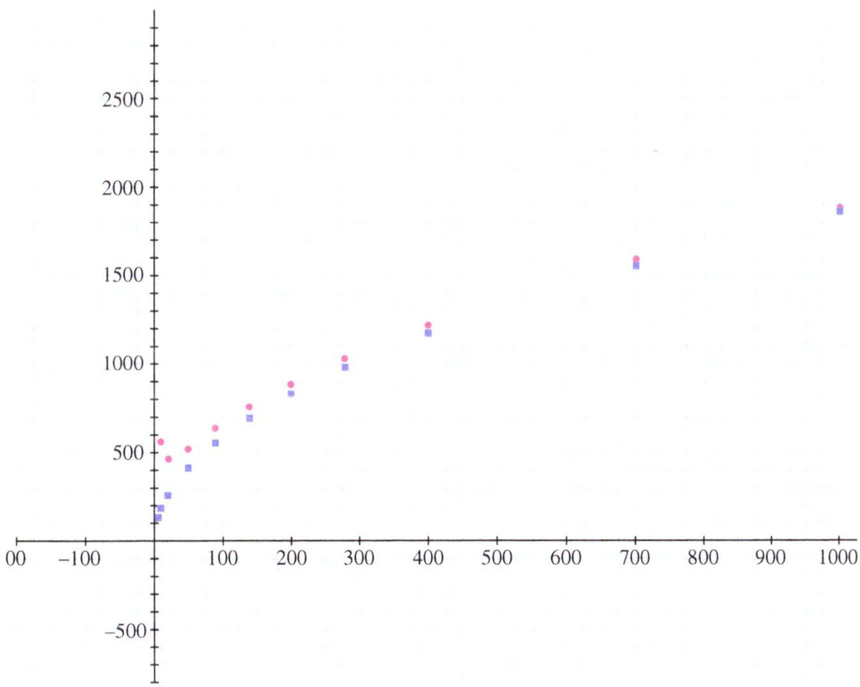

Fig. 4.6.2. Helium sound speed as calculated using Eq. (4.6.1).

The calculation is straightforward (see below), using the equation of continuity and the integrated form of the Bernoulli equation, as in the case of an ideal gas; the square of the speed is the adiabatic derivative of the pressure with respect to the density, where "adiabatic" now means that the temperature is evaluated according to

$$\frac{\partial(f+\rho ST)}{\partial T}\bigg|_{\rho,S}=0,$$

i.e., $(V-b)T^n=$ constant. This use of the word adiabatic is standard and explains why the last equation is referred to as the adiabatic relation.

The equations that govern a first-order, adiabatic perturbation of a pure phase in equilibrium, in which the perturbed fields depend on only one spatial coordinate, are the linearized equations of motion,

$$\frac{d\ddot{\rho}}{\rho}=-\dot{v}'=\frac{\partial^2\rho}{\partial x^2}\frac{\partial}{\partial\rho}(f+sT)\bigg|_S$$

(see Section 4.1). The derivative is one of the definitions of the chemical potential. As always, since the specific entropy is uniform, this leads to

$$c^2=\frac{\partial p}{\partial\rho}\bigg|_S.$$

In the present case,

$$T=\left(\frac{\rho}{(1-b\rho)k}\right)^{1/n},\quad p=\frac{\mathcal{R}}{k^{1/n}}\left(\frac{\rho}{(1-b\rho)}\right)^{1+1/n}-a\rho^2$$

and

$$c^2=\frac{\gamma\mathcal{R}T}{(1-b\rho)^2}-2a\rho. \tag{4.6.1}$$

At the critical point, this reduces to

$$c^2=\frac{0.8314\times10^8\times5.19}{4}\frac{5/3}{(1-17.4\times0.0697)^2}-2\times0.0697^2\times2162.5\times10^6.$$

The result is $c=134.5\,\mathrm{m/s}$. The numerical value applies to helium, with $n=3/2$; it is a small deviation from Laplace's formula $\sqrt{\gamma\mathcal{R}T}=134.1\,\mathrm{m/s}$.

Table 4.6.1 gives a range of values. Comparison with Table 4.6.2 shows little deviation from experiments, or from measured values, at low pressures, but at the highest pressure the fit to experiment is very bad, showing no improvement with respect to the ideal gas model. A better fit is achieved with the general cubic model discussed at the end of the chapter.

Table 4.6.1. Liquid–vapor critical temperature and pressure.

Substance	Critical temperature	Critical pressure
Argon	−122.4°C (150.8 K)	48.1 atm (4870 kPa)
Ammonia	132.4°C (405.6 K)	111.3 atm (11,280 kPa)
Bromine	310.8°C (584.0 K)	23,102 atm (10,300 kPa)
Cesium	1664.85°C	94 atm (9500 kPa)
Chlorine	143.8°C (417.0 K)	76.0 atm (7700 kPa)
Ethanol	241°C (514 K)	62.18 atm (6300 kPa)
Fluorine	−128.85°C (144.30 K)	51.5 atm (5220 kPa)
Helium	−267.96°C (5.19 K)	2.24 atm (227 kPa)
Hydrogen	−239.95°C (33.20 K)	12.8 atm (1300 kPa)
Krypton	−63.8°C (209.4 K)	54.3 atm (5500 kPa)
CH_4 (Methane)	−82.3°C (190.9 K)	45.79 atm (4640 kPa)
Neon	−228.75°C (44.40 K)	27.2 atm (2760 kPa)
Nitrogen	146.9°C (126.3 K)	33.5 atm (3390 kPa)
Oxygen	−118.6°C (154.6 K)	49.8 atm (5050 kPa)
CO_2	31.04°C (304.19 K)	72.8 atm (7380 kPa)
N_2O	36.4°C (309.6 K)	71.5 atm (7240 kPa)
H_2SO_4	654°C (927 K)	45.4 atm (4600 kPa)
Xenon	16.6°C (289.8 K)	57.6 atm (5840 kPa)
Lithium	2950°C (3220 K)	652 atm (66,100 kPa)
Mercury	1476.9°C (1750.1 K)	1720 atm (174,000 kPa)
Sulfur	1040.85°C (1314.00 K)	207 atm (21,000 kPa)
Iron	8227°C (8500 K)	
Gold	6977°C (7250 K)	5000 atm (510,000 kPa)
Aluminum	7577°C (7850 K)	
Water	373.946°C (647.096 K)	217.7 atm (22.06 MPa)

Source: Wikipedia.org.wiki/Critical point.

Table 4.6.2. Some van der Waals **parameters**.

	a (L^2atm/mol^2)	b (L/mol)
Acetic acid	17.71	0.1065
Anhydride	20.158	0.1263
Acetone	16.02	0.1124
Acetonitrile	17.81	0.1168
Acetylene	4.516	0.0522
Ammonia (NH_3)	4.225	0.0371
Argon	1.355	0.0320
Benzene	18.24 (0.82)	0.1154 (0.1193)
Bromobenzene	28.94	0.1539
Butane (C_4H_{10})	14.66 (13.89)	0.1226 (0.1164)
Carbon dioxide	3.640 (0.658)	0.04267 (0.0429)
Carbon disulfide	11.77 (0.25)	0.07685 (0.0726)

(*Continued*)

Table 4.6.2. (*Continued*)

	a (L^2 atm/mol^2)	b (L/mol)
Carbon monoxide	1.505 (0.472)	0.03985 (0.0395)
Carbon tetrachloride	19.7483	0.1281
Chlorine (Cl_2)	6.579 (0.343)	0.05622 (0.0542)
Chlorobenzene	25.77	0.1453
Chloroethane	11.05 (0.66)	0.08651 (0.0903)
Chloromethane	7.570	0.06483
Cyanogen (C_2N_2)	7.769	0.06901
Cyclohexane	23.11 (21.92)	0.1424 (0.1411)
Diethyl ether	17.61	0.1344
Diethyl sulfide	19.00	0.1214
Dimethyl ether	8.180 (0.690)	0.07246 (0.0774)
Dimethyl sulfide	13.04	0.09213
Ethane (C_2H_6)	5.562 (0.580)	0.0638 (0.0651)
Ethanethiol	11.39	0.08098
Ethanol (C_2H_5OH)	12.18 (0.56)	0.08407
Ethyl acetate	20.72	0.1412
Ethylamine	10.74	0.08409
Fluorobenzene	20.19	0.1286
Fluoromethane	4.692	0.05264
Freon	10.78	0.0998
Germanium tetrachloride	22.90	0.1485
Helium	0.03457	0.0237
Hexane (C_6H_{14})	24.71	0.1735
Hydrogen (H_2)	0.2476 (0.2452)	0.02661 (0.265)
Hydrogen bromide	4.510	0.04431
Hydrogen chloride	3.716	0.04081
Hydrogen selenide	5.338	0.04637
Hydrogen sulfide	4.490	0.04287
Iodobenzene	33.52	0.1656
Krypton	2.349 (5.193)	0.03978 (0.0106)
Mercury (Hg)	8.200	0.01696
Methane (CH_4)	2.283 (0.303)	0.04278 (0.0431)
Methanol	9.649 (0.476)	0.06702 (0.0659)
Neon	0.2135 (0.208)	0.01709 (0.0167)
Nitric oxide	1.358	0.02789 (0.0289)
Nitrogen (Ni_2)	1.408 (0.370)	0.03913 (0.0387)
Nitrogen dioxide	5.354	0.04424
Nitrous oxide	3.832	0.04415
Oxygen	1.378	0.03183
Pentane	19.26 (0.09)	0.146 (0.1449)
Phosphine (PH_3)	4.692	0.05156
Propane (C_3H_8)	8.779 (9.39)	0.08445 (0.0905)

(*Continued*)

Table 4.6.2. (*Continued*)

	a (L^2 atm/mol^2)	b (L/mol)
Silane (SiH_4)	4.377	0.05786
Silicon tetrafluoride	4.251	0.05571
Sulfur dioxide	6.803	0.05636
Tin tetrachloride	27.27	0.1642
Toluene (C_7H_8)	7 24.38 (0.86)	0.1463 (0.1497)
Water	5.536	0.03049
Xenon	4.250 (0.192)	0.05105 (0.0516)

Source: Wikipedia.org.wiki/Van der Waals.

4.7. Refinements of the Expression for the Internal Energy

The heat capacity

We consider the region of one phase, at or above the critical point. We shall attempt to account for the anomalies in this region by variation of the heat of evaporation, more precisely the parameter ϵ, with temperature. The potential, with this term now included, is

$$V_1(T,\rho) = \mathcal{R}\rho T \ln\frac{\rho}{(1-b\rho)T^n k_0} - \kappa(\rho) - \rho\epsilon_1(T).$$

The pressure is not affected. We shall assume that the function ϵ takes the form

$$\epsilon(T) = \alpha(T-T_c)^{1-\beta}.$$

We expect that α is positive and that β is small, positive, to make the evaporation energy zero at the critical point. Then

$$C_V = n\mathcal{R} + \alpha\beta(1-\beta)t^{-\beta}, \quad t := T - T_c$$

is infinite at the critical point. Experiments with helium indicate that $\alpha < 0$ and that $\beta \approx 0.04$.

The speed of sound

With $\tilde{\psi} = \epsilon$ there is a correction to Eq. (4.3.3),

$$\frac{d\ddot{\rho}}{\rho} = \frac{T''}{\rho}\frac{\partial p}{\partial T} + \frac{\rho''}{\rho}\frac{\partial p}{\partial \rho} + \epsilon_1''(T).$$

This is supplemented by the on-shell condition

$$\ln\frac{\rho}{(1-b\rho)T^n k_0} - n = \epsilon_1'/\mathcal{R}, \qquad (4.7.1)$$

which implies that, to first order,

$$(n+\epsilon_1''T/\mathcal{R})\frac{dT}{T} = \frac{1}{1-b\rho}\frac{d\rho}{\rho}.$$

Hence

$$\frac{d\ddot{\rho}}{\rho} = \mathcal{R}T''\frac{1}{1-b\rho} + \frac{\rho''}{\rho}\frac{\partial p}{\partial \rho} + \cdots,$$

which gives for the speed c of sound propagation

$$c^2 = \frac{\mathcal{R}T}{n(1-b\rho)^2} + \frac{\partial p}{\partial \rho} + \cdots. \tag{4.7.2}$$

What follows is the general third-order equation of state. Evaluating the second term in (4.7.2), we find

$$c^2 = \frac{\gamma\mathcal{R}T}{(1-b\rho)^2} - \frac{a}{\rho^2}\frac{2(\mathcal{V}-b)+x}{((\mathcal{V}-b)^2+x(\mathcal{V}-b)+y)^2}.$$

Experiment shows a pronounced dip in the speed of sound as a function of density or temperature, in a small region above the critical point. To first order in the perturbation, it is equal to

$$\frac{\rho''}{\rho}\left(\frac{\mathcal{R}T}{(1-b\rho)^2} - 2a\right) + T''\frac{\mathcal{R}}{1-b\rho}.$$

The last term was simplified with the help of the modified on-shell condition, namely

$$\mathcal{R}\left(\ln\frac{\rho}{(1-b\rho)T^n k_0} - n\right) + \epsilon' = 0.$$

This also gives us

$$dT = \frac{1}{1-T\epsilon''/n\mathcal{R}}\frac{T}{n}\frac{1}{1-b\rho}\frac{d\rho}{\rho}.$$

Thus,

$$c^2 = \frac{d\rho''/\rho}{dT''/T} = \mathcal{R}dT''\left(n + \frac{1}{1-b\rho} + \frac{\epsilon' - T\epsilon''}{\mathcal{R}}\right) + d\rho''\left(\frac{b\rho T}{(1-b\rho)^2} - 2a\right).$$

To first order in ϵ, it gives the squared speed

$$c^2 = \frac{\mathcal{R}T}{(1-b\rho)^2}\left(\gamma + \frac{T/n\mathcal{R}\epsilon''}{1-T/n\mathcal{R}\epsilon''}\right).$$

4.8. The Cubic Equation of State

It is sometimes of interest to reduce the errors inherent in the van der Waals expression for the pressure, and many suggestions have been made. A generalization developed by Patel and Teja takes the form

$$p = \frac{\mathcal{R}T}{1-b\rho} - \frac{a\rho^2}{(1-c\rho)(1-d\rho)}, \tag{4.8.1}$$

with a certain relation between the parameters b, c, d. The associated free energy density is

$$f(\rho, T) = \rho\mathcal{R}T \ln\frac{\rho}{(1-b\rho)T^n} + \frac{a\rho}{c-d}\ln\frac{1-c\rho}{1-d\rho} + \rho\psi(T),$$

with k_0 constant, the function ψ arbitrary so far. There are arguments to show that this function can best be taken to be zero; in that case, the internal energy density is

$$U = n\mathcal{R}T + \frac{a}{c-d}\ln\frac{1-c\rho}{1-d\rho}.$$

The van der Waals gas is the case in which $c = d = 0$.

Provisionally, we shall set $\psi = 0$, deferring refinements until they are needed, as when we take up the study of the neighborhood of the critical point, then

$$f(T, \rho, S) = \mathcal{R}T\rho\ln\frac{\rho}{(1-b\rho)T^n} + \kappa(\rho), \tag{4.8.2}$$

$$\kappa(\rho) := \frac{a\rho}{c-d}\ln\frac{1-c\rho}{1-d\rho}.$$

The adiabatic condition (4.2.4) (from variation of T) is

$$\ln\frac{\rho}{(1-b\rho)T^n k_0} = n, \tag{4.8.3}$$

with k_0 constant, and the on-shell internal energy density at equilibrium is

$$u = n\mathcal{R}\rho T\rho + \rho\kappa(\rho) + \text{constant } \rho.$$

We use the expression (4.6.1) for the pressure,

$$p = \frac{\mathcal{R}T}{(\mathcal{V}-b)} - \frac{a}{(\mathcal{V}-c)(\mathcal{V}-d)}, \tag{4.8.4}$$

but specialize to the case considered by Patel and Teja. We introduce $\mathcal{V}' = \mathcal{V} - b$, and define x, y by writing

$$p = \frac{\mathcal{R}T}{\mathcal{V}'} - \frac{a}{\mathcal{V}'^2 + x\mathcal{V}' + y}. \tag{4.8.5}$$

The special case considered by Patel and Teja is the case that $y = 2b^2$.

An extremum of this function of $\mathcal{V}'$ occurs at

$$\frac{\mathcal{R}T}{a} = \frac{\mathcal{V}'^2(2\mathcal{V}' + x)}{(\mathcal{V}'^2 + x\mathcal{V}' + y)^2},$$

where

$$\frac{\mathcal{R}T}{p} = \frac{\mathcal{V}'^2(2\mathcal{V}' + x)}{\mathcal{V}'^2 - y}. \tag{4.8.6}$$

At the critical point

$$\mathcal{V}'^3 - 3y\mathcal{V}' - xy = 0, \quad x = y^{-1}\mathcal{V}'(\mathcal{V}'^2 - 3y).$$

Finally, at the critical point, the value of b is given by the experimental values of the compressibility $Z_{\rm cr} = \mathcal{R}T/p$ and the density $\rho_{\rm cr}$ by

$$\left.\frac{\mathcal{R}T}{p\mathcal{V}}\right|_{\rm cr} = \frac{(\mathcal{V} - b)^3}{y\mathcal{V}}.$$

For He_4, the experimental values are

$$T_{\rm cr} = 5.19, \quad \rho_{\rm cr} = 0.0697, \quad p_{\rm cr} = 0.227 \times 10^7, \quad Z_{\rm cr} = 0.301, \quad \mathcal{V}_{\rm cr} = 14.35,$$

and the parameters are

$$a = 2.378 \times 10^9, \quad b = 3.61, \quad x = 15.31, \quad y = 26.06.$$

These parameters were used to calculate the speeds given in lines labeled "cubic" in Table 4.8.1.

Table 4.8.1. Helium sound speed.

$p = 0.01$			$p = 1$			$p = 50$		
T	ρ	v	T	ρ	v	T	ρ	v
5.19	0.0001	134.0	5.19	0.001	133.3	5.19	0.1417	780.0
10	0.5E–4	186.1	10	0.5E–3	186.1	10	0.118	569.5
20	0.23E–4	263.2	20	0.24E–3	263.4	20	0.08	462.5
50	0.1E–4	416.2	50	0.1E–3	416.5	50	0.04	524.7
90	0.55E–5	558.4	90	0.55E–4	558.5	90	0.021	642.4
140	0.34E–5	696.4	140	0.34E–4	696.5	140	0.016	764.5
200	24E–5	832.5	200	0.24E–4	832.5	200	0.0113	889.1
280	0.17E–5	984.9	280	0.17E–4	985.0	280	0.0082	1033
400	0.12E–5	1177	400	0.12E–4	1177	400	0.0058	1217
700	0.7E–6	1557	700	0.7E–5	1557	700	0.0034	1589
1000	0.5E–6	1861	1000	0.5E–5	1884	1000	0.00237	1887

Source: Reprinted with permission from: Sychev, V.V. and Selover, T.B. (1987) *Thermodynamic Properties of Helium*, Hemisphere Publishing Corporation.

4.9. Stability

This is a concept that will become very difficult later, as we come to deal with more complicated problems. We have already studied the van der Waals gas; it will provide the first illustration.

Suppose that the velocity is zero, then the Bernoulli equation show that

$$\rho\dot{\vec{v}} = -\vec{\nabla}p.$$

In simple systems, the temperature can be eliminated with the help of the adiabatic condition; then the pressure is a function of the density and the specific entropy, supposed constant, hence

$$\rho\,\dot{\vec{v}} + \left.\frac{\partial p}{\partial \rho}\right|_S \vec{\nabla}\rho = 0.$$

If the velocity is to the right, say, and the acceleration is in the same direction, then the space between the particles will increase and the density must decrease; this implies that, for stability, $dp/d\rho$ is positive. In more detail, suppose the acceleration is in the positive x direction and imagine a virtual change that makes the acceleration increase. Then by the equation of continuity, the gradient of the density must decrease. So again, the coefficient must be positive.

An equivalent argument is to calculate the speed of propagation of sound. We found that its square is the adiabatic derivative of the pressure with respect to the density and learned that if this is positive, we will have wave propagation, while if it is negative, there will be decay.

In the case of the van der Waals fluid, this condition is violated in the rising part of the curve that shows pressure against volume. Where $dp/d(1/\rho)$ is positive, the configuration is unstable. If we approach this region along an isotherm, from right to left, then after passing the point of coexistence we enter a metastable region. In the absence of a disturbance, as the pressure is increased slowly, condensation can be avoided, but a small disturbance may initiate a catastrophic condensation. If the top of the curve is reached, condensation becomes inevitable. Similarly, the region below the bubble point may be entered without actually creating evaporation, even so far as to make the pressure negative, but upon a catalytic perturbation evaporation will be rapid.

Chapter 5

A Theory of Fluid Mixtures

The traditional treatment of **mixtures** often relies on rules concerning the additivity of properties; one of the most important ones is the additivity of the energies of the components of a composite system. Most fundamental of them all is the postulate of additivity of Lagrangians:

$$\begin{gathered}\mathcal{L} = \sum_{i=1,2} \mathcal{L}_i = \sum_i \rho_i(\dot{\Phi}_i - \vec{v}_i{}^2/2) - f - sT, \\ f = \sum_i f_i, \quad s = \sum_i s_i;\end{gathered} \tag{A}$$

it incorporates the additivity of energies and the additivity of entropies. The algebraic sum of the Lagrangians of two adiabatic systems describes an adiabatic system with two non-**interacting components**.

Realistically, coexistent systems are never non-interacting and ways have to be found to take the interaction into account. The strategy will be *to begin* by joining the Lagrangians of components with the expectation that corrective terms, "interactions", will have to be added. The systematic application of this point of view is not widespread in thermodynamics. (See, for example, many studies of critical phenomena in mixtures.) But it is a standard, and successful, strategy in more developed branches of theoretical physics — and it incorporates a general concept that helps us understand mixtures:

> **In the limit when interactions are weak, each constituent behaves as if the others were absent.**

An overstatement; it overlooks the fact that the two temperatures are replaced by just one and the important role that is consequently played by the adiabatic condition, about which much more below.[50]

Recall that in unary thermodynamics, it is axiomatic that the free energy (or any other fundamental potential expressed in terms of the natural variables) tells us all that we need to know about a substance. It follows that a mixture in which each constituent behaves as if it were alone is described by variables that satisfy the usual equations of motion of a unary system. The natural variables for the mixture are thus the collection of natural variables for the ingredients. And if this mixture is characterized by an action principle, then the action must be a sum of individual actions in the limit of no interactions.

Although the attitude that has just been expressed is quite traditional, it is decidedly less so when seen in the context of the formula

$$s = \rho_1 S_1 + \rho_2 S_2, \tag{B}$$

for it implies the relation

$$\rho_1 s_1 + \rho_2 S_2 = -\frac{\partial f}{\partial T},$$

which says more than $s = -\partial f/\partial T$.

The principles expressed in Eqs. (A) and (B) are characteristic of the Lagrangian approach; much more consequential than the idea that each gas behaves as if it were alone.

The persistent use of a **Lagrangian** does much to distinguish our approach to thermodynamics, the choice of variables is just as important. I confess to being perplexed by some common usage, especially the occasional suppression of degrees of freedom and the preference for molar fractions over partial densities. Here is the addition formula for free energy of a mixture of ideal gases,

$$F = \frac{f_1 + f_2}{\rho} = N_1 F_1 + N_2 F_2 + \mathcal{R}T(N_1 \ln N_1 + N_2 \ln N_2).$$

In terms of ordinary densities, this statement takes the form $f = f_1 + f_2$.

Until further notice, it will be assumed that no chemical reactions are taking place in the mixtures under consideration, and that the equilibria are uniform. In this case, an important idea is the Gibbs–Dalton hypothesis developed in this chapter and used in Chapter 6.

---

[50]Cf. the Gibbs–Dalton hypothesis in Section 5.4.

Chemical interactions change the picture materially; they will occupy us in Sections 5.8 and 5.9. Immiscible fluids form another category of mixtures, the subject of Chapter 6.

5.1. Variables and Potentials

A description of two non-interacting systems requires two sets of variables. In particular, each system has a temperature or a temperature field. This is natural if the two systems occupy different, non-overlapping regions of space. Situations exist where two systems occupy the same space and for which it has nevertheless been found useful to maintain a notion of separate temperatures, for example, in plasma physics and in Cosmology. But in this book, the temperature shall always be taken to be a unique field extended over the total vessel. Two components may be present in the same region but we shall try to get by with only one temperature field. If this strategy fails, then it shall be abandoned, reluctantly. That has not come to pass so far.

Taken for granted is the idea that each distinct component has an independent flow vector field. Otherwise, there would be no way to formulate the separate **conservation of masses** and separate equations of continuity, for each component (and no way to visualize the "mixing"). More precisely, we hold that, *if two mass densities are separately conserved*, then one needs two independent velocity fields $\vec{v}_1$ and $\vec{v}_2$ in order to be able to assert that $\dot{\rho}_1 + \text{div}(\rho_1\vec{v}_1) = 0$ and $\dot{\rho}_2 + \text{div}(\rho_2\vec{v}_2) = 0$.[51] But if only the total mass is conserved, then there can be only one velocity and only one continuity equation. The analysis becomes simpler if one assumes that the mixture behaves as a one-component fluid[52] with properties defined by interpolation between those of the components. This drastic assumption, first used by van der Waals, is tempting in the case of mixtures of many components; it is common, for example, in calculations relating to the interior of the Sun. But it ignores physical degrees of freedom and cannot account for all observed phenomena. Others introduce individual densities and velocities without comment, as in Bois and Kubicki (2000). Let us keep an open mind on this question.

By adopting a two-velocity approach to binary mixtures without chemical reactions, we depart in an essential way from what is common practice.

[51] As in Landau's 2-fluid theory of liquid helium with its two equations of continuity.
[52] "Air is a mixture that behaves, under normal conditions, like an ideal gas". See Müller and Mueller (2009).

The most important 2-fluid theory is Landau's theory of superfluidity. But Landau's equations are not derived from an action principle and consequently, even in that case, we shall propose a new direction (see Chapter 8). For an example of a treatment of mixtures in terms of multiple velocities, see Glavatsky (2009).

In the ideal case, when there is (if there were) no interaction between the components, there can be no objection to two separate and independent velocities; they are unaware of each other and they pass through each other. But this simple scenario is not realistic, for the insistence on a single temperature amounts to an interaction between the two components. There is only one adiabatic condition.

The imperfect transparency of each component with respect to the other will be further affected when we allow for an explicit interaction between the components, the simplest being an addition to the total free energy density,

$$f = f_1 + f_2 + f_{\rm int}, \quad f_{\rm int} = \alpha\rho_1\rho_2, \quad \alpha \text{ constant}.$$

An interaction of this type was used with some success by van der Waals. With α positive, it has the effect of discouraging the overlap of the densities of the two species and the interaction is a repulsion.

From now on, it is taken for granted that the dynamical variables of mixtures at rest are the **individual densities** ρ_i, individual specific entropies S_i (Section 5.2), and T. The variables chosen by Gibbs, and used almost universally ever since, are the total volume or density and the **molar fractions**, instead of the individual densities. This leads to a definition of the Gibbs function as a function of pressure, temperature, and molar fractions, related to the free energy by a Legendre transformation, $G(P,T,x) = F(V,T,x) + VP$. In our approach, using densities, the **Gibbs potential** density is not related to the free energy density by a Legendre transformation. The **Legendre transformations** of densities are natural generalizations of those listed in Section 3.7, including the following:

$$p(T,\mu_1,\mu_2,\ldots) = \sum_i \rho_i\mu_i - f, \tag{5.1.1}$$

with

$$\mu_i = \frac{\partial f}{\partial\rho_i}\Big|_T = \frac{\partial u}{\partial\rho_i}\Big|_s$$

and

$$dp = sdT + \sum_i \rho_i d\mu_i.$$

The natural variables of the internal energy are s and $\rho_1, \rho_2, \ldots$:

$$u(\rho_1, \rho_2, \ldots, s) = f(\rho_1, \rho_2, \ldots, T) + sT,$$
$$du = \sum \mu_i d\rho_i + T ds, \quad df = \sum \mu_i d\rho_i - s dT.$$

All the potentials have $n+1$ variables (n is the number of components). The Clapeyron equation, that is not presented in the literature in differential form, arises from the fixing of the molar fractions. Instead, the correct equation is

$$\left.\frac{\partial p}{\partial T}\right|_{\mu_1, \mu_2, \ldots} = s.$$

The interpretation of the pressure as a thermodynamical density potential was stressed in Section 3.5. The thermodynamic variables are the entropy density s or its conjugate T and each partial density ρ_i with its conjugate μ_i.

For a multicomponent system, with no interaction between the constituents, we shall make the natural assumption that the free energy is

$$f = \sum_i f_i,$$

where $f_1, f_2, \ldots$ are the same expressions for the free energy of the components, in terms of their natural variables, that apply when they are by themselves.

The principal density potentials for multicomponent thermodynamics are

$$p(T, \vec{\mu}) \quad f(T, \vec{\rho}), \quad u(s, \vec{\rho}), \quad h(s, \vec{\mu}).$$

The introduction, by Gibbs, of molar fractions as principal variables, the associated approach to chemical potentials and the complications that this brings to analysis (Gibbs–Duhem relation, a separate treatment of open systems) will be eschewed; the variables are T, the densities, and the chemical potentials defined above.

Molar fractions and "free energy of mixing"

With the introduction of molar fractions and the associated chemical potentials, Gibbs arrives at expressions for the differentials of thermodynamical potentials. When those relations are rewritten in terms of densities, all reference to molar fractions drops out and one ends up with the simpler set of differentials given above. We leave the verification of this to the skeptical reader.

We must not neglect to point out that the above formula for pressure can be expressed in a more familiar fashion as

$$p(T,\rho,N_1,\ldots) = -f + \rho \frac{\partial f}{\partial \rho}\bigg|_{T,N_1,N_2,\ldots}, \tag{5.1.2}$$

where $T, N_1, N_2, \ldots$ are the molar fractions. However, the formula $f = f_1 + f_2$ may seem controversial. The standard approach is to express the free energy for N moles of the liquid as NF, thus F is the free energy of one mole of the pure gas when it is alone. Then the usual formula for the mixture is

$$F = f/\rho = N_1F_1 + N_2F_2 + F_{\text{mix}}.$$

We can compare the two expressions in the case of a mixture of two ideal gases. In terms of molar densities, the free energy of an ideal gas is

$$f(\rho_i, T) = \mathcal{R}\rho_i T \ln \frac{\rho_i}{T^{n_i}}.$$

In the mixture, if the molar fraction is $N_1 = \rho_1/\rho$, one writes

$$f_i(\rho_i, T) = \rho_i \mathcal{R}T \ln \frac{\rho_i}{T^{n_i}} = \rho N_i \mathcal{R}T\left(\ln \frac{\rho}{T^{n_i}} + \ln N_i\right). \tag{5.1.3}$$

We hold that this separation is wholly unwarranted. It results in the unwieldy expression

$$F = \frac{f_1 + f_2}{\rho} = N_1F_1 + N_2F_2 + \mathcal{R}T(N_1 \ln N_1 + N_2 \ln N_2). \tag{5.1.4}$$

The last term is F_{mix}; it is regarded as a "mixing correction" and in the general case it is replaced by a phenomenological correction.[53] When the densities are used as fundamental variables, all the awkward mixing terms disappear.

For non interacting gases, the more natural formula $f = f_1 + f_2$ will be used in this book; there is no mixing term. Interactions will be introduced by adding a term that depends on both sets of variables,

$$f = f_1 + f_2 + f_{\text{int}'}.$$

The same remarks apply to entropy; the concept "**entropy of mixing**" does not appear again in this book. The densities f and s are additive in the absence of interaction.

Remark. The **Gibbs-Duhem** relation is the statement that the function f depends on the $n+1$ variables $\rho_i = \rho N_i$ and T, and not separately on ρ.

[53] See Lemmon and Jacobsen (1999), where we find $F = N_1F_1 + N_2F_2 + F_{\text{mix}} + F_{\text{int}}$.

5.2. Entropy and Equations of Motion

The following expression for the Lagrangian density is appropriate when there are no chemical reactions and all the densities are independent variables. Note that each individual kinetic energy term $-\rho_i\vec{v}_i^2/2$ must be accompanied by the term $\dot{\Phi}_i$ for Galilei invariance (see Section 3.11). Clearly, this implies a set of canonical pairs of basic variables $(\rho_i, \dot{\Phi}_i)$. It was argued, and postulated in the introduction to this chapter, that the action for an ideal composite system — one without interactions between the constituents — must be additive; consequently, before including an interaction term, we have the tentative Lagrangian,

$$\mathcal{L} = \sum_{i=1,2} \mathcal{L}_i = \sum_i \rho_i(\dot{\Phi}_i - \vec{v}_i{}^2/2) - f - sT \tag{5.2.1}$$

with

$$f = \sum_i f_i \quad \text{and} \quad s = \sum s_i.$$

Variation of the temperature gives the adiabatic condition

$$s = -\frac{\partial f(T, \rho_1, \rho_2)}{\partial T} = \frac{\partial f_1(T, \rho_1)}{\partial T} + \frac{\partial f_2(T, \rho_2)}{\partial T}; \tag{5.2.2}$$

in the absence of interactions, it expresses a weak form of the Gibbs–Dalton hypotheses, to be discussed below. This formula is one of the most important consequences of formulating thermodynamics as an **`action principle`**.

The structure of the kinetic term is appropriate in the absence of chemical reactions. It is strongly motivated by the need to preserve the equations of continuity. We are very reluctant to modify the structure of this term; thermodynamic interactions will be taken into account by modifications of the potential. Changes that are appropriate in the case of chemical reactions will be discussed in Section 5.8.

We shall suppose that particles of distinct components are distinguishable; whence the additivity of the entropy. More precisely, for the same reasons that have made us limit attention to the case of uniform specific entropy densities in a one-component system, we shall assume that, interactions present or not, the entropy density of a multicomponent system, the density s that appears, multiplied by T, as the last term in the above Lagrangian, is a linear function of the densities, *with constant coefficients*

$$s = \sum \rho_i S_i. \tag{5.2.3}$$

This assures that the identification of the pressure with the on-shell value of the Lagrangian density extends to mixtures. That S_1 and S_2 are uniform in the absence of interactions have been our working assumptions; no exceptions have been encountered so far. Obedience to this command is presumed to have been achieved during mixing, by a non-adiabatic development, but we are concerned only with the final result, and the adiabatic development that may follow. Equation (5.2.3) remains true when interactions are present is, at the time of writing, a working assumption. It implies the absence of unexpected entropic forces (forces proportional to the gradient of the specific entropies) in the Bernoulli equation, and that all the other fields are also uniform in the absence of external forces.

This is a place to recall what is one of the most important aspects of Gibbs approach, the role of entropy in the formulation of the variation principle. In the case of a simple system, the entropy is fixed. In general, it is agreed that an adiabatic system is one with fixed entropy; we have followed the lead of Prigogine in generalizing that:

> **A unary adiabatic system has a fixed entropy distribution; the entropy density is $s = \rho S$ with a fixed, uniform value of S.**

Here we are going further, to postulate, as a working assumption, that

> **In an adiabatic Lagrangian the entropy density is a fixed linear function of the densities, with constant coefficients.**

In the case of a binary mixture, $s = \rho_1 S_1 + \rho_2 S_2$. Note the inevitable appearance of two **entropy parameters**; the allocation of entropy between the two components is now an issue. The explicit modeling of the entropy density is an important feature of our procedure; it is not standard. It aims at an improved **predictive power** and the possibility of doing without the ubiquitous "**reference state**". A little reflection will show that this approach is new. There may be antecedents, but discussions of the structure of the entropy density are extremely rare. We mean by that the proposal of an expression for the entropy density of mixtures in terms of the mass densities, that goes into the formula for $f + sT$, that is, the formula $s = \rho_1 S_1 + \rho_2 S_2$ and the assumption that S_1, S_2 are uniform. It is not that this assumption is inevitable, but that, if there are exceptions then that would be worthy of notice. It is a working assumption (Fig. 5.2.1).

The appearance of two entropy parameters, adiabatic invariants, in the expression for the total Lagrangian of a two-component system, invites

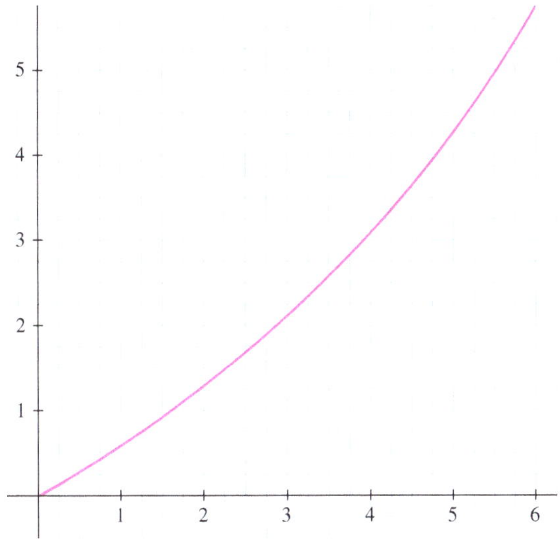

Fig. 5.2.1. **Entropy path.** The coordinates are the specific entropies S_1, S_2. To each adiabatic configuration there corresponds a point in the plane; addition of heat makes it trace out a path.

a re-examination of the concept of two time scales in thermodynamics. Experience shows that the addition of heat to any system eventually leads to an adiabatic system with well-defined entropy; so if the formula $s = \sum_i \rho_i S_i$ is valid, then the ultimate adiabatic system is characterized by a particular partition of the total entropy among the two components; that is, by a one-dimensional subset of the entropy plane, the plane with coordinates S_1, S_2. This subspace defines a path along which the system moves as the entropy is changed by the addition of heat. But what determines the path, and how does the system reach a point on this path?

Suppose the system is initially in a state of absolute equilibrium, reached after being isolated long enough for dissipation to have run its course. Suppose further that a manipulation of the system leads to a Lagrangian in which $s = \sum \rho_i S_i$ with fixed but randomly chosen values of S_1, S_2. Mathematically, it appears to be a well-defined adiabatic system with adiabatic processes that, by definition, do not modify the entropy parameters. The motion that brings the system back to the point of absolute equilibrium is a process of another kind; that is, it is not adiabatic and the most economical standpoint is to characterize it as one of dissipation or degradation driven by a search for maximal entropy. But the choice of the actual path may have a different explanation: It may be driven by the search for maximum entropy, but another factor may be the condition of adiabatic stability. We can offer no general insight into this problem as yet; the best we can do is to examine particular processes and the way that the problem has been approached in the past.

From now on, we shall assume that Eq. (5.2.3) holds and that the coefficients S_i are often uniform. It is proposed as a useful working assumption; to be honored until it is proved wrong.

Variation of the action with respect to the densities gives, when the entropy parameters are uniform, a pair of **Bernoulli equations**

$$\frac{Dv_i}{Dt} = -\mu_i - S_i T, \quad \mu_i := \frac{\partial f}{\partial \rho_i}\bigg|_T, \quad i = 1, 2. \tag{5.2.4}$$

This too is hardly conventional. Landau's two-flow theory recognizes two independent velocities and incorporates similar equations, but instead of individual chemical potentials μ_i on the right-hand side, there appears the gradient of the total pressure p in both equations. That is, the two fluids interact with each other by an unusual mechanism that is not found in other physical systems. Certainly, mutual interactions must be included. But we shall take a suggestion from the field theories of elementary particles, to introduce the interaction in a different way, by means of an interaction term $f_{\rm int}$, a function of the two densities, in the total Lagrangian density. Landau assumes that each fluid feels the pressure of the other, and that is reasonable; but we shall obtain the same effect by including an interaction (attractive or repulsive) in the Lagrangian.

The concept of the entropy of the mixture as a linear function of the densities leads to a new view of the problem of mixtures and a new way to analyze experiments. A principal problem is to determine, for any mixture, the path that it chooses as it winds its way through the entropy plane.

We may be opening up a new direction of development of thermodynamics. It is based on what we believe to be the most important concept introduced by Gibbs: the concept of "**Assigned Entropy**". In global, adiabatic thermodynamics, a numerical value is assigned to the entropy; then the energy functional is minimized. In the local theory of mixtures, the entropy density has to be given as a linear function of the densities, with the values of the coefficients, to define an adiabatic system. Linearity is a working assumption. It is strange that no previous attempt to model the assigned entropy could be found; perhaps, a clear statement of the concept was not widely known.

Partial pressures

The (total) pressure was defined in (5.1.1). As in the case of a unary system, this is justified by showing that it coincides with the pressure on the walls. "Partial pressures" p_i can be defined by

$$p = \sum_i p_i, \quad p_i := \rho_i \frac{\partial f_i}{\partial \rho_i} - f_i,$$

but their significance is unexpectedly complicated.

In Chapter 3, we derived the Bernoulli equation, Eq. (3.6.1), for a unary system. To express the Euler–Lagrange equation, in which a principal term is $\vec{\nabla}(f + sT)$, in terms of the gradient of the pressure, we used the identity

$$\rho\vec{\nabla}\left(\mu + T\frac{\partial s}{\partial \rho}\right) = \vec{\nabla}p + \left(\rho\frac{\partial s}{\partial \rho} + \frac{\partial f}{\partial T}\right)\vec{\nabla}T + \rho T\vec{\nabla}\frac{\partial s}{\partial \rho} = 0.$$

Then we used the adiabatic relation $\partial f/\partial T = -s$ to simplify, to obtain the familiar Bernoulli equation of motion for a unary system in terms of $\vec{\nabla}p$,

$$\dot{\vec{v}} + \vec{\nabla}(\vec{v}^2/2 - \Phi) = \frac{1}{\rho}\vec{\nabla}p.$$

For a binary system, we get the same equation for each component,

$$\rho_i\vec{\nabla}\left(\mu_i + T\frac{\partial s_i}{\partial \rho_i}\right) = \vec{\nabla}p_i + \left(\rho_i\frac{\partial s_i}{\partial \rho_i} + \frac{\partial f_i}{\partial T}\right)\vec{\nabla}T + \rho_i T\vec{\nabla}\frac{\partial s_i}{\partial \rho_i}$$
$$= 0 \qquad (5.2.5)$$

but now there is an additional difficulty; we have only one adiabatic condition, $\partial f/\partial T = -s$, and no way to reduce the quantities $\partial f_i/\partial T$. The trouble with this approach is that we have no physical interpretation of the "partial pressures" and no formula for $\partial f_i/\partial T$. This development leads nowhere, in general, but the situation changes somewhat in the case that the Gibbs–Dalton hypothesis is valid, as we shall see.

All this is just to show that the partial pressure may be a useless concept. The Euler–Lagrange equations have no need to be expressed in terms of partial pressures. They are explicit expressions for the accelerations in terms of the densities and the temperature.

There is, perhaps, one exceptional case: superfluid Helium is a mixture in which the superfluid component S_2 of the specific density is zero. In this case, $\partial f_1/\partial T = -S_1$ and $\partial f_2/\partial T = 0$ and this gives rise to an entropic force $\rho_1 S_1 \vec{\nabla}T$. Such a term appears explicitly in Landau's equations of motion and is used to explain the fountain effect. The problem with the interpretation of the partial pressures is circumvented by the replacement of both by p.

5.3. Mixture of Two Ideal Gases

This example will throw some light on the question of allocation of entropy.

Imagine a vessel, divided into two parts by a wall, the two parts filled with an ideal gas of type 1 and an ideal gas of type 2, respectively. Assume that the wall is impenetrable to either gas but that it admits the transfer of heat. The Lagrangian density for this system is

$$\mathcal{L} = \sum_{i=1,2} \mathcal{L}_i = \sum_i (\rho_i(\dot{\Phi}_i - \vec{v}_i{}^2/2) - f_i - s_i T),$$

where $s_i = \rho_i S_i$ and S_1, S_2 are adiabatic invariants. It is postulated that there is a unique, continuous **temperature field**, extending over both vessels, continuous at the wall (see Section 3.8).

This Lagrangian describes two separate systems that interact with each other solely by sharing the boundary condition implied by the continuity of the temperature at the wall. The feasibility of this setup is amply confirmed by experiments.

What happens if the **dividing wall** is removed? The most common observation is that **mixing** takes place, the particles of gas number 1 begin to "diffuse" into the space formerly reserved for the other gas, and vice versa. If no external forces are acting, the mixture often becomes uniform. At this stage of our study, we shall not inquire into the physical, probably non-adiabatic process that causes the mixing, but only try to describe the final states.

The process aside, we need to find the Lagrangian of the system after mixing. One result of mixing is a reassignment of entropies. After mixing, each of the originally separated gases has expanded to a larger volume, with attendant changes in the entropy parameters; what are the final values? A law discovered by Dalton states as follows:

Dalton's law. At equilibrium, the total pressure of the mixture of two ideal gases (the pressure against the walls) is the sum of the pressures that would have been obtained if each gas would have occupied the total volume in the absence of the other at the same temperature.

Important as it is, this information is a very small part of what we wish to know: what are the dynamical properties of this mixture, that is, what is the Lagrangian for this new gas? This need to know more led some people to overstate Dalton claims. Dalton was said to have held that each gas in a mixture behaves as if it were alone. He did not. Some of the controversy is discussed in Dalton's book (Dalton, 1802).[54]

If indeed it were so, that each gas in the mixture behaves as if it were alone, then we should obtain the Hamiltonian density for the mixture by simply adding the two Hamiltonian densities and integrating over the vessel filled by both, assigning a different temperature to each gas. This concept of dual temperatures is unphysical except in certain cases where a method is devised to measure the two separate temperatures. We postulate a single temperature field and we are led to consider the following Lagrangian density for the mixture,

$$\mathcal{L} = \sum_i (\rho_i(\dot{\Phi}_i - \vec{v}_i{}^2/2) - f_i - s_i T), \tag{5.3.1}$$

where $f_1(T, \rho_1)$ and $f_2(T, \rho_2)$ are the expressions for the free energy appropriate for each gas when alone. We repeat that this is undertaken merely as a first step towards the eventual construction of a realistic Lagrangian. The adoption of this Lagrangian is in accord with the spirit of Dalton's law. The thermodynamic pressure is

$$p = \sum \rho_i \frac{\partial f}{\partial \rho_i} - f = \sum \mathcal{R}_i \rho_i T \tag{5.3.2}$$

and the entropy density is

$$s = \sum s_i = \sum \rho_i S_i.$$

The formula for the total pressure expresses Dalton's law; the inclusion of interactions will invalidate it, unless it is of total degree one. This suggests that we keep in mind a particularly interesting interaction,

$$f_{\rm int} \propto \sqrt{\rho_1 \rho_2}. \tag{5.3.3}$$

The first question that needs to be discussed arises from the appearance of two entropy parameters when only one was expected. The total entropy density is a linear function of the two densities depending on two parameters S_1, S_2. Exploring the merits of *any* explicit expression for the Lagrangian of a binary or multicomponent system is a departure from common practice. Attempts to model the Gibbs function are common, but they involve a "reference pressure" p_0 (usually one atmosphere) and a "`reference state`" with an unspecified Gibbs function $G(T, p_0)$ consecrated in tables. Confrontation

[54]Here is a sample statement of the law in a modern textbook: "...the assumption is made that each gaseous component of the mixture behaves as if it existed at the temperature of the mixture and filled the entire volume ..." (Holman, 1974).

with observation would seem to point to a determination of the correct form for the unknown function, but we have seen no attempt to do this. The special, linear form that is being postulated for the **assigned entropy** gives a structure to the modeling that can easily be tested by experiments. The overall plan is to test naive expressions for the Lagrangian, including the entropy, and make improvements as necessary. The most important advantage of this program is that each attempted application will suggest a refinement of the Lagrangian and, especially, the fact that one Lagrangian has to accommodate a variety of effects and experiments. The search is on for the best possible Lagrangian capable of accounting for all properties of the mixture of the two gases.

Additivity of the Lagrangian for a system of two non-interacting components embodies additivity of the Hamiltonians, which is natural, and also additivity of the thermodynamic potentials, even when that is only as a benchmark, subject to modifications as required by observation. Though it seems natural, this attitude is not standard. Consequently, from this point onwards, there will be a significant divergence from the usual approach. To defend this deliberate departure from tried and tested methods in the treatment of mixtures, it may be said that, certainly, there can be nothing wrong with it in the limit of a very weak interaction between the components. And this cannot be said with as much assurance of some other approaches.[55] Our reward will be a theory with a very much enhanced predictive power. The predictions are, of course, to be tested.

5.4. Gibbs–Dalton Hypothesis for Ideal Gases ...

The additivity of energies that is implied by the addition of Lagrangians is an attractive assumption, and probably a must in the absence of interactions, but what can be done about the two entropy parameters? Addition of heat to the system must change the parameters and if the system is to end up in a well-defined configuration, then it has to follow a fixed path in the "entropy plane" with coordinates S_1, S_2. A partial answer to this question is provided

[55] This criticism, if that is what it is, can be countered, to some extent, by making the observation that, realistically, no interaction implies vanishing density. A problem would arise, however, with the discovery of gases that are ideal at a finite density.

by a literal interpretation of the **Gibbs-Dalton hypothesis**, developed by Gibbs from Dalton's law, namely,

> **Gibbs–Dalton hypothesis. When the mixture of ideal gases reaches a final equilibrium each gas has the entropy that it would have if it were alone.**

Discussion. In our context, Dalton's law appears to be expressed by Eq. (5.3.2) and to be a direct consequence of the additivity of Lagrangians. But this is not what Dalton's law means. Dalton's law $p = \sum p_i$ has content only after we decide what the symbols "p_i" stand for. To Dalton, they are the pressures of a pure gas with temperature T and density ρ_i; in the case of an ideal gas, $p_i = \mathcal{R}_i \rho_i T$. But the summands $p_i = \rho_i \partial f_i / \partial \rho_i - f_i$ in Eq. (5.3.2) are expressed in terms of the temperature, the densities and the entropy. To define them, we need to eliminate the entropy; hence, Dalton's law follows from (5.3.2) only if the entropy is a sum of partial entropies and each partial entropy is related to density and temperature by the same expression as it is in the case of the components in isolation. This is what leads to the Gibbs–Dalton hypothesis.

One interpretation of the Gibbs–Dalton hypothesis is the adiabatic relation that is one of the Euler–Lagrange equations,

$$s = -\frac{\partial f}{\partial T}.$$

We shall call this the weak form of the hypothesis. It is of little use without a prescription for assigning a value to the entropy parameters. We have insisted on maintaining Gibbs' point of view, that the variational principle minimizes the energy for a fixed value of the entropy. In extended, one-component thermodynamics, this was interpreted in the sense of the specific entropy density, and the latter has been found to be uniform in a wide context. In the case of a heterogeneous system, Gibbs does not offer guidance. The last formula serves only to calculate the entropy for a given configuration of the system, which is of little use.

To be useful, the adiabatic relation needs to be supplemented by a rule for assigning entropy. A first step, and a major one, is to demand additivity, in the strong form

$$s = \sum s_i = \sum \rho_i S_i,$$

with uniform values of S_1 and S_2. That leaves only the problem of tracing the path followed by the system through the entropy plane, a problem for the solution of which we defer to experiment.

A stronger interpretation of the Gibbs–Dalton hypothesis is possible, namely,

> **Strong Gibbs–Dalton hypothesis. When the mixture of ideal gases reaches a final equilibrium, each gas has the entropy that it would have if it were alone at the same density and temperature.**

When each gas is alone, these relations hold:

$$S_i = -\mathcal{R}_i\left(\ln\frac{\rho_i}{T^{n_i}} - n_i\right), \quad i = 1, 2 \text{ separated gases.}$$

According to the strong version of the hypothesis, it holds for a mixture but only at equilibrium:

$$S_i = -\mathcal{R}_i\left(\ln\frac{\rho_i}{T^{n_i}} - n_i\right), \quad i = 1, 2 \text{ at equilibrium.}$$

This is what we shall mean by "the strong Gibbs–Dalton hypothesis" (for a mixture of ideal gases). It is expected to hold in the absence of any chemical reactions or external forces.[56] The virtue of the statement is that, as a statement about the value of the entropy, it applies to the adiabatic system in general, not just to equilibrium states; it removes the ambiguity in the Lagrangian by fixing a relation between the two (or several) entropy parameters defined in (5.2.1).

Consider two ideal gases with

$$f_i + s_i T = \mathcal{R}_i\rho_i T\ln\frac{\rho_i}{T^{n_i}} + \rho_i S_i T, \quad i = 1, 2.$$

If each gas were alone, we should have the two equations of motion (adiabatic relations),

$$\mathcal{R}_i\rho_i\left(\ln\frac{\rho_i}{T^{n_i}} - n_i\right) + \rho_i S_i = 0, \quad i = 1, 2, \tag{5.4.1}$$

from variation of each Lagrangian with respect to the temperature, but variation of the proposed total Lagrangian (5.3.1) with respect to the common temperature yields only one adiabatic equation:[57]

$$\sum_{i=1,2}\left(\mathcal{R}_i\rho_i\left(\ln\frac{\rho_i}{T^{n_i}} - n_i\right) + \rho_i S_i\right) = 0. \tag{5.4.2}$$

---

[56] Please remember that these relations are restrictions on the values of T and the densities. The numerical values assigned to S_1 and S_2 are **`adiabatic invariants`**.

[57] This is enough to give the formula for the internal energy density, $u = \sum_i n_i\mathcal{R}_i\rho_i T$.

The strong Gibbs–Dalton hypothesis requires that both of Eqs. (5.4.1) hold for the unified system, *but only at equilibrium.* If the equilibrium is characterized by uniform densities, then this amounts to the required relation between S_1 and S_2, a relation that provides the required path in the entropy plane. Suppose that there is a value T_0 of T such that both terms vanish, then for all T

$$\mathcal{R}_i\rho_i\left(\ln\frac{\rho_i}{T^{n_i}} - n_i\right) - \rho_i S_i = n_i\mathcal{R}_i\rho_i \ln\frac{T_0}{T}$$

and the equation of motion (5.3.2) reduces to

$$\left(\sum n_i R_i \rho_i\right)\ln\frac{T_0}{T} = 0.$$

This can be true only at $T = T_0$. That is, if such a value of T exists when the densities take equilibrium values, then the equation of motion implies that this is the value of T at equilibrium and that both terms vanish separately at equilibrium. The strong Gibbs–Dalton hypothesis can thus be stated within the present context as follows:

Gibbs–Dalton hypothesis, strong version:

> **For correctly assigned values of the entropy parameters S_1 and S_2, the equilibrium temperature is uniquely fixed by the requirement that both terms in (5.4.2) vanish separately.**

Away from equilibrium, only (5.4.2) holds and each component gas no longer behaves as if the other component were absent! The two gases at equilibrium may not be "aware of each other", but in configurations other than equilibrium, the pair of adiabatic conditions that characterize the pure gases is replaced by the single adiabatic condition (5.4.2).

For a system of ideal gases with fixed total quantities of each component and with uniform equilibrium densities, the Gibbs–Dalton hypothesis predicts the following entropy path:

$$\frac{S_1}{n_1\mathcal{R}_1} - \frac{S_2}{n_2\mathcal{R}_2} = \ln\frac{\rho_2^{1/n_2}}{\rho_1^{1/n_1}}, \tag{5.4.3}$$

where ρ_1, ρ_2 are the densities at equilibrium. It agrees with Boltzmann's distribution law if $n_1 = n_2$ (see Section 5.4).

5.5. ... and What Lies Behind It

We shall attempt to determine what it is that would drive the mixed gas to satisfy the strong version of the Gibbs–Dalton hypotheses.[58] Let

$$x = \frac{\rho_1}{k_{01}(eT)^{n_1}}, \quad y = \frac{\rho_2}{k_{02}(eT)^{n_2}}, \quad S_i = -\mathcal{R}_i \ln k_{0i}, \quad i = 1, 2,$$

so that the Gibbs–Dalton hypothesis makes $x = y = 1$, and the adiabatic relation (5.4.2) takes the form

$$x \ln x + ay \ln y = 0, \quad a = (\mathcal{R}_2 k_{02}/\mathcal{R}_1 k_{01})(eT)^{n_2 - n_1}. \tag{5.5.1}$$

Let us fix S_1 and S_2, and consider an equilibrium configuration in which all the fields are uniform. Figure 5.5.1 shows the locus of points in the x, y plane at which Eq. (5.5.1) holds for a particular value of T. Since the variables x, y are constant multiples of the densities these variables are in fact scaled densities and we may refer to the plane of the figure as the density plane. There is a family of such curves, one for each value of T. The figure also shows several lines of constant total pressure,

$$p = \left(\rho_1 \frac{\partial}{\partial \rho_1} + \rho_2 \frac{\partial}{\partial \rho_2} - 1\right)(f_1 + f_2) = T(\mathcal{R}_1 \rho_1 + \mathcal{R}_2 \rho_2). \tag{5.5.2}$$

The loci of p — the isobars — are straight lines. The isotherm intersects the isobar at two points, at one point, or at no point in the ρ-plane. If they intersect at no point, then there is no such configuration.

If the curves touch at one point then the densities are uniquely determined. The tangents of the two curves at any point are given by

$$\mathcal{R}_1 d\rho_1 (\ln x + 1) + \mathcal{R}_2 d\rho_2 (\ln y + 1) = 0,$$

and

$$\mathcal{R}_1 d\rho_1 + \mathcal{R}_2 d\rho_2 + (\mathcal{R}_1 \rho_1 + \mathcal{R}_2 \rho_2)\frac{dT}{T} = 0,$$

respectively; they coincide, for a fixed value of T, only if

$$\ln x = \ln y. \tag{5.5.3}$$

By virtue of Eq. (5.4.2), this implies that both logarithms are zero, as required by the Gibbs–Dalton hypothesis.

[58] The answer given by Boltzmann to this question is the foundation of statistical mechanics. Here, we are looking for an answer within our chosen context of thermodynamic field theory.

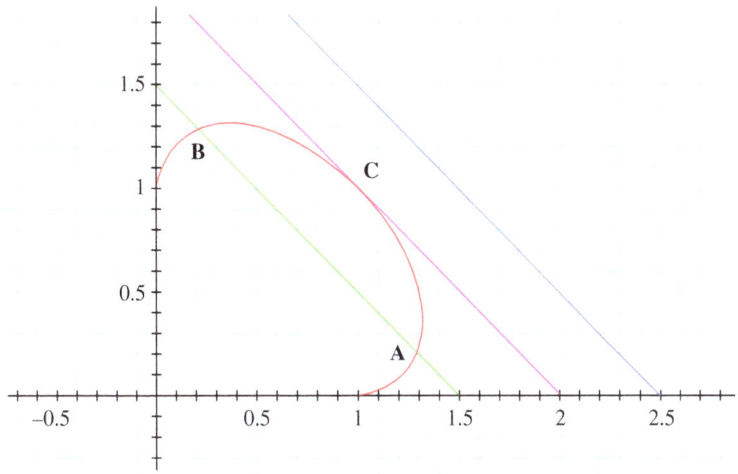

Fig. 5.5.1. Gibbs–Dalton. The coordinates are the (scaled) densities x and y. The oval is the locus of (5.4.2) determined by fixing T, S_1 and S_2. The straight lines are isobars.

There remains the possibility of two intersections at points A and B in the figure. To justify the Gibbs–Dalton hypothesis, we have to throw some aspersions on this case. Well, if the gas at equilibrium is homogeneous, then the ratio $r = \rho_1/\rho_2$ is given in advance by the ratio of total masses. This number need not coincide with the value of either of the ratios $r(A)$ or $r(B)$. If it is intermediate between those values, then the gas must separate into two parts, one part in which $r = r(A)$ and another part in which it is $r = r(B)$. Later, we shall see that mixtures exist, where this does indeed happen, but in the case of ideal gases, or at high temperature or low pressure for most gases, we know that it does not; so there is no stationary state available. This is the same as to say that such states are adiabatically unstable to **convection**. Flow will develop and eventually the kinetic energy will degenerate to heat. The final state will be stationary but the values of the entropy parameters must have changed. If an equilibrium is finally reached, then it must satisfy the Gibbs–Dalton hypothesis.

The only stable, homogeneous equilibrium available is point C, where the two curves have the same tangent. At that point, both terms in (5.4.2) vanish separately, as we have shown.

If the values of the densities and of the temperature are given, then this is a restriction on the values of the two entropy parameters S_1 and S_2. For a given system with fixed mass and with uniform equilibrium densities, there will be a curve $T \mapsto (S_1(T), S_2(T))$, the image of which is the path followed by the system as the temperature is varied so as to move it through a succession of equilibrium configurations (see Fig. 5.5.1).

We seem to be able to conclude that the Gibbs–Dalton hypothesis finds the only stable mixture and that this is what determines how the entropy is apportioned between the two components. An argument based on the maximal entropy principle was expected. Instead, we have here an analogue of **free expansion**; the assumed initial point is unstable and goes into an agitated state, out of equilibrium. The final equilibrium is reached by degradation of the kinetic energy, leading to a state of maximal entropy.

Thus, it seems that the Gibbs–Dalton distribution of entropy is not reached by a purely dissipative process but because it is the only one that allows for the existence of equilibrium states, or more generally for any stable adiabatic system. (The fact that values of pressure and entropy cannot be assigned at random is easily proved by more elementary examples (see Section 3.5 and Fig. 3.5.2.) If this conclusion is correct, it shows that some important questions about equilibrium thermodynamics are answered only by the extension to hydro-thermodynamics.

It seems that the entropy chosen by the Gibbs–Dalton hypothesis is a kind of **critical point**. Recall that, in general, a critical point in thermodynamics is one in which a Legendre transformation is singular (Tisza, 1970). For example, consider the Legendre transformation

$$p(T, q_1, q_2) = \sum \rho_i \mu_i - f, \quad \mu_i = \frac{\partial f}{\partial \rho_i}, \qquad i = 1, 2.$$

An example of a critical point is a point where

$$\det(\partial^2 p / \partial \rho_i \partial \rho_j) = 0.$$

The critical point at hand is defined by the vanishing of the Jacobian of the transformation T, $p \to \rho_1$, ρ_2,

$$J = \left| \frac{\partial T \partial p}{\partial \rho_i \partial \rho_j} \right|;$$

that is, a point where the two gradients in the density plane

$$\left(\frac{\partial T}{\partial \rho_1}, \frac{\partial T}{\partial \rho_2} \right), \quad \left(\frac{\partial p}{\partial \rho_1}, \frac{\partial p}{\partial \rho_2} \right)$$

are parallel. The derivations are to be taken with S_1, S_2 fixed.

With this interpretation, we can ask for a generalization of the hypothesis to the case of a mixture of real, interacting gases. This question will be taken up again in Section 6.1.

The suggested form of the Lagrangian for a mixture, with the concept of an entropy space and the need to find the path through this space chosen by the system, is a central feature of the treatment proposed in this book.

In the case of a mixture of real gases, we do not expect the original Gibbs–Dalton hypothesis to apply; this includes van der Waals fluids as well as ideal fluids with other types of mutual interaction. Both of these cases will be studied in Section 5.6. It also does not apply to the case of two ideal gases in the case when dissociation and recombination take place. This case shall be studied in Section 5.8. The condition for equilibrium, Eq. (5.7.3), is not compatible with the strong Gibbs–Dalton condition unless $R_2(n_2+1) = R_1(n_1+1)$.

5.6. Gibbs–Dalton Hypothesis for Some Real Gases

In the case of two fluids of very different nature, the Dalton law of partial pressures is not expected to be valid, and the Gibbs–Dalton hypothesis even less. The Gibbs–Dalton hypothesis applies to ideal gases, and real gases are not ideal gases. Can anything be said about the distribution of entropy in a mixture of real gases? We shall answer this question in the affirmative.

One expects that there is an **interaction** between two dissimilar atoms. As was first shown by van der Waals (1873/1899), this interaction can sometimes be simulated by adding an interaction term to the free energy. Because we are interested in the equilibrium states the density gradients will be small and it is enough to consider an interaction energy that is an algebraic function of the densities, not involving their gradients. As a preliminary test of this idea, we merely add the following term to the free energy density:

$$f_{\text{int}} = \alpha(\rho_1\rho_2)^k, \tag{5.6.1}$$

with α and k constant. There is direct experimental evidence in favor of this expression, since the measured "excess free energy" (Rowlinson, 1959, p. 138) sometimes has exactly this shape, with k in the region of 0.5 to 1 or higher (see Fig. 5.6.1). The forces are often repulsive, which makes α positive. The total free energy is $f_1 + f_2 + f_{\text{int}}$. For f_1 and f_2 we shall use, in a first instance, the expressions that pertain to the one-component ideal gases. Thus, at equilibrium, the Hamiltonian density is

$$\begin{gathered}
\rho_1\vec{v}_1^{\,2}/2 + \rho_2\vec{v}_2^{\,2}/2 + f + sT, \quad f = f_1 + f_2 + \alpha(\rho_1\rho_2)^k, \\
s = \sum \rho_i S_i, \\
f_1 = \mathcal{R}_1 T\rho_1 \ln\frac{\rho_1}{T^{n_1}}, \quad f_2 = \mathcal{R}_2 T\rho_2 \ln\frac{\rho_2}{T^{n_2}}.
\end{gathered} \tag{5.6.2}$$

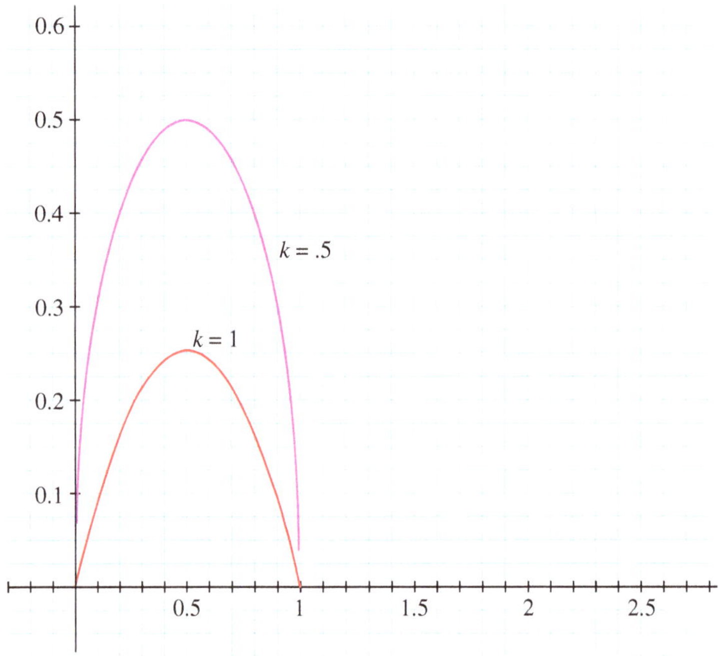

Fig. 5.6.1. Illustration of the dependence of the excess free energy of a mixture on the concentration for a fixed value of the total density. It is not always positive.

The Gibbs–Dalton hypothesis specifies a relation between S_1 and S_2 by focusing on the unique equilibrium configuration. We are not admitting external forces, so the densities are uniform and determined by the total volume and the masses; the temperature is also uniform. The statement of the strong Gibbs–Dalton hypothesis is a relation between the entropy parameters that is related to the equilibrium values of the densities.

Now, we ask in what way this rule is expected to be modified by the interaction. A simple and possibly universal hypothesis is available: that the stability interpretation discovered in Section 5.2 applies, namely,

$$\det \frac{\partial(p,T)}{\partial(\rho_i,\rho_j)} = 0. \tag{5.6.3}$$

The partial derivatives are adiabatic; that is, the entropy parameters are kept fixed. We have

$$\begin{aligned} dp = {} & [\mathcal{R}_1 T + k(2k-1)\alpha(\rho_1\rho_2)^k/\rho_1]d\rho_1 \\ & + [\mathcal{R}_2 T + k(2k-1)\alpha(\rho_2\rho_2)^k/\rho_2]d\rho_2 \end{aligned}$$

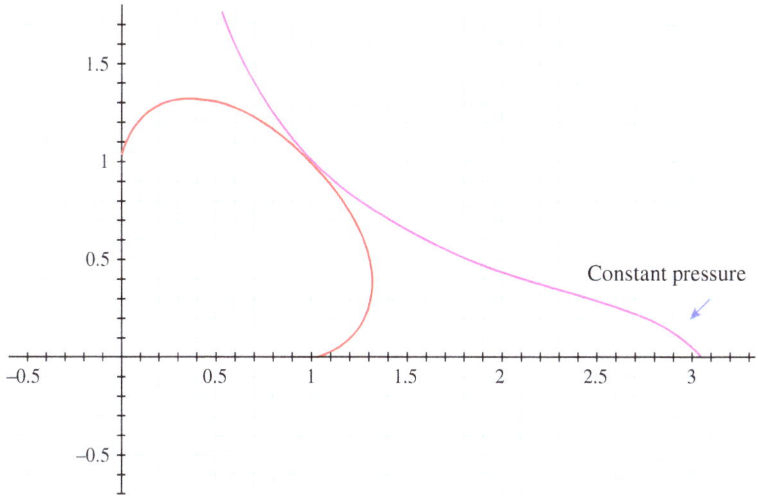

Fig. 5.6.2. Gibbs–Dalton. The effect of interaction on the Gibbs–Dalton critical point. The oval is the locus of solutions of the adiabatic condition for fixed values of T, S_1, S_2, the coordinates x, y are the densities, scaled: $x\ln x + y\ln y = 0$. The other curve is a locus of constant pressure, $x + y + (xy)^{4.2} = 3.05$.

and the adiabatic condition

$$\mathcal{R}_1\rho_1 \ln x + \mathcal{R}_2\rho_2 \ln y = 0,$$

where

$$x = \frac{\rho_1}{k_{01}(eT)^{n_1}}, \quad y = \frac{\rho_1}{k_{02}(eT)^{n_2}}$$

gives

$$dT \propto [\mathcal{R}_1(\ln x + 1)]d\rho_1 + [\mathcal{R}_2(\ln y + 1)]d\rho_2.$$

The condition for tangency takes the form

$$\frac{1+\ln x}{1+\ln y} = \frac{1+\alpha'(\rho_1\rho_2)^k/R_1\rho_1 T}{1+\alpha'(\rho_1\rho_2)^k/R_2\rho_2 T}, \quad \alpha' := k(2k-1)\alpha.$$

The Gibbs–Dalton relation $x = y = 1$ is thus modified unless the molar density is the same for both gases. This is illustrated in Fig. 5.6.2 for $k = 1/2$, $\alpha' > 0$.

There remains an interesting possibility that α' may in special cases be negative. Then the situation illustrated in Fig. 5.6.3 may be realized. Here, the generalized Gibbs–Dalton hypothesis is satisfied at two points, with different concentrations. Before we conclude that two different concentrations may coexist, we have to investigate the condition of equilibrium.

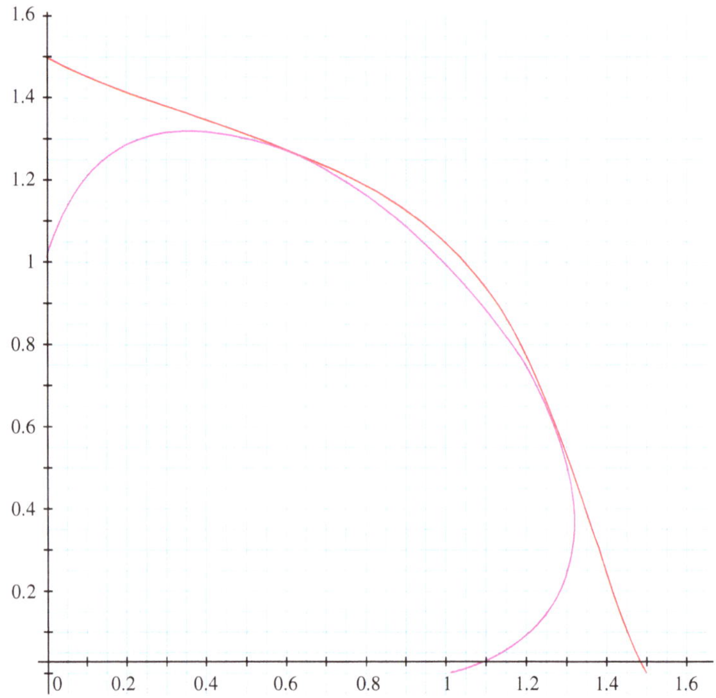

Fig. 5.6.3. Gibbs–Dalton. An interesting possibility. Same as the preceding equation but the pressure is $x + y - 0.37(2k - 1)(xy)^k$, $k = 1.2$ or $x + y + 2.6(2k - 1)$ with $k = 0.4$.

5.7. The Simplest Mixtures: Boltzmann Theory

Consider a gas with a ground state with free energy $f = 0$ consisting of molecules with two quantum levels. More precisely, consider a molecule that converts freely between two energy levels, a lowest level 1 with vanishing free energy $f_1 = 0$ and an "excited" level 2 with internal energy density $f_2 = \epsilon\rho_2$. Suppose that

$$\epsilon = (S_2 - S_1)T. \tag{5.7.1}$$

This is analogous to what we found to be an expression of Maxwell's rule.

Now, let us take a route that may be risky: irrespective of the properties that we have imputed to the gas, assume (with Gibbs–Dalton, strong version) that the familiar expression for the entropy of an ideal gas applies to each component of the mixture:

$$S_i = -\mathcal{R} \ln \frac{\rho_i}{(eT)^{n_i}}, \quad i = 1, 2. \tag{5.7.2}$$

Thus,

$$\frac{\rho_2}{\rho_1} = (eT)^{n_2 - n_1} e^{-\epsilon/\mathcal{R}T}. \tag{5.7.3}$$

This is Boltzmann's distribution law for the population of the two levels. Equations (5.7.1) and (5.7.3) are standard lore and not to be tampered with, this in spite of any doubts that may arise concerning the actual direct, experimental verification of Eq. (5.7.3).

Consider now the Lagrangian density

$$\dot{\Phi}\rho - f - sT = \dot{\Phi}\rho - \sum\Big(\mathcal{R}_i T\rho_i \ln\frac{\rho_i}{T^{n_i}} + \rho_i S_i T\Big).$$

Note that only the total density is conserved; therefore, there is only one velocity potential and there is no way to include the binding energy directly though it is accounted for via Eq. (5.7.1). The new equations of motion are the adiabatic relation (variation of T)

$$\sum \rho_i\left(\mathcal{R}_i \ln\frac{\rho_i}{(eT)^{n_i}} + S_i\right) = 0,$$

the equation of continuity (variation of Φ) and the hydrostatic condition (variation with $d\rho_1 = -d\rho_2$)

$$\ln\frac{\rho_2^{1/n_2}}{\rho_1^{1/n_1}} = -(S_2 - S_1), \qquad (^*)$$

where, as before, S_i is the entropy per mol of the ith component. I have followed the same method as for saturation. The two cases are very different, for in one case the two components are separated and in the other case they are mixed, so that we cannot be sure that this is right.

5.8. Chemical Reactions: Dissociation

We consider the simplest example of a chemical reaction involving ideal gases, the transformation between atomic and molecular **hydrogen**,

$$H_2 \leftrightarrow 2H.$$

At sufficiently low temperatures, the gas is almost purely molecular with molecular weight and adiabatic index

$$m_1 = 2, \quad n_1 = 5/2.$$

At equilibrium, we shall have $\vec{v} = 0$ and

$$\rho_1 = \rho = M/V, \quad \ln\frac{\rho}{T^{n_1}} - n_1 = -s_1/R_1.$$

At sufficiently high temperatures, the gas is almost purely atomic, with $m_2 = 1$, $n_2 = 3/2$. At equilibrium,

$$\rho_2 = \rho = M/\mathcal{V}, \quad \ln\frac{\rho}{T^{n_2}} - n_2 = -s_2/\mathcal{R}_2.$$

Consider a mixture of two gases, similar in all respects to H and H_2, except that the reaction analogous to $H_2 \leftrightarrow 2H_1$ does not take place. The Lagrangian density would be

$$\mathcal{L}' = \sum \Big((\rho_i \dot\Phi_i - \vec v_i^{\,2}/2 - \phi) - \sum \Big(\mathcal{R}_i T \rho_i \ln \frac{\rho_i}{T^{n_i}} + T\rho_i S_i \Big)\Big).$$

A hypothesis that can be tried is that this expression retain some validity for the real hydrogen problem. But it must be changed in essential ways. In $\mathcal{L}'$, there are two independent velocity fields and therefore two independent potentials Φ_1 and Φ_2. Variation leads to independent conservation laws, atoms and molecules are independently conserved. The difference between this hypothetical system and real Hydrogen is that in the real system only the total density (grams/cm^3) is preserved, so we have only one conservation law and we can admit only one variation, hence only one independent potential velocity field,[59]

$$\vec v_1 = \vec v_2 = \vec v, \quad \dot\Phi_1 - \dot\Phi_2 = \epsilon = \text{constant}. \tag{5.8.1}$$

The **Lagrangian** density becomes

$$\mathcal{L} = \rho_1 \dot\Phi_1 + \rho_2 \dot\Phi_2 - \rho(\vec v^{\,2}/2 + \phi) - \sum \Big(\mathcal{R}_i T \rho_i \ln \frac{\rho_i}{T^{n_i}} + T\rho_i S_i \Big), \tag{5.8.2}$$

with $\rho = \rho_1 + \rho_2$. The distinction between one molecule and two atoms is not entirely lost, for we are able to maintain two independent densities; indeed, this must be so for otherwise the problem of two kinds of gases, atomic and molecular, would not exist. The densities are separately observable and we shall derive a formula for the concentrations in the mixture. Generalizations, to the case of more than two component fluids with chemical interactions, will be considered in Section 5.9. Some formal considerations are being relegated to the end of this section.

Let us study the equations of motion in the case that the fluid is at rest, $\vec v = 0$ everywhere. The effect of gravity will be neglected. Variation with respect to T yields the adiabatic relation

$$\sum \Big(\mathcal{R}_i \rho_i \Big(\ln \frac{\rho_i}{T^{n_i}} - n_i \Big) + \rho_i S_i \Big) = 0. \tag{5.8.3}$$

Independent, local variation of the two densities shows that all the fields are uniform. The actual values of ρ_1 and ρ_2 are not obtained, as each variational equation gets a contribution from a $\rho\dot\Phi_i$-term. But in variation of the densities with ρ fixed, $\delta\rho_1 = -\delta\rho_2$, the contributions of these terms cancel out

[59] In this section, the densities are in grams, not mols.

because $\Phi_1 - \Phi_2$ is fixed and we obtain an additional relation,

$$R_1T\left(\ln\frac{\rho_1}{T^{n_1}}+1\right)+S_1T-\mathcal{R}_2T\left(\ln\frac{\rho_2}{T^{n_2}}+1\right)-S_2T=0.$$

Now, it is at once clear that the question that has occupied us in the first four sections of this chapter has a different dimension for this, very different, type of mixture. In the condition of equilibrium, we have an additional relation, for fixed temperature, between densities and entropies. Equations (5.8.1) and (5.8.3), together, define a mapping,

$$(\rho_1,\rho_2)\leftrightarrow(S_1,S_2).$$

If this mapping is one-to-one, is more information needed? Certainly, it is locally one-to-one almost everywhere. We may ask if the Gibbs–Dalton hypothesis is valid if it is compatible with the equation of motion. Contrary to the case of the unary fluid, the answer is that it is most likely not, and that if it happens to be, then this circumstance gives no useful information.

Gibbs' phase rule

For a mixture of C components and P phases, this famous rule says that the number of degrees of freedom is $C+2-P$, a non-negative number. It is based on counting the number of relations at equilibrium and comparing with the number of independent field variables. In other words, $C+2$ is the number of *a priori* field variables and P is the number of conditions for equilibrium. In the case of a single component, this rule predicts that there is a unique equilibrium for any pair of values of p and T. In the binary mixture with a single phase, we have only one condition for equilibrium and one more condition is needed to reach the same result, either (1) Gibbs–Dalton hypothesis or (2) the second condition for equilibrium that comes with a second phase. In the second case, the Gibbs–Dalton hypothesis is not satisfied.

Solving the equations

The densities are the normal kind, in grams/cm^3. The particular linear combination that appears here reflects the fact that it is mass that is preserved by the reaction, and not (for example) the number of particles. Since for hydrogen $\mathcal{R}_2 = 2\mathcal{R}_1$, Eq. (5.5.2) reads

$$\ln\frac{\rho_1}{\rho_2^2}-(n_1-2n_2)\ln T-1=\frac{S_2-S_1}{\mathcal{R}_1}.$$

In general, with $\mathcal{R}_i = \mathcal{R}/m_i$, where m_1, m_2 are the atomic molecular weights,

$$\ln \frac{\rho_1^{m_2}}{\rho_2^{m_1}} - (n_1 m_2 - n_2 m_1)(\ln T + 1) = \frac{m_1 m_2}{\mathcal{R}}(S_2 - S_1). \tag{5.8.4}$$

The right-hand side is usually expressed in terms of the binding energy, by means of Boltzmann's formula

$$\epsilon = (S_1 - S_2)T. \tag{5.8.5}$$

We shall try to derive this formula.

Minimal energy

As in the case of condensation, we need to understand what is meant by "energy" in this context. The boundary conditions are the same, the Hamiltonian is still the free energy; we only need to include the binding energy in the Hamiltonian, thus

$$h = \epsilon\rho_1 + f_1 + f_2.$$

Probably, this cannot be strictly justified, but it is in fact standard, see, for example, Landau and Lifshitz (1959). To do this without affecting the action — we already have the correct equations of motion — all that is needed is a slight change of vocabulary. The Lagrangian density is

$$\dot{\Phi}\rho + \epsilon\rho_1 - sT - h, \quad h = \epsilon\rho_1 + f_1 + f_2,$$

the Euler–Lagrange equations are the same as before and, at equilibrium, minimization of the Hamiltonian implies the Boltzmann relation (5.8.5).

What we have gained is the proper definition of "energy" as understood by Boltzmann, and Boltzmann's formula (5.8.5). This result confirms the validity of our entropy postulate for the mixture, namely,

$$s = \rho_1 S_1 + \rho_2 S_2;$$

cf. the discussion that follows Eq. (4.4.1). To sum up, the thermodynamical action principle, knowing nothing about binding energies, gives (5.8.4); at equilibrium, the principle of minimum energy, without invoking the concept of entropy, gives the same result but with $(S_2 - S_1)T$ replaced by ϵ. Together, they yield **Boltzmann's formula**

$$T(S_2 - S_1) = \epsilon. \tag{5.8.5'}$$

This relation is usually taken for granted; see, for example, Prigogine (1965, p. 41). Thus, finally,

$$\ln \frac{\rho_1}{\rho_2^2} - (n_1 - 2n_2)\ln T - 1 = \epsilon/R_1 T. \tag{5.8.6}$$

To compare this with the result of Eckert (1920) that was famously applied by Saha (1921) to stellar atmospheres, we replace the partial densities by the partial pressures, defined here by $p_i = \mathcal{R}_i \rho_i T$, to get

$$\ln \frac{p_1}{p_2^2}\frac{\mathcal{R}_2^2}{\mathcal{R}_1} - ((n_1 + 1 - 2(n_2 + 1))\ln T - 1 = \frac{\epsilon}{\mathcal{R}_1 T}$$

or, finally,

$$\ln \frac{p_1}{p_2^2} = \frac{C_{P1} - 2C_{P2}}{\mathcal{R}}\ln T + \frac{\epsilon}{\mathcal{R}_1 T} + 1 - \ln 2\mathcal{R}, \tag{5.8.7}$$

where $C_{Pi} = (n_i + 1)\mathcal{R}_i$.

This is the result contained by Saha from kinetic theory.

Remark. The **Euler-Lagrange equations** (5.8.2) and (5.8.3) allow us to determine how the total entropy is distributed among the two species. Both approaches agree that the total entropy density before mixing is

$$-\sum \mathcal{R}_i \rho_i \left(\ln \frac{\rho_i}{T^{n_i}} - n_i\right),$$

and there may be an expectation that the two summands represent the share of entropy allocated to each species as predicted by the Gibbs–Dalton hypothesis. Instead, from the condition of equilibrium and the adiabatic condition, we obtain

$$S_1 = -R_1\left(\ln \frac{\rho_1}{T^{n_i}} - n_1\right) - \frac{\rho_2}{\rho}[\mathcal{R}_1(1 + n_1) - \mathcal{R}_2(1 + n_2)].$$

This is not in contradiction with the classical theory, since that theory does not deal with the question, it can be understood as follows. "**Entropy production**" refers to the difference between the entropies in two hypothetical, equilibrium situations, at the beginning and at the completion of a process. During the process, the traditional position is that entropy is undefined. Our understanding is that, at equilibrium, the total entropy is determined (in terms of the temperature) by the adiabatic condition and that the allocation of entropies at equilibrium (the difference $S_1 - S_2$) was calculated with the same result as in the classical theory. "Entropy production" can be calculated by comparing the total entropy at final equilibrium with the total equilibrium before mixing, provided that we are given the

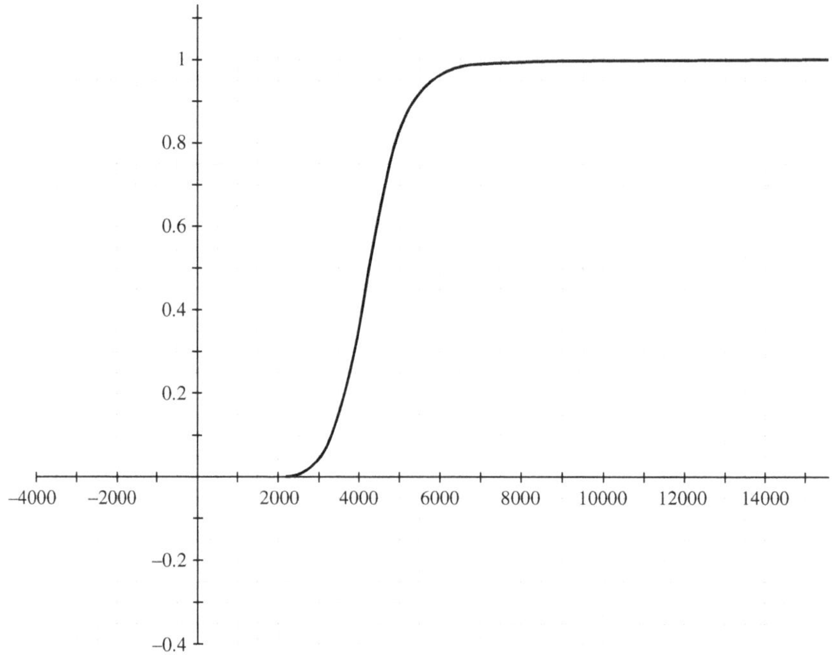

Fig. 5.8.1. The ratio r against T, Eq. (5.8.4).

temperatures. In the case that no chemical reaction takes place, we know that the temperature does not change. We do not try to determine whether this entropy production is due to mixing, to the progress of the chemical reaction or to the removal of the wall of separation. The classical theory does not say, nor does it ask.

De Groot and Mazur (1962) have proposed that, once mixing is complete, entropy production can be understood in terms of the progress of the chemical reaction. In that theory, $\dot{\Phi}$ is the rate of specific heat flow.

Numerical calculations produce Figs. 5.8.1 and 5.8.2.

The molecular binding energy is about 4.5 ev, or in relation to the rest energy, in energy units $\epsilon = (4.5/1876 \times 10^6)c^2 \approx 2.15 \times 10^{12}$. Thus,

$$\epsilon/\mathcal{R}_1 = 4.3 \times 10^{12}/0.83214 \times 10^8 = 51674.$$

Figure 5.8.1 shows r versus T, for $n_1 = 5/2$, $n_2 = 3/2$ with $1/ep = 1000$.

The measurements are usually performed under conditions of constant pressure. Setting $\rho_1/2 + \rho_2 = p/\mathcal{R}T$ we get, when $n_2 = 5/2$ and $n_2 = 3/2$,

$$\frac{\rho_2^2}{\rho\rho_1} = \frac{r^2}{1-r^2} = \frac{\mathcal{R}}{2ep}T^{3/2}e^{-\epsilon/R_1T}; \quad r = \rho_2/\rho. \tag{5.8.8}$$

This relationship is plotted in Fig. 5.8.2 for $\mathcal{R}ep = 1000$.

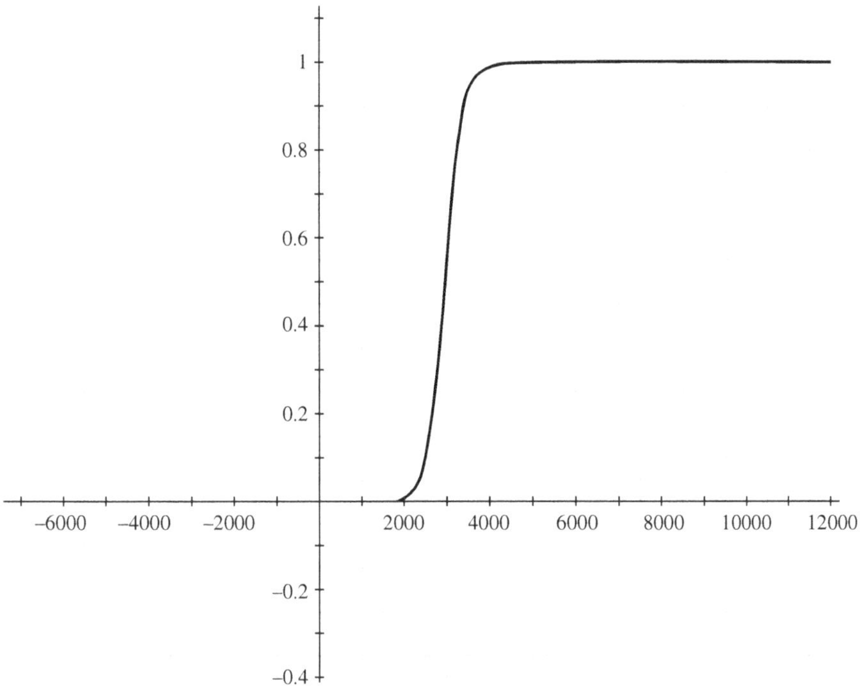

Fig. 5.8.2. The ratio r against T, Eq. (5.8.5).

We must return to the beginning to answer a reasonable question. If we have only one velocity then we have only one velocity potential, but two distinct densities. The system is no longer Hamiltonian. The best reply that we can make is to describe the procedure in slightly different terms. We conserve the full set of variables, retain the Hamiltonian structure, and work out all the Euler–Lagrange equations. Then we impose the constraint (5.8.1). The only thing that can go wrong is that the constraint is inconsistent with the equations of motion. The result is exactly the same. Below, we shall propose a generalization of this constraint and it will then be shown that it responds favorably to some tests. In the meantime, we are encouraged by the results obtained here.

5.9. Other Chemical Reactions

More complicated reactions raise new questions. Consider the choice of variables in the slightly more complicated case of a reaction of the type

$$A \leftrightarrow B + C, \tag{5.9.1}$$

where ν_A molecules of type A change into ν_B molecules of type B and ν_C molecules of type C. Let ϵ be the excess energy produced. The densities (in mols) of components A, B, C are ρ_A, ρ_B, ρ_C. We shall convey these data more concisely as follows. Define

$$\nu_A A - \nu_B B - \nu_C C =: \vec{\nu} \cdot \vec{A}, \quad \vec{\nu} := (\nu_A, -\nu_B, -\nu_C), \quad \vec{A} := (A, B, C)$$

and combine the formula (5.9.1) with the information about the energy in this form

$$\vec{\nu} \cdot \vec{A} = \epsilon.$$

Two densities are independently conserved, for example,

$$\nu_A \rho_A + \nu_B \rho_B \quad \text{and} \quad \nu_B \rho_B - \nu_C \rho_C.$$

We therefore need two Lagrange multipliers and two velocities. The space of conserved densities is a two-dimensional linear vector space and any two independent densities form a basis for it. A sum of squared velocities is not linear, so what is the correct expression for the kinetic energy? Any attempt to eliminate one of the velocities leads to disorderly formulas; it is better to impose a constraint.

We take the complete Lagrangian density, regardless of interactions, to be, as in the case that there are no chemical reactions,

$$\mathcal{L} = \rho_A(\dot{\Phi}_A - \vec{v}_A^2/2) + \rho_B(\dot{\Phi}_B - \vec{v}_B^2/2) + \rho_C(\dot{\Phi}_C - \vec{v}_C^2/2) - f_A - f_B - f_C - sT, \tag{5.9.2}$$

with

$$s = \rho_A S_A + \rho_2 S_B + \rho_C S_C \tag{5.9.3}$$

and with uniform coefficients S_A, S_B, S_C. The densities are in grams per cm^3 and the kinetic energies are in common, cgs units.

Conjecture. *The effect of the reaction is to impose the* **`linear constraint`**

$$\nu_A \dot{\Phi}_A - \nu_B \dot{\Phi}_B - \nu_C \dot{\Phi}_C = \epsilon. \tag{5.9.4}$$

Immediate support for this idea comes from the equations of motion. Variation with respect to the constrained potentials,

$$\vec{\nu} \cdot \delta \dot{\vec{\Phi}} = 0 \tag{5.9.5}$$

gives the two conservation laws,

$$\delta \mathcal{L} = \sum_i (\dot{\rho}_i + \vec{\nabla} \cdot (\rho_i \vec{v}_i)) \delta \Phi_i = 0$$

or

$$\dot{\rho}_i + \vec{\nabla} \cdot (\rho_i \vec{v}_i) = \nu_i Z(x), \quad i = A, B, C. \tag{5.9.6}$$

This is a succinct, and I think, a beautiful expression for the equations of continuity that hold during a chemical reaction.

This account of the independent variables and the conservation laws is an alternative to the treatment by Prigogine (1949) and his school.

The conjecture (5.9.4) is based on the significance of the velocity potential, more precisely the time derivative $\dot{\Phi}$, as an energy parameter. We have already remarked, in the simpler case of dissociation, as well as in the case of saturation, that it includes the latent heat. See Section 4.3, "Interpretation of the velocity potential".

At equilibrium, all the fields are uniform and the remaining information that comes from the Euler–Lagrange equations includes the adiabatic relation

$$\frac{\partial}{\partial T}(f_A + f_B + f_C + sT) = 0 \tag{5.9.7}$$

and stationarity of (5.8.2) with respect to the local variations of the densities. A principal result is that, at equilibrium, in the absence of external forces, the densities are uniform. We shall assume that each component is an ideal gas, then (5.9.7) takes the familiar form

$$\sum_{i=A,B,C} \left(R_i \rho_i \left(\ln \frac{\rho_i}{T^{n_i}} - n_i \right) + s_i \rho_i \right) = 0.$$

Additional information comes from an independent global variations of the densities that leave invariant the constraint (5.9.4), namely,

$$\delta\vec{\rho} \propto \vec{\nu} := (\nu_A, -\nu_B, -\nu_C).$$

Thus,

$$\vec{\nu} \cdot \vec{\mu} = T\vec{\nu} \cdot \vec{S}, \quad \vec{\mu} = (\mu_A, \mu_B, \mu_C).$$

In the case of three ideal gases, the full expression for this is

$$\frac{(\rho^{\nu/m})_A}{(\rho^{\nu/m})_B (\rho^{\nu/m})_C} (eT)^{-(n/m)_A + (n/m)_B + (n/m)_C} = e^{-\vec{\nu}\cdot\vec{S}/\mathcal{R}T}. \tag{5.9.8}$$

Here, m_A is the molecular weight of fluid A and $(n/m)_A := n_A/m_A$. The affinity to the Saha equation is evident. For other fluids, the right-hand side depends on the interaction between the components. It is a function of the densities alone, independent of the temperature. The chemists resort to accumulated tables.

As in the case of dissociation, we claim that maximization of the entropy makes the Hamiltonian stationary or equivalently that

$$T\vec{\nu} \cdot \vec{S} = \epsilon, \tag{5.9.9}$$

and here it is becoming manifest that the constraint (5.9.1) is pertinent. The last equation governs the partitioning of the entropy. At equilibrium, the condition that the Hamiltonian is stationary is equivalent to the condition

$$\vec{\nu} \cdot (\vec{\Phi} - T\vec{S}) = 0. \tag{5.9.10}$$

In other words, when the entropy is distributed according to Eq. (5.9.9), then the constraint (5.9.4) is a property of the equilibrium state.

Equation (5.9.9) leads to the alternative statement

$$\frac{(\rho^{\nu/m})_A}{(\rho^{\nu/m})_B(\rho^{\nu/m})_C} = (eT)^{(n/m)_A-(n/m)_B-(n/m)_C} e^{-\epsilon/\mathcal{R}}. \tag{5.9.11}$$

Either form gives a value for the "**reaction constant**"

$$K_c = \frac{(\rho^{\nu/m})_A}{(\rho^{\nu/m})_B(\rho^{\nu/m})_C},$$

as a function of the temperature. While the specialization to ideal gases limits the practical value of this result, it is nevertheless a little strange that this function is always taken from experiment and that no attempt is made to predict it, with the exception of the work of Saha (1921).

5.10. Taking Stock

The approach to binary mixtures through the addition of Lagrangians furnishes a two-parameter family of model Lagrangian densities. The two parameters are interpreted in terms of the entropies assigned to each type of molecule. As heat is added to the system, the entropy must change and a decision has to be made about the allotment of additional entropy to each gas; this gives us the concept of a two-dimensional entropy space and the path in this space that is chosen by the system. We had expected that the key to the determination of the path is entropy maximization, but this expectation has not been confirmed.

We have studied some mixtures of gases that are free to transmogrify, and we have seen two examples of a rule that would have the physical path through entropy space reduce to relations of the type $\epsilon = (S_2 - S_1)T$. In the case of ideal gases that do not make chemical reactions, the stronger statement furnished by the strong Gibbs–Dalton hypothesis may apply.

When we begin to consider mixtures of van der Waals gases in the next chapter, we shall find that condensation is a major part of it, so that both items of insight summarized here will be put to severe tests.

So far, there was no attempt to predict **rates** of chemical reactions. In the case that the reactions are driven by input of energy, the rates will be determined by the rate of input of heat or other forms of energy, as in electrolysis. In the case of spontaneous reactions that produce heat, the difficulty lies in the almost impossible task of preparing suitable initial conditions. A few very special circumstances may offer some hope for fruitful analysis.

1. Many systems have metastable configurations that can be reached with care, such as is the case with supercooled vapors or superheated liquids. Then an initial state can be prepared that is characterized by simple properties, such as uniformity. Possibly, the decay of such states may take place without convection setting in or with convection with a high degree of symmetry.
2. Solids may offer possibilities for analysis because of the simplification that is the absence of convection. The simplest instance is offered by the slab when an initial uniform configuration is disturbed by suddenly bringing one of the bounding plates to a higher temperature. There may be an opportunity of deriving Fourier's heat transfer law from thermodynamics.

Project 1. Study speed of sound in a mixture over a wide range of densities and temperatures. Collect data and test models. Then use the same models to study critical phenomena.

Project 2. Adapt the Gibbs–Dalton hypothesis to real gases.

Project 3. Use the entropy model to study chemical reactions. Holman gives several examples that may be useful.

Chapter 6

Immiscible Fluids

We shall continue our study of real fluids with an interesting phenomenon that is strangely absent from some textbooks on thermodynamics and the main subject in others. Here, in the first place, we shall introduce interactions between the components of a **mixture**. After all, the form of the free energy of a real, pure gas reflects the influence that each particle has on its neighbors; so it is reasonable to expect that unlike particles also interact with each other.

6.1. Ideal Gases with Mutual Interaction

When two fluids are mixed, and afforded enough time to settle down, one of the two things may happen. (1) The system approaches an equilibrium state in which the mixture is uniform, the concentration $\rho_1/(\rho_1 + \rho_2)$ having the same value throughout the vessel. Such mixtures are said to be **miscible**. (2) But sometimes the two fluids may not mix so readily. A small amount of each fluid can be dissolved in the other, either way, within limits. But at a range of temperature and pressure, if the average concentration is too high, then a different type of equilibrium may be observed, typically containing bubbles of one fluid mixture floating in the background of another fluid mixture with very different concentration. Because this chaotic configuration, with variable droplet size and a major influence of surface tension, is difficult to analyze, let us admit the presence of the standard gravitational field. Then a likely result is that the heavier fluid occupies the lower part of the vessel (a small amount of the lighter fluid dissolved in it) while the lighter fluid occupies the rest (a small amount of the heavier fluid dissolved in it).

Examples involving two liquids, a liquid and a gas or a gas and a solid, are common, but it is reported that the combination of any pair of "gases" is completely miscible.[60] The processes by which such things may come about are almost beyond analysis, but we can try to understand the final, equilibrium states. Gravitation is used only to explain the observation that the fluid separates into distinct regions with smooth boundaries and our studies concern the final, equilibrium configurations.

In the case of two fluids of very different nature, the Dalton law of partial pressures is not expected to be valid, and the strong Gibbs–Dalton hypothesis must at least be modified (see Section 5.5). One expects that there is an interaction between two dissimilar atoms. As before, in Section 5.6, we add a van der Waals interaction term to the free energy. Because we are interested in the equilibrium states, the density gradients will be small and it is enough to consider an **interaction energy** that is an algebraic function of the densities, not involving their gradients. We shall include the following term in the free energy density of the mixture of two ideal gases:

$$f_{\text{int}} = \alpha(\rho_1\rho_2)^k, \tag{6.1.1}$$

with α and k constants. There is direct experimental evidence in favor of this expression (see Section 5.6). The total free energy is $f_1+f_2+f_{\text{int}}$. For f_1 and f_2, we shall use, in a first instance, the expressions that pertain to the respective one-component ideal gases. Thus, at equilibrium, the Hamiltonian density is

$$\rho_1\vec{v}_1^2/2 + \rho_2\vec{v}_2^2/2 + f, \quad f = f_1 + f_2 + \alpha(\rho_1\rho_2)^k, \quad s = \sum \rho_i S_i, \tag{6.1.2}$$

where, *for the time being*, the individual energy densities are those of ideal gases,

$$f_1 = \mathcal{R}_1 T\rho_1 \ln\frac{\rho_1}{T^{n_1}}, \quad f_2 = \mathcal{R}_2 T\rho_2 \ln\frac{\rho_2}{T^{n_2}}. \tag{6.1.3}$$

A great deal of attention is paid to the "**excess free energy (density)**" of a real mixture (Fig. 6.1.1). It is defined as $f_{\text{excess}} = f - f_1 - f_2$, where f_1 and f_2 refer to situations where the individual fluids are alone with the same density and temperature (cf. Rowlinson, 1959). The densities f_1 and f_2 are known in the case of ideal gases, and more or less known in other cases; yet, the idea of modeling the excess, as a correction to the sum $f_1 + f_2$ is not

[60] This is what we read in many places, but if a van der Waals fluid above the critical point is "a gas" then there are numerous cases of immiscible gases (Scott and van Konynenburg, 1980, p. 503).

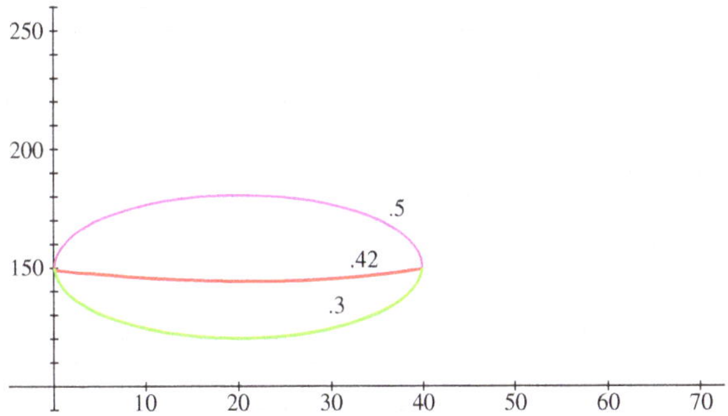

Fig. 6.1.1. Free energy. The function f with $\alpha = 0.5$, 0.42, 0.3, $k = 0.56$ and fixed total density, as a function of the concentration of the mixture. The middle curve has the characteristic shape with two extrema. However, they are maxima, not minima, for either sign of α. Ideal gases are perfectly miscible and this type of interaction does nothing to alter that property.

popular (see, however, Lemmon, 1996; Lemmon and Jacobsen, 1999). For an accumulation of free energy data, see Hicks and Young (1975).[61]

Following van der Waals, an alternative procedure has been to model the mixture as a single component fluid with properties that are a compromise between those of the components, with an explicit dependence of the van der Waals parameters on the molar fractions.[62] This is true, in particular, of an influential paper by Scott and van Konynenburg (1970). Note especially an expression for the excess free energy, valid in the limit of infinite pressure (Scott and van Konynenburg, 1980, p. 499), agrees with Eq. (6.1.1), with $k = 1$.[63]

Our point of view is that a mixture, having more degrees of freedom than the one-component fluid, must exhibit features that cannot be simulated by a single component, either physically or mathematically. A single-component model uses an incomplete set of dynamical variables and therefore cannot obey the fundamental relations of thermodynamics.

[61]Unfortunately, their definition of free energy excess includes the "free energy of mixing" so that a detailed confrontation could not be made.

[62]Rowlinson voices some degree of skepticism: "Such attempts rest on the fallacy that the forces $(\alpha - \beta)$ between two molecules of species α and β are always determinable from the strengths of the forces $(\alpha - \alpha)$ and $(\beta - \beta)$". Rowlinson does not offer an alternative but see the quoted papers by Lemmon and Jacobsen.

[63]One aspect of this approach is particularly reasonable; it is assumed that the "displaced volume" of gas number 1, represented by the coefficient b in the expression for the pressure, must be affected by the presence of gas number 2, and vice versa. I have not tried to include this effect in the calculations.

Lemmon *et al.* (2000) include a factor T in their expression for the excess free energy to make it look like a contribution to the entropy. A contribution to the free energy that is linear in T can always be compensated by a shift in the entropy. We shall make it a definition that the interaction strength is independent of the temperature.

Unfortunately, some of the concepts that help us understand non-interacting gases do not apply when interactions are present. The total pressure is well defined; it coincides with the on-shell value of the Lagrangian density,

$$p = \sum \rho_i \frac{\partial f}{\partial \rho_i} - f \quad (= \mathcal{L} \text{ on shell}). \tag{6.1.4}$$

In the present case,

$$p = \sum \mathcal{R}_i \rho_i T + (2k-1)\alpha(\rho_1 \rho_2)^k. \tag{6.1.5}$$

This is illustrated in Fig. 6.1.2 for the case that $(2k-1)\alpha > 0$. The dependence of p on the concentration (at constant temperature and volume) is linear in the absence of interaction; it dips below the straight line when $(2k-1)\alpha$ is negative and it rises above it if it is positive.

But there is no natural definition of **partial pressures** and Dalton's law cannot be applied; that is one reason why the Gibbs–Dalton hypothesis is

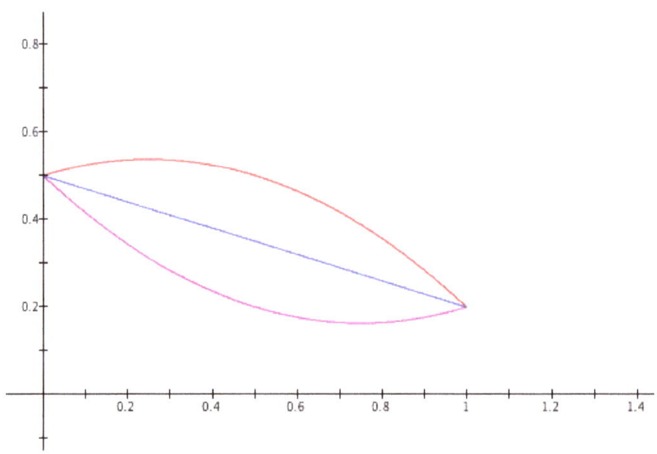

Fig. 6.1.2. Pressure at constant temperature, showing the effect of interaction on the pressure profile. The abscissa is the concentration. When α is positive the upper curve illustrates the case that the exponent k is greater than 0.5, the lower curve the case that $k < 0.5$.

preferable; it can be formulated and suitably modified to take the interaction into account, as we have seen in Chapter 5. The equations of motion, from variation of the densities, give (6.1.2) or

$$\dot{\Phi}_i = \frac{\partial(f + sT)}{\partial \rho_i}\bigg|_T .$$

In the case of a simple system, we can work with the pressure or with the chemical potential as we like, but in the case of mixtures we had better get used to using the latter. In general, the pressure is one of the potentials (potential densities), not a variable.

At constant pressure, the volume as a function of concentration is often less than the sum of the volumes of the separate components. Observation of excess free energy and the deficit of volume are ready sources of information about the parameters α and k.

We shall consider the possible existence and stability of a two-phase equilibrium. After completing this simplest instance of interacting fluids with a brief summary, we shall have enough confidence to attack van der Waals' fluids in Section 6.2.

Data

There exist collections of experimental data that can be related to the interaction density in mixtures. Before we can use it, we must be sure to know what it is that is measured. A large compilation of measured values of excess energies (Hicks and Young, 1975) qualifies them as values of the enthalpy (density). This choice is based on the idea that the experiments were done under circumstances where the system was in diathermic contact with the surroundings and under constant (atmospheric) pressure. That the pressure rather than the volume was fixed is easy to understand, for pressures equalize almost instantaneously. Whether there was enough time for changes in the temperature to adjust may seem doubtful, although, if that was the express intention of the experimenters, then there is no reason why it should not be. But the action principle that has been developed, so far, is adiabatic, and we shall continue to apply it to isolated systems exclusively. The fluctuations of shortest time parameter are the adiabatic changes and they are the variations that determine the instantaneous equilibrium configuration. The postulate of minimum energy therefore refers to either the Helmholtz free energy $\int f$ (if the volume is fixed) or the Gibbs free energy $\int f + p$ (if the pressure is fixed). For simplicity, we shall deal with the former case only.

The difficulty with direct experimental sources is the telegraphic nature of information, for example, "The critical temperatures were observed by allowing the tubes to warm slowly in an environment of measured uniform temperature". Neither "uniform temperature" nor "observed" are defined. Comments provided by the compiler is sometimes useful.

Conditions for equilibrium

According to the action principle, a static configuration is characterized by the condition that the action must be stationary with respect to variations of the temperature and the densities. Local variations of the temperature give the adiabatic condition. The principal statement that comes from variation of the densities is that the fields must be piecewise uniform. If the fluid separates into two parts with different concentrations, then additional, global variations that remain to be investigated are variations that hold the total masses fixed, piecewise uniform variations that satisfy

$$\delta\rho_1^u = -\delta\rho_1^v, \quad \delta\rho_2^u = -\delta\rho_2^v,$$

where the superscripts u, v refer to the two parts of the fluid. This implies continuity of the pressure and the temperature and (the Euler–Lagrange equations)

$$\text{Disc}(\dot\Phi_i - TS_i) = \text{Disc}\ \mu_i := \mu_i^u - \mu_i^v, \quad i = 1, 2.$$

The principle of maximum entropy (= minimum free energy) makes both sides equal to zero.

That both chemical momenta are continuous is the condition that a particle, of either type, at the dividing surface is subject to the same total force whether it is considered to belong to one side of the boundary or the other; the continuity of $\dot\Phi_i - S_iT$ means that **`entropic forces`** are balanced as well.

To sum up, equilibrium of the divided fluid implies continuity of temperature, pressure and $\dot\Phi_i - TS_i - \mu_i$, $i = 1, 2$. Minimum energy implies that $\dot\Phi_i - TS_i$ and μ_i are separately continuous.

To solve these equations, several methods can be used.

The method of the `common tangent`

The conditions of equilibrium are

$$T^u = T^v, \quad p^u = p^v, \quad \mu_i^u = \mu_i^v, \quad i = 1, 2. \tag{6.1.6}$$

We shall regard the densities as coordinates in a two-dimensional Euclidean space and use a geometric notation. The equations of equilibrium shows that the tangent planes to the equipotential surface f = constant, at the two points, are parallel (see Section 4.4). Now,

$$p = (\mathcal{R}_1\rho_1 + \mathcal{R}_2\rho_2)T + \alpha(2k-1)(\rho_1\rho_2)^k. \tag{6.1.7}$$

Since $p = \rho_1(\partial f/\partial\rho_1) + \rho_2(\partial f/\partial\rho_2) - f$, it follows that

$$f(\rho_1,\rho_2)^u - f(\rho_1,\rho_2)^v = (\rho_1^u - \rho_1^v)\mu_1^v + (\rho_2^u - \rho_2^v)\mu_2^v.$$

That is, the two tangent planes actually coincide. This is exactly the same situation that occurs in saturation, as illustrated in Fig. 4.3.1 and subsequent figures, but in a higher dimension, making illustration more difficult.

As an example, we may take $\beta = 2$, $k = 1$ and thus $\alpha = 2\mathcal{R}T/\rho$. The concentrations at the minima of f in Fig. 6.1.3 are $x = 0.24853$ and $y = 0.75147$. The pressure is $\mathcal{R}\rho T + \alpha\rho_1\rho_2 = \mathcal{R}$, so that the excess pressure at fixed volume must be more than twice the pressure of the corresponding ideal, non-interacting mixture. It is likely that, when this condition is met, one or both of the real gases will be in the domain where they must be described by the van der Waals equation of state.

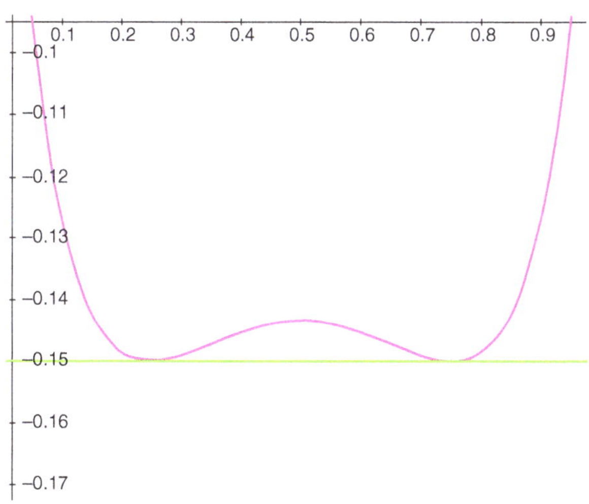

Fig. 6.1.3. **Common tangent**. Example, $\beta = 2$, $k = 1$; the dependence of f on the concentration, on the line $\rho_1 + \rho_2 = 1$. The free energy density has been normalized to give the tangent zero slope.

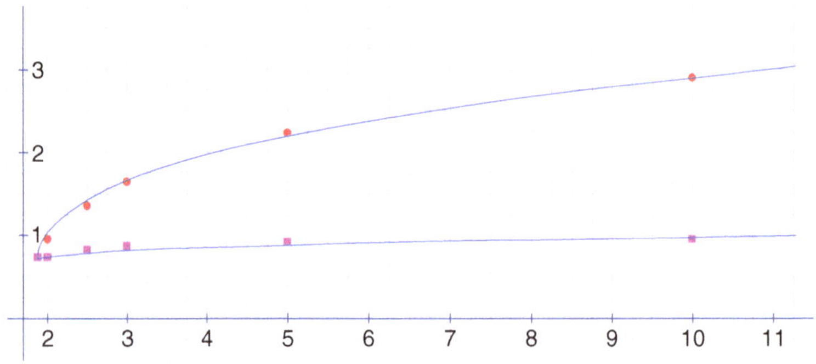

Fig. 6.1.4. The range of values of the parameters k (abscissa) and β (ordinate) for which the free energy density exhibits two minima is the interval between the two lines.

For assigned values of the parameters α, k, there is a finite range of values of α, approximately,

$$2 < \frac{\alpha}{\mathcal{R}T}\rho^{2k-1} < 10.$$

This is the range of temperatures for which separation occurs (Fig. 6.1.4). The upper limit is referred to as UCST or upper critical solution temperature. At this temperature, any decrease in the temperature converts the system from a homogeneous liquid to a two-phase system. The lower limit, L(ower)CST, the same transformation takes place when the temperature is increased (Scott and van Konynenburg, 1980).

Symmetry

If it happens that the formula for the pressure, when expressed in terms of molar densities, is symmetric with respect to the interchange of the two densities, the fact that $p^u = p^v$ strongly suggests that we look for solutions that satisfy

$$(\rho_1^u, \rho_2^u) = (\rho_2^v, \rho_1^v). \tag{6.1.8}$$

For a determination of equilibrium using (6.1.8), see Fig. 6.1.5. This method is seldom applicable.

Direct search

Both methods used so far to find the coexistent mixtures depend on symmetry, on the assumption that the two coexistent mixtures have conjugate concentrations, $(\rho_1, \rho_2)^u = (\rho_2, \rho_1)^v$. But there are other solutions, and to

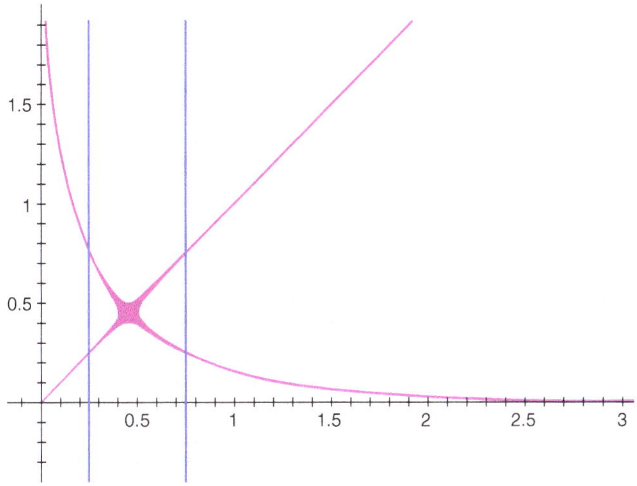

Fig. 6.1.5. Determination of concentrations using symmetry. The curve is the locus of pairs of densities that satisfy the symmetry condition (6.1.8). The two vertical blue lines mark the values of x with the selected value of the pressure. The coexistent densities are the coordinates of the intersections.

find them all we know of no method better than the following. We produce plots of the loci defined by the three equations

$$\mu_1(x, y) = A, \quad \mu_2(x, y) = B, \quad p(x, y) = C,$$

where A, B, C stand for real numbers. These numbers are adjusted so that the first two curves intersect at two points; then they are adjusted further until an isobar passes through both of them. For an illustration, see Fig. 6.1.6.

This method is the more rewarding, for it will give us a visual representation of the nature of the physical phenomenon. We shall refine this scheme and carry out the calculations in several particular cases later.

Summary

We have explored a simple phenomenon that involves a mixture of two ideal gases, slightly modified by a simple interaction term added to the free energy. In this case, we have verified, under certain circumstances, the separation of the mixture into two parts with quite different concentrations. What makes real gases more complicated, and more interesting, is the fact that interactions between like particles also come into play, in a regime not too far from saturation, where the individual fluid components behave, approximately, like van der Waals fluids. In fact, it is commonly believed that the phenomenon of immiscibility cannot occur in fluids above the condensation

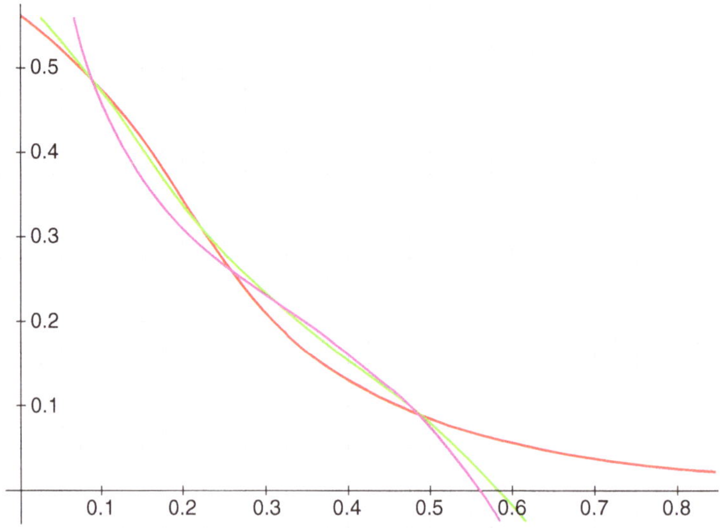

Fig. 6.1.6. A typical determination of **coexistence**. The ordinates are the two densities. The red lines are loci of the chemical potentials, the green line is an isobar. The two triple intersections are the densities of two coexistent densities. The geometry is typical of immiscibility, as opposed to condensation of one of the constituents.

point. It is even said that it does not happen with gases; what lies behind this is a too restricted definition of "gas".

What about the entropies of these mixtures? It would be naive to suppose that the Gibbs–Dalton law applies, since we expect that it is affected by the interaction. Fortunately, as we have seen, and as we shall confirm below, it is not necessary to know the entropies in the present investigation. What we need to know about the entropy was discussed above; it led to the statement that the chemical potentials are continuous.

6.2. Mixture of Two van der Waals Fluids

The spinodal of unaries

Refer to Fig. 6.2.1; it is the same as Fig. 4.3.1, with additions. The curve is the van der Waals potential, the function $p(\rho, T)$ for a fixed value of T. This function is sometimes referred to as an equation of state, which is misleading since ρ, T is not the natural set of variables for the pressure — the natural variables are μ and T.

The points A and B in the figure are the points of coexistence for the chosen temperature; both p and μ have the same values at both points. They define a pair of stable configurations in equilibrium with each other. A very

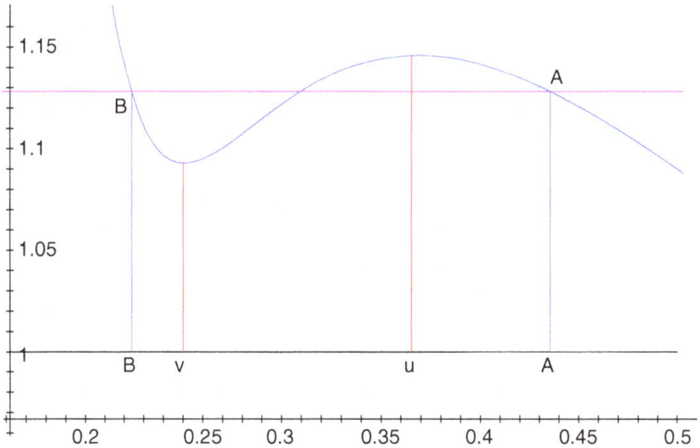

Fig. 6.2.1. From Chapter 4; with the points of the spinodal added; it is the manifold that consists of the two points u and v on the density axis defined by (6.2.1). The abscissa is the inverse density. The temperature is fixed below critical.

useful concept is the "**spinodal**"; in the present case, it is the set of densities labeled u, v, points at which $\partial p/\partial\rho|_T = 0$. A better characterization of this manifold is

$$\left.\frac{\partial^2 f}{\partial \rho^2}\right|_T = 0 \tag{6.2.1}$$

(note that $\partial p/\partial\rho|_T = \rho\partial^2 f/\partial\rho^2$); it has the advantage of being defined by the free energy as a function of its natural variables and it has a simple and natural generalization to mixtures.

The most important property of the spinodal is the following. The two points are bracketed by the points A, B; the unstable point between A and B lies between the spinodal points and this gives a fairly accurate indication of the location of the points of coexistence once the spinodal is known. As the temperature is raised, points A and B approach each other and coalesce with the spinodal (points u and v) at T_{cr}.

The mapping $p \mapsto \rho$ is three-valued when the pressure is between the values $p(u)$ and $p(v)$, single-valued when the density is outside this interval.

The **spinodal of binaries**

The set of 2-phase configurations of a typical binary system, for each pair of values p, T, is a one-dimensional subset of the density plane. Since, in the case of van der Waals fluids, $x = b_1\rho_1$ and $y = b_2\rho_2$ take values between 0 and 1, we shall use coordinates x, y with this range, $0 < x$, $y < 1$ and refer to $\vec{\rho} = (x, y)$ as a point in density space.

For fixed T below a certain limit $T_{\max}$, the density space of a typical binary system is divided into three or more distinct regions, separated and sometimes bounded by points of the spinodal, this being the locus of points on which

$$\det\left(\frac{\partial f}{\partial\rho_i\partial\rho_j}\right)_T = 0,$$

the natural generalization of (6.2.1). These are points where the mapping $\vec{\mu} \to \vec{\rho}$ is two-valued. They form lines that separate a single-valued domain from a three-valued domain. The spinodal is not the line of coexistence, but it gives a good indication of where the coexistent mixtures are, and it has the advantage of being easy to calculate and to plot.

Here is an example of what is observed in a typical experiment. The experimenter usually begins with a homogeneous mixture at low pressure, a gaseous system with fixed concentration. Because the concentration cannot be easily manipulated, it is natural to use the concentration as one of the variables, another being the total density or its inverse, the volume. The temperature may be kept fixed while the pressure is slowly increased to take the system through a sequence of adiabatic equilibria.

As the pressure is increased (with T fixed), the system typically traces a radial line in **density space**. Eventually, as the density approaches the spinodal, condensation sets in; this point is referred to as the **dew point**. At best, the droplets fall like rain and accumulate at the bottom of the vessel.

Figures 6.2.2–6.2.4 show what happens to a model of an ethane–propane mixture in the two-dimensional density space. The oval (purple) is the spinodal of the mixture. As the temperature is raised, the oval contracts and reduces to a point at a temperature $T_{\max} = 370.5$, above which no phase separation occurs. When T is reduced to 369.5, which is the critical temperature of propane, the oval touches the vertical axes; at lower temperatures, the oval is a horizontal band across density space.

The slope of the radial line in Fig. 6.2.3 is the concentration ρ_2/ρ_1. In the most standard type of experiment, the system moves out along a radial line as the pressure is increased. If the radial line intersects the spinodal oval, then the system moves along the radial until, before quite touching the spinodal, it reaches a point of condensation, the dew point, shown in Fig. 6.2.3 as A, with density $\vec{\rho}^u = (x^u, y^u)$.

What happens next, as the pressure is further increased, is this. In addition to the gas at the point $A = \vec{\rho}^u$, there appears a liquid at the point $B = \vec{\rho}^v$ on the same isobar. Initially, the population of this point, that is,

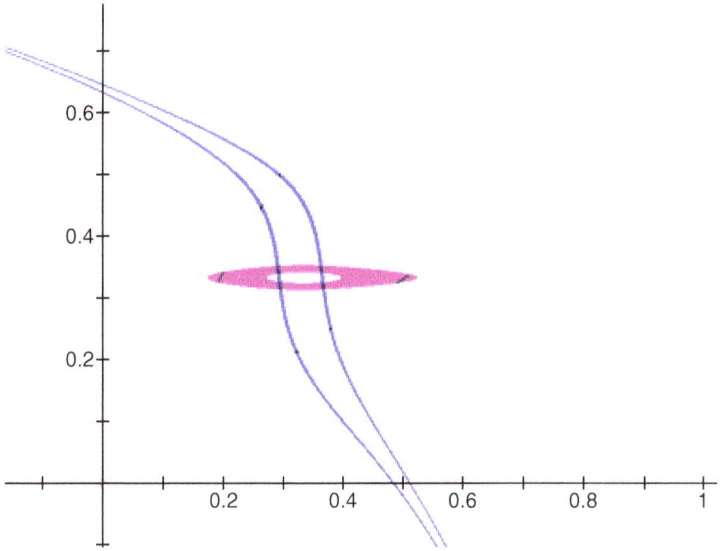

Fig. 6.2.2. Density diagram for the model of the mixture of methane (abscissa) and propane (ordinate) at $T = 370.4$, $\alpha = 5$, showing the spinodal (purple) and two (blue) **isobars**. The spinodal is a closed line bordering an oval. Radials indicate critical concentrations for this value of T.

the total amount of liquid with this density, is negligible, but with increasing pressure the system gradually transfers, by condensation, from A to B. To maintain the average concentration of the total sample, this transfer is accompanied by a movement of both points towards the right along "lines of coexistence" at a nearly constant distance from the spinodal. This continues until the upper point reaches the radial of the original concentration (the continuation of the original radial line) at which point the condensation is complete and the whole system is at the point B', the bubble point. The lower point of coexistence has moved from the original point A to reach the point A' (on the same isobar as B') with the occupation diminishing to zero. Further increasing the pressure does not lead to any unusual phenomena, the liquid remains homogeneous, moving along the original radial line, upwards from B'.

At lower temperatures, the spinodal reaches the vertical axis, at $T = 369.5$, the critical temperature of propane, as well as the vertical line $x = 1$. At still lower temperatures, it approaches the horizontal axis at $T = 306$, the critical temperature of ethane, as shown in Fig. 6.2.4.

Experimental reports often give much less information, most experiments are stopped on reaching the dew point, to record the dew point pressure

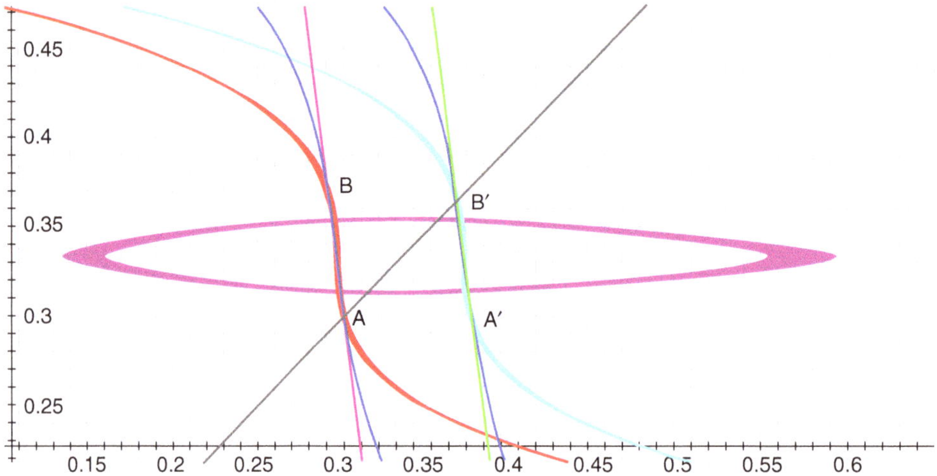

Fig. 6.2.3. Density diagram for (the same model of) the mixture of methane and propane at $T = 370$, $\alpha = 5$. If the temperature is lowered to 369.5, which is the **critical temperature** of propane, then the spinodal reaches the vertical axis and below that temperature it is a nearly horizontal band that reaches across the entire interval $0 < x < 1$. The blue lines are isobars, the red ones and the light blue ones are loci of the chemical potentials.

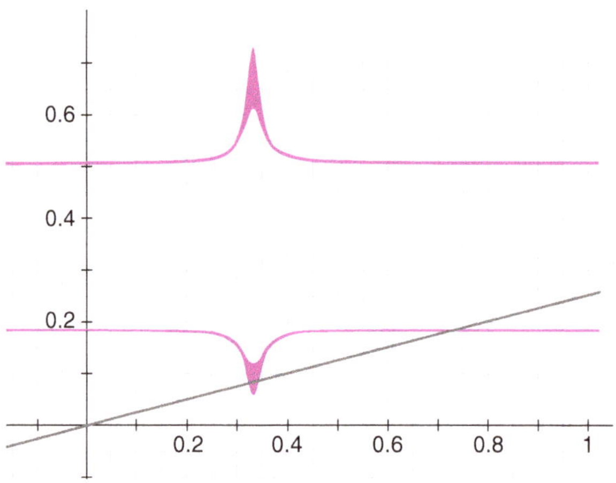

Fig. 6.2.4. As the two preceding figures but the temperature is slightly above the critical temperature of ethane.

at the chosen temperature. In a companion experiment, one starts with a homogeneous liquid at high pressure and traces out the approach to the point B', **the bubble point**. The actual separation is often not observed. Results are presented as a T, p graph.

Calculation of the line of coexistence

The theoretical calculation of the points of **coexistence** is painful without the assistance of an appropriate computer. These points are defined by the conditions of equilibrium,

$$T^u = T^v, \quad p^u = p^v, \quad \vec{\mu}^u = \vec{\mu}^v.$$

The best strategy that was found is the following. Choose a value of T and an isobar and select a point $\vec{\rho}_0$ on it. Experience allows to select a point fairly close to the line of coexistence. Draw the loci (all for fixed T and $\vec{\rho}_0$)

$$p(\vec{\rho}) = p(\vec{\rho}_0), \quad \mu_1(\vec{\rho}) = \mu_1(\vec{\rho}_0), \quad \mu_2(\vec{\rho}) = \mu_2(\vec{\rho}_0).$$

The last two should have two further intersections, one inside the spinodal (ignore it) and one on the upper side. This latter will probably not lie on the same isobar; in that case, slide the point $\vec{\rho}_0$ along the chosen isobar until the required triple intersection is achieved on the upper side. This locates a pair of coexistent concentrations, the points A and B. The points A' and B' are easily determined since they are known to be on the radial line of the original, average concentration.

6.3. Sample Applications: First Example

This is all about mixtures of van der Waals fluids. The experimental literature is vast, but much of it reports only partial results; there is very little observation of actual separation. Theoretical attempts are, for the most part, applications of van der Waals' one-fluid method that we have already mentioned; more shall be said about this below. There have been many attempts to account for the data using variations of the van der Waals approach. Because these theories are phenomenological, and consequently subject to few constraints, they are often successful in reproducing selected data. The attempts reported here are also less than completely successful but the work is far from complete.

In this section, we continue to base the approach on the action principle, with a complete set of thermodynamical variables, with the approach to mixtures of Chapter 5 and with the component unaries modeled as van der Waals fluids. Because we aim at making predictions, we must restrict the number of free parameters; consequently, our first attempt uses the following expression for the free energy density,

$$f = f_1 + f_2 + f_{\text{int}}, \quad f_{\text{int}} = \alpha \rho_1 \rho_2.$$

There is only one free parameter, α, and the van der Waals parameters in f_1 and in f_2 are determined by observations of the pure components. The interaction term is of the type used by others, but much more restricted; cf. Scott and van Konynenburg (1980) pose

$$f_{\text{int}} = Q(c)\rho^2,$$

where ρ is the total density, c is the concentration and Q is a polynomial — and these are not the only adjustable parameters.

Here, we shall examine some mixtures of the simplest **organic fluids**.

"Experimental value of the pressure". For propane, these are very difficult to find; and most of the discrepancy arises from the inconsistency of the few experimental figures. For consistency, the numbers 4.55 and 4.09 in the first column should have been equal (Pressure in MPa). As an approximation to the real, unary fluids, the van der Waals model is successful in the qualitative sense. Quantitatively, it usually reproduces the experimental properties to better than 10 percent accuracy near the critical temperatures, better at higher temperatures but progressively less well at lower temperatures. Our hope is to develop a model of the mixtures with a comparable degree of agreement with experiments (Table 6.3.1).

The **vapor pressures** of ethane and propane are calculated from the van der Waals formula and collected in Table 6.3.2.

Table 6.3.1. Critical parameters.

Methane	16	190.6	4.63	7.83	0.229	0.0427
Ethane	30	306	4.96	5.21	0.550	0.0640
Propane	44	369.5	4.52	3.95	0.877	0.0844
Butane	58	25.1	3.58	2.72	1.466	0.123

Source: From Hicks and Young (1975, Appendix A). Units are MPa for pressure, MPa(Lit/mol)^2 for a and (mol/Lit) for b. The figures are consistent with van der Waals' formula within 1 percent.

Table 6.3.2. Vapor pressure.

Ethane	T	306	290	270	250	220	200
p	4.97	4.00	2.96	2.12	1.14	0.7	
Exp.	4.96	3.51	2.20	1.30	0.49	0.22	
Propane	T	369.5	360	340	320	300	280
p	4.55	4.1	3.23	2.49	1.87	1.35	
Exp.	4.09	3.5	2.42	1.58	0.98	0.624	

Binary van der Waals mixtures: Formulas

For a mixture of any two van der Waals fluids, the assumed free energy density is

$$f = f_1 + f_2 + \alpha\mathcal{R}(\rho_1\rho_2)^k, \tag{6.3.1}$$

$$f_1 = \mathcal{R}T\rho_1 \ln\frac{\rho_1}{(1-b_1\rho_1)T^{n_1}} - a_1{\rho_1}^2, \quad f_2 = \mathcal{R}T\rho_2 \ln\frac{\rho_2}{(1-b_2\rho_2)T^{n_2}} - a_2{\rho_2}^2, \tag{6.3.2}$$

with the constant α of order unity and $k = 1$. To simplify, we set $x = b_1\rho_1$, $y = b_2\rho_2$. We also drop the factor T^n in the logarithm since it makes no contribution to the determination of the equilibrium,

$$f_1 = \frac{\mathcal{R}}{b_1}\left(Tx\ln\frac{x}{1-x} - \frac{27}{8}T_1x^2\right), \quad f_2 = \frac{\mathcal{R}}{b_2}\left(Tx\ln\frac{y}{1-y} - \frac{27}{8}T_2y^2\right),$$

and

$$f = \frac{\mathcal{R}}{b_1}\left(Tx\ln\frac{x}{1-x} - \frac{27}{8}T_1x^2\right) + \frac{\mathcal{R}}{b_2}\left(Tx\ln\frac{y}{1-y} - \frac{27}{8}T_2y^2\right) + \alpha\mathcal{R}\left(\frac{xy}{b_1b_2}\right)^k;$$

T_1 and T_2 are the critical temperatures of the constituents. The derivatives of the free energy density are, when the exponent k in (6.2.1) is unity,

$$\mu_i(x,y) = \mathcal{R}T\left(\ln\frac{\rho_i}{(1-b_i\rho_i)T^{n_i}} + \frac{1}{1-b_i\rho_i}\right) - 2a_i\rho_i + \alpha\mathcal{R}\rho_j, \quad i,j = 1,2 \text{ or } 2,1,$$

and

$$f_{ii} = \frac{\mathcal{R}T}{\rho_i(1-b_i\rho_i)^2} - 2a_i, \quad i = 1,2, \quad f_{12} = \alpha\mathcal{R}.$$

The thermodynamic pressure is

$$\frac{p}{\mathcal{R}} = \frac{1}{b_1}\left(T\frac{x}{1-x} - \frac{27}{8}T_1x^2\right) + \frac{1}{b_2}\left(T\frac{y}{1-x} - \frac{27}{8}T_2y^2\right) + \frac{\alpha}{b_1b_2}xy.$$

The spinodal, a two-dimensional manifold in the 3-space with coordinates T and the two densities, is defined by

$$\det\left(\frac{\partial^2 f}{\partial\rho_i\partial\rho_j}\right) = 0.$$

Table 6.3.3. Maximal temperature.

α	1	2	5	10	50
$T_{\max}$	369.6	369.8	371.1	375.4	443.6
$p/\mathcal{R}$	1661	1685.5	1765	1930	2860
p	13.81	14.01	14.67	16.05	23.78

In a form suitable for the computer,

$$f_{11} = \mathcal{R}b_1\left(T\frac{1}{x(1-x)^2} - \frac{27}{4}T_1\right),$$

$$f_{22} = \mathcal{R}b_2\left(T\frac{1}{y(1-y)^2} - \frac{27}{4}T_2\right),$$

and $f_{12} = f_{21} = \alpha$. The spinodal is thus the locus

$$b_1 b_2\left(\frac{T}{x(1-x)^2} - \frac{27}{4}T_1\right)\left(\frac{T}{y(1-y)^2} - \frac{27}{4}T_2\right) = \alpha^2. \tag{6.3.3}$$

For a fixed value of T, it is empty unless T is less than a "maximal" value that, for the ethane/propane mixture with $\alpha = 10$, is

$$T_{\max} = 375.4\,\mathrm{K}$$

The variation of this with the strength of the interaction is shown in Table 6.3.3.

Table 6.3.3 is the **maximal temperature**, being the highest temperature at which separation occurs, for ethane/propane with varying interaction strength.

At this value of T, the spinodal is a single point at $x = y = 1/3$. Below this value, the single point opens to a closed oval, as shown in Figs. 6.2.2 and 6.2.3. Both figures also show two isobars, each of which cuts the oval. Straight lines from the origin are lines of constant concentration; in Fig. 6.2.2, they are tangent to the ovals.

Coexistence

The conditions for coexistence of two mixtures are the continuity of T, p and

$$\mu_i(x,y) = \mathcal{R}T\left(\ln\frac{\rho_i}{(1-b_i\rho_i)T^{n_i}} + \frac{1}{1-b_i\rho_i}\right) - 2a_i\rho_i + \alpha\rho_j, \quad i,j = 1,2 \text{ or } 2,1.$$

An efficient way to perform the calculation was already presented.

The dynamical variables have the following values at points A, A', B, B', in Fig. 6.2.3.

Table 6.3.4. Maximal temperature.

Point	x^u	y^u	p/R
A	0.37	0.3	1658
B	0.291	0.370	
A'	0.373	0.294	1847
B'	0.364	0.364	

Table 6.3.4 is the two pairs of coexistent mixtures for given initial low density, 50–50, gaseous, model mixture of ethane and propane. Most experiments measure only the pressure.

Points that are `critical` to the experimenter

In a region of temperatures just below $T_{\max}$, the spinodal is an oval, simply connected and convex. In a typical experiment, one starts with a low pressure and low densities, x and y normalized molar densities less than 1/3. The mixture is homogeneous. The pressure is increased, or the temperature is increased, or both, but the concentration is kept fixed. One is moving out from the origin of the density diagram along one of the radials. This radial may cross the spinodal or it may pass to the left or to the right of it. In the first case, separation will occur at a point below the spinodal; what happens here was described in Section 6.2. Figures 6.2.2 and 6.2.3 illustrate this in the case of a model.

As the temperature is changed, the intersection of any radial line eventually becomes tangent to the spinodal. The meniscus disappears if this point is very close to the spinodal, as in Fig. 6.2.2; at this point, the densities of liquid and gas become equal; these are the critical points reported by the experimenter. As the temperature descends to T_2, the critical temperature of the least volatile gas, the oval reaches the vertical axis. Below this temperature, there is an interval of temperatures where the oval still may have a radial tangent at the high end, at very high pressures, then this also disappears. At still lower temperatures, as we approach the critical point of the other gas, it returns (see Fig. 6.2.4).

This behavior of the model is in contradiction with experiments. We must conclude that our first choice of f_{int} is inappropriate, but before looking for alternatives, we shall look briefly at another example, the argon–oxygen system.

The critical points to the experimenter are the configurations at which the radials determined by the concentration are tangent to the spinodal oval. They are uniquely determined by the temperature, at least for temperatures close to $T_{\max}$. Figures 6.2.2–6.2.4 show the shape of the spinodal for a descending series of temperatures for a model that has $f_{\text{int}} \propto \rho_1\rho_2$.

Remark. The studies of mixtures in terms of expressions for the free energy are legion in the literature. It is unexpected, however, to encounter an approach that is phenomenological to the point that additional input, an "equation of state" or more precisely an expression for the pressure, is needed (Hicks and Young, 1975); this even in the case where the inspiration comes from van der Waals and the free energy had already been used. In thermodynamics, the pressure is determined by the free energy density, $p = \vec{\rho}\cdot\vec{\nabla} f - f$. But the "free energy" of the one-fluid model is not the thermodynamic free energy and the one-field model is not thermodynamics. Consequently, the formula that is used to calculate the critical points is without justification. Separate phenomenological input (with additional adjustable parameters) is invoked in the form of an independent relation between pressure and temperature.

6.4. Second Mixture of van der Waals Fluids

The plan for this section is not just to study another mixture, but to consider another type of interaction.

Here, we examine mixtures of argon and oxygen. These two gases have approximately the same van der Waals parameters.

Gas number 1, argon, $\mu_1 = 40$, $n_1 = 3/2$, $T_1 = 150\,\text{K}$, $p_1 = 4.9\,\text{MPa}$, $a_1 = 1.355$, $b_1 = 0.03201$.

Gas number 2, oxygen, $\mu_2 = 32$, $n_2 = 5/2$, $T_2 = 155\,\text{K}$, $p_2 = 5\,\text{MPa}$, $a_2 = 1.382$, $b_2 = 0.03186$.

The units are bar.Lit^2/mol^2 for a and Lit/mol for b.

We begin by examining the spinodal isobars, Eq. (6.4.1):

$$b_1 b_2 \left(\frac{T}{x(1-x)^2} - \frac{27}{4}T_1\right)\left(\frac{T}{y(1-y)^2} - \frac{27}{4}T_2\right) = \alpha^2. \tag{6.4.1}$$

This relates α directly related to $T_{\max}$, the highest temperature at which separation can occur at any concentration, *the* critical temperature. We should like to begin by choosing the best value of the coupling strength α. Unfortunately, this is rarely measured. Jones and Rowlinson tell us only that $T_{\max}$ is higher than 149.4 K; our model says that it is higher than 155. Table 6.4.1 relates $T_{\max}$ to α.

Table 6.4.1. Effect of strength of interaction.

T_{max}	α
155.2	0.001
156.0	0.0032
160.0	0.0098
164.2	0.0160
171.2	0.0260
179.6	0.0380
186.7	0.048

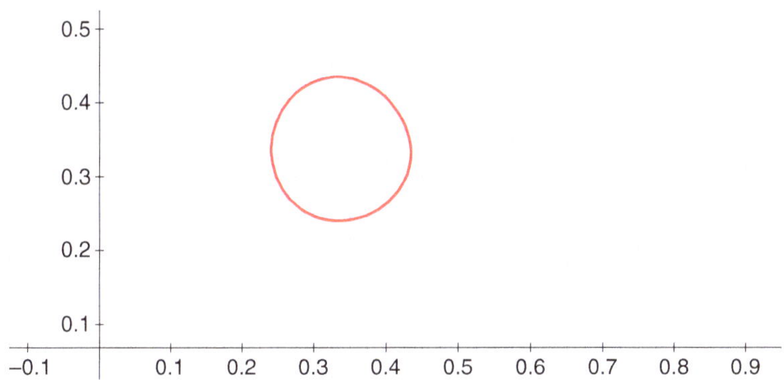

Fig. 6.4.1. The spinodal for a temperature slightly lower than critical.

Table 6.4.1 is the highest temperature at which separation occurs, as a function of the strength of the interaction $f_{\text{int}} = \alpha\rho_1\rho_2$.

Figure 6.4.1 shows the spinodal for $\alpha = 0.016$, $T = 160\,\text{K}$. In Fig. 6.4.2, $T = 155\,\text{K}$, which is the critical temperature of oxygen, the spinodal oval has grown to reach the vertical axis. In Fig. 6.4.3, the temperature is 149.4 K, the highest temperature recorded by Jones and Rowlinson. The blue line is the locus of the concentration, value 0.202 as in the experiment. According to observation, this is a critical point so we expect it to be nearly tangent to the spinodal. Since the temperature is below the critical temperature (155 K) of oxygen, this is in clear disagreement with the model.

Later, we shall attempt to understand the propagation of sound in mixtures with the same model and the same interaction as we have used here, $f_{\text{int}} = \alpha\rho_1\rho_2$, to find that it cannot account for sound propagation either. Instead, what works very well for the speed of sound is

$$f_{\text{int}} = \alpha\sqrt{\rho_1\rho_2}.$$

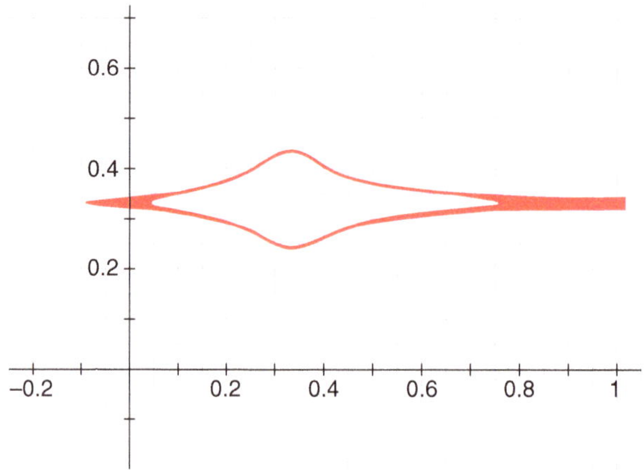

Fig. 6.4.2. As the preceding figure, but $T = T_2 = 155\,\text{K}$.

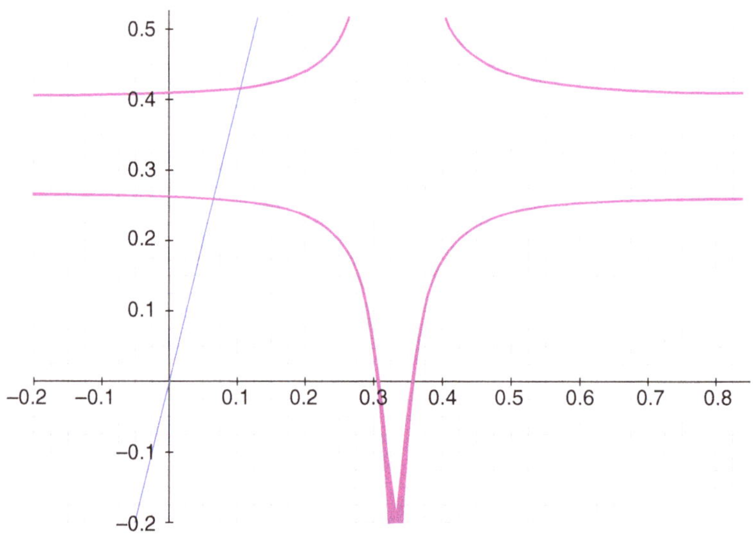

Fig. 6.4.3. Here $T = 149.4\,\text{K}$. The radial line indicates the concentration in the experiment of Jones and Rowlinson.

Although the success of this model in accounting for the speed of sound refers to experiments at room temperature, it is natural to ask what it can do for the critical phenomena being discussed here. Indeed, our approach obliges us to use the same Lagrangian for both phenomena. As we have stressed, this is an essential feature of our method, the main source of a greatly enhanced predictive power.

The square root interaction density for mixtures

It was natural to expect that the expression for the interaction energy take the same form as the self-interaction energy in van der Waals formula. That is why we have given this idea a fair hearing, to see it fail. We already had some experience with sound propagation that suggested a different form for $f_{\rm int}$ (Fronsdal, 2011). To test the idea that the same Lagrangian density must serve for both phenomena, we now investigate the viability of

$$f_{\rm int} = \mathcal{R}\alpha\rho_1\rho_2 + \mathcal{R}\alpha'\sqrt{\rho_1\rho_2},$$

with α provisionally equal to zero. The chemical potentials

$$\mu_1 = f_1 = \mathcal{R}T\left(\ln\frac{\rho_1}{(1-b_1\rho_1)T^{n_1}} + \frac{1}{1-b_1\rho_1}\right) - 2a_1\rho_1 + \mathcal{R}\frac{\alpha'}{2}\sqrt{\rho_2/\rho_1},$$

$$\frac{f_{11}}{\mathcal{R}} = b_1\left(\frac{T}{x(1-x)^2} - \frac{27}{4}T_1\right) - \frac{\alpha'}{4}\sqrt{b_1^3/b_2}\sqrt{y/x^3},$$

$$\frac{f_{12}}{\mathcal{R}} = \frac{\alpha'}{4}\sqrt{b_1b_2}/\sqrt{xy},$$

and the spinodal is the locus

$$f_{11}f_{22} = \left(\frac{\alpha'}{4}\right)^2\frac{b_1b_2}{xy}$$

or

$$\left(\frac{T}{x(1-x)^2} - \frac{27}{4}T_1 - \sqrt{\frac{b_1}{b_2}}\frac{\alpha'}{4}\sqrt{\frac{y}{x^3}}\right)$$
$$\times\left(\frac{T}{y(1-y)^2} - \frac{27}{4}T_2 - \sqrt{\frac{b_2}{b_1}}\frac{\alpha'}{4}\sqrt{\frac{x}{y^3}}\right) = \left(\frac{\alpha'}{4}\right)^2\frac{1}{xy}.$$

To interpret it, we need isobars. The pressure is independent of the interaction, simply

$$\frac{p}{\mathcal{R}} = \frac{1}{b_1}\left(\frac{x}{1-x}T - \frac{27}{8}T_1x^2\right) + \frac{1}{b_2}\left(\frac{y}{1-y}T - \frac{27}{8}T_2x^2\right).$$

The expressions for the μ_i are

$$\frac{\mu_1}{\mathcal{R}T_1} = \frac{T}{5.19}\left(\ln\frac{x}{1-x} + \frac{1}{1-x}\right) - \frac{27}{4}x + 36.0321\sqrt{\frac{20}{50}}y + 0.03112\frac{\alpha'}{\mathcal{R}}\sqrt{\frac{y}{x}},$$

$$\frac{\mu_2}{\mathcal{R}T_2} = \frac{T}{290}\left(\ln\frac{y}{1-y} + \frac{1}{1-y}\right) - \frac{27}{4}y + 1.38765\sqrt{\frac{20}{50}}x + 0.001334\alpha'\mathcal{R}.$$

Table 6.4.2. Results from the square root model.

T	p	c	Exp.
132.0	520.77	0.798	
138.0	760	0.70	0.599
143.8	865	0.34	0.401
149.4	1100	0.19	0.202

The formulas that we need are, the free energy,

$$\hat{f} = \frac{fb_1}{\mathcal{R}T_{c1}} = T_r x \ln \frac{x}{1-x} - \frac{27}{8}x^2 + b\left(T_r y \ln \frac{y}{1-y} - c\frac{27}{8}y^2\right) + \beta xy,$$

the chemical potentials,

$$\mu_x(x,y) = T_r\left(\ln \frac{x}{1-x} + \frac{1}{1-x}\right) - \frac{27}{4}x + \beta y,$$

$$\mu_y(x,y) = bT_r\left(\ln \frac{y}{1-y} + \frac{1}{1-y}\right) - bc\frac{27}{4}y + \beta x,$$

the second derivatives,

$$f_{xx} = \frac{T_r}{x(1-x)^2} - \frac{27}{4}, \quad f_{yy} = b\frac{T_r}{y(1-y)^2} - bc\frac{27}{4}, \quad f_{xy} = \beta,$$

and the pressure,

$$\frac{pb_1}{\mathcal{R}T_{c1}} = T_r\left(\frac{x}{1-x} + b\frac{y}{1-y} - \frac{27}{8}(x^2 + bcy^2)\right) + \beta xy.$$

Reasonable agreement with the observations of Jones and Rowlinson is obtained with $\alpha = 120$, as shown in Table 6.4.2 and in Figs. 6.4.4–6.4.7.

The four experiments are illustrated by the plotting of the spinodal (purple) and the relevant isobar (blue) in the density plane (see Figs. 6.4.4–6.4.7). Each radial (green) line is the path followed during one of the experiments in which the concentration is constant. The experiment started at low density and pressure and was terminated at the dew point.

Table 6.4.2 is the square root model with $\alpha = 120$. The third column is the (molar) concentration $c = x/x + y$. The temperatures are those of the experiment and the last column shows the experimental concentrations. Experimental values of the pressure are not available.

Quantitatively, the model agrees very well with the measurements. The shape of the spinodal is quite different in this model and this is the type of structure that is needed to account for the experiment of Jones and Rowlinson. It may be noted that the geometry exhibited at the first two

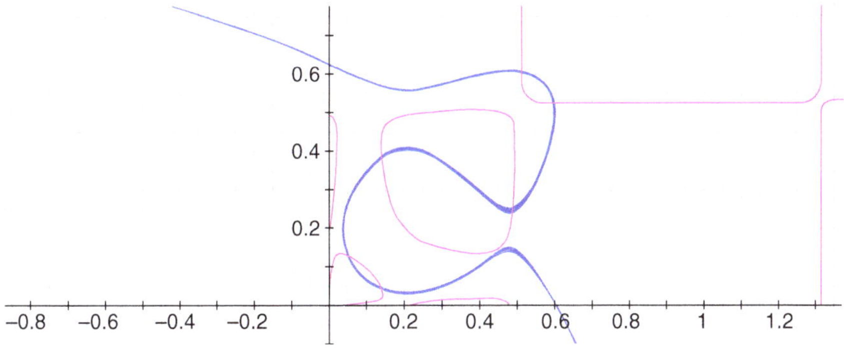

Fig. 6.4.4. Here, $T = 149.4\,\mathrm{K}$.

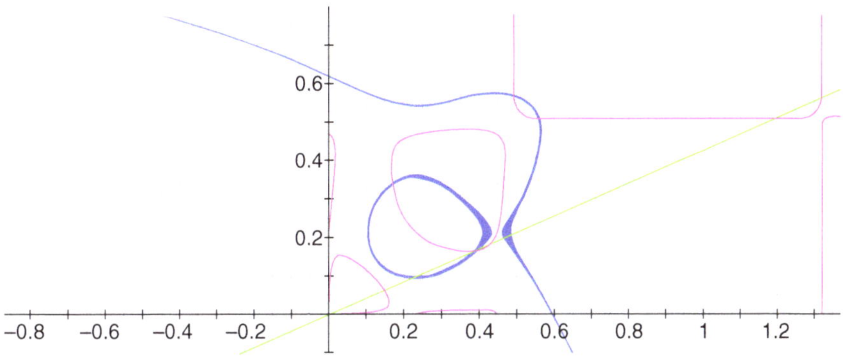

Fig. 6.4.5. Here, $T = 143.8\,\mathrm{K}$.

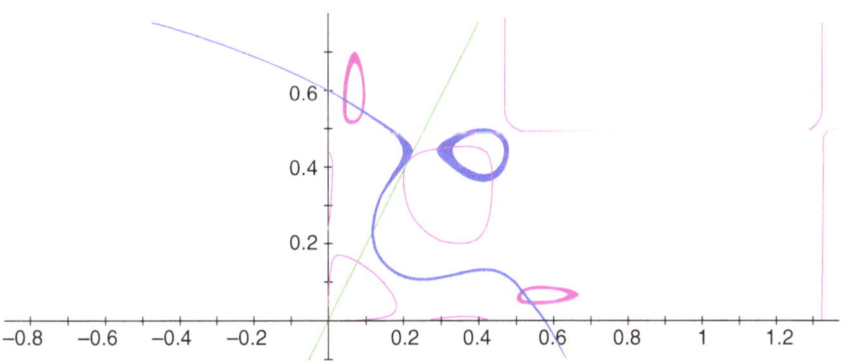

Fig. 6.4.6. Here, $T = 138\,\mathrm{K}$.

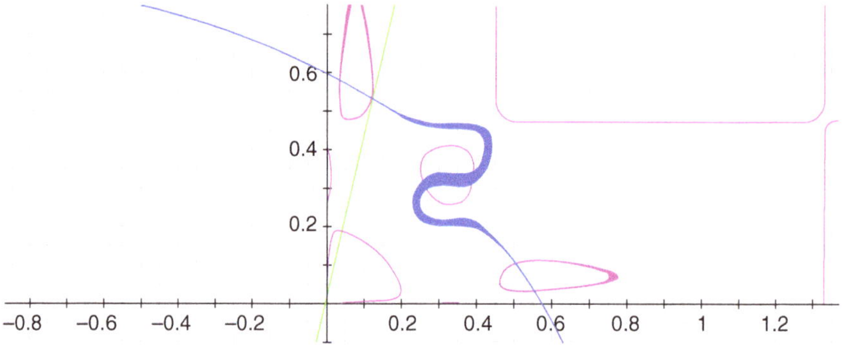

Fig. 6.4.7. Here, $T = 132\,\mathrm{K}$.

Table 6.4.3. Nitrogen and argon's square root model.

T	p	c	Experiment
131.0	580	0.58	0.8
134.5	600	0.42	0.6
141.0	865	0.4	
146	1100	0.30.2	

points is quite different from the last two points; something that cannot be revealed by the experiment.

Nitrogen and Argon

This case is very similar but there is less experimental information available.

Gas number 1, nitrogen, $\mu_1 = 34$, $n_1 = 5/2$, $T_1 = 126.3$, $b_1 = 0.03913$.
Gas number 2, argon.

Taking $\alpha' = 100$, we get a good fit to three out of four experimental points.

Table 6.4.3 shows the nitrogen and argon square root model. The first three columns in the table give the result of our model; the last is experimental.

These bare numbers do not give an idea of the complicated picture that is illustrated in Figs. 6.4.8–6.4.10. The picture changes very rapidly with the input parameters, so that experimental uncertainties play a role.

We conclude that the Lagrangian with a square root interaction is in qualitative agreement with measurements if the interaction strength α' is near $100(\mathrm{cm/sec})^2$. Later, we shall see that the same Lagrangian can account for the speed of sound in the mixture.

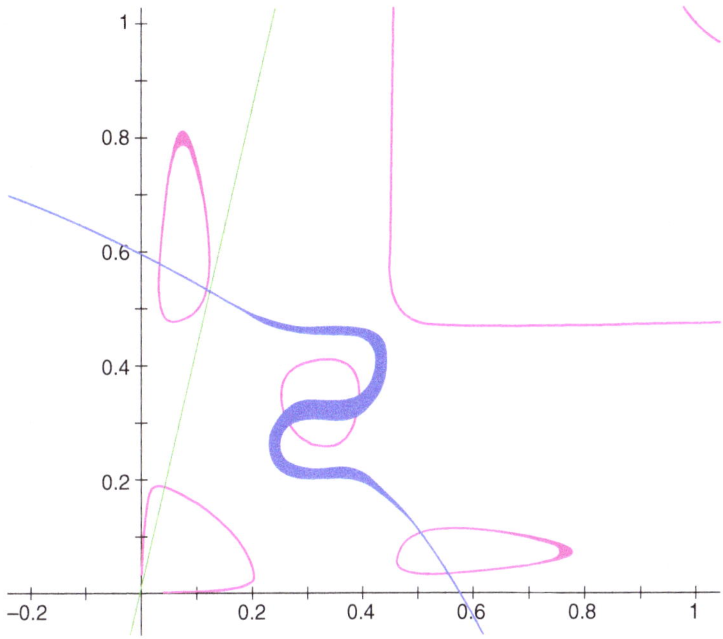

Fig. 6.4.8. The singular point for $T = 146\,\mathrm{K}$ and $\beta = \pm 3$.

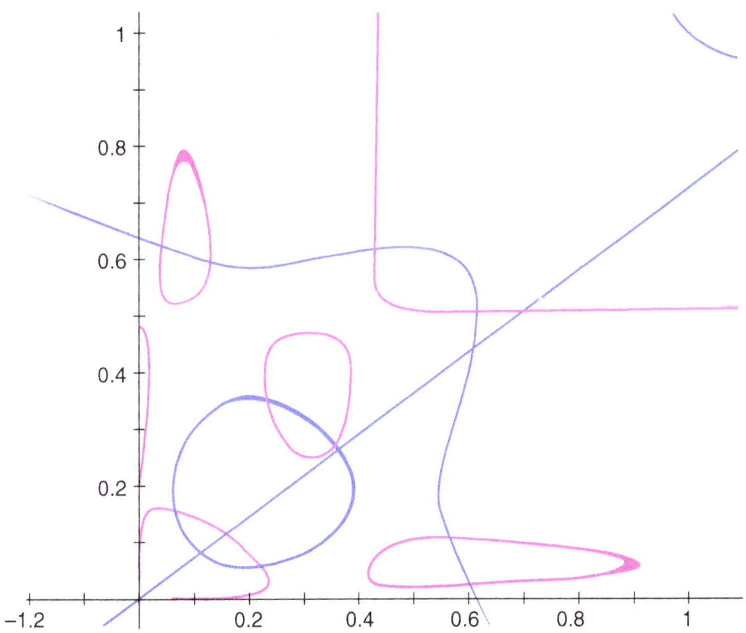

Fig. 6.4.9. The singular point for $T = 134.5\,\mathrm{K}$ and $\beta = \pm 3$.

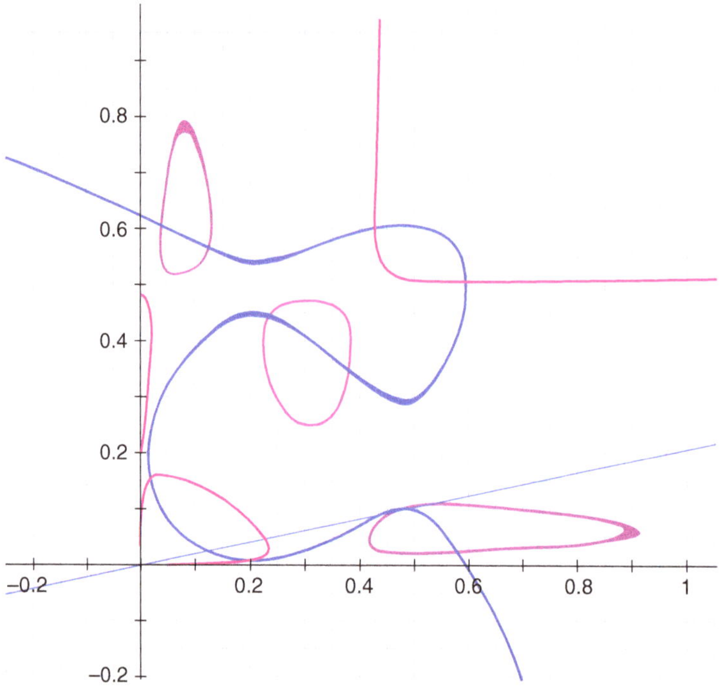

Fig. 6.4.10. The singular point for $T_r = 146\,\mathrm{K}$ and $\beta = \pm 3$.

Figure 6.1.6 is a clear example to show what happens near a critical point. We plot loci of μ_1 (red curves) and μ_2 (blue curves) for the same value of T, α and densities $(x, y) = (a, 0.6a)$. This is the value of x at the point where the line $y = 0.6x$ nearly touches the closed line. It is the case shown in Fig. 6.1.6, where the three loci of μ_1, μ_2 and p are seen to touch, and in Fig. 6.4.3. It is a critical point of the set of configurations that is projected out by fixing the relative concentration to $y/x = 0.6$.

This scenario is close to the experimental situation. One begins with a homogeneous mixture at the higher temperature and reduces the temperature very slowly. At the critical temperature, condensation sets in, the 10-to-6 homogeneous mixture seizes to be stable.

6.5. Helium and Xenon

Critical parameters:

$$\text{He}: \; T_{\text{cr}} =: T_1 = 5.19\,\mathrm{K}, \quad p_{\text{cr}} = 2.24\,\mathrm{atm} = 227\,\mathrm{kPa},$$

$$\text{Xe}: \; T_{\text{cr}} =: T_2 = 289.8\,\mathrm{K} = 16.6^\circ\mathrm{C}, \quad p_{\text{cr}} = 57.6\,\mathrm{atm} = 5840\,\mathrm{kPa}.$$

Van der Waals parameters:

$$\text{He}: \ a_1 = 0.0346\,\text{barLit}^2/\text{mol}^2, \quad b_1 = 0.0237\,\text{Lit/mol},$$
$$\text{Xe}: \ a_2 = 4.250\,\text{barLit}^2/\text{mol}^2, \quad b_2 = 0.05105\,\text{Lit/mol}.$$

Interaction: $f_{\text{int}} = \mathcal{R}\alpha\rho_1\rho_2$.

This is a mixture that has been difficult to deal with; the van der Waals approach has not been successful. We begin by briefly confirming this within our framework.

The free energy density is thus $f = f_1 + f_2 + \mathcal{R}\alpha\rho_1\rho_2$.

The maximal (critical) temperature

The maximal temperature of the mixture is the lowest temperature for which the determinant of the matrix of second partial derivatives of the free energy is positive definite, at this point

$$b_1 b_2 \left(\frac{T}{x(1-x)^2} - \frac{27}{4}T_1\right)\left(\frac{T}{y(1-y)^2} - \frac{27}{4}T_2\right) = \alpha^2. \tag{6.5.1}$$

We list the critical temperature for some values of the coupling strength.

$$\alpha^2 = 507, \quad T_{\text{cr}}, = 340,$$
$$\alpha^2 = 50, \quad T_{\text{cr}} = 295.7,$$
$$\alpha^2 = 40, \quad T_{\text{cr}} = 294.6,$$
$$\alpha^2 = 30, \quad T_{\text{cr}} = 293.5,$$
$$\alpha^2 = 20, \quad T_{\text{cr}} = 292.3,$$
$$\alpha^2 = 8.5, \quad T_{\text{cr}} = 291.$$

These values are taken from a graph, more accurate values can be obtained digitally, for example, at $\alpha^2 = 50$, $T_{\text{cr}} = 295.686$. From now on, this is the value of α that has been used. The associated value of the normalized densities is 1/3; this gives the exact expression

$$T_{cr} = \frac{T_1 + T_2}{2} + \sqrt{\left(\frac{4}{27}\right)^2 \alpha^2 T_1 T_2 + \left(\frac{T_1 - T_2}{2}\right)^2}. \tag{6.5.2}$$

For $\alpha^2 = 50$, it gives $T_{\text{cr}} = 295.686$, both Mathematica (File Untitled-2) and Graphing Calculator.

The densities are molar, so $\mathcal{R}$ is the universal gas constant.

I calculated the spinodal for $\alpha^2 = 50$ and temperatures ranging from 292.5 K to 150 K. As expected, the spinodal oval exists for $T \leq 295$ K and reaches the vertical axis when $T = 289.8$ K.

Again as expected, each pair of coexistent densities bracket the spinodal at a distance is nearly constant except when the isobar is tangent to the spinodal; these are the configurations that are labeled "critical" in the sense of the experimenter. Confrontation with experiments makes it very clear that this model is not realistic. The conclusion so far is that the interaction $\alpha\rho_1\rho_2$ does not give results that agree with experiments. I have done some calculations with the alternative interaction proportional to $\beta\sqrt{\rho_1\rho_2}$ but I have not gone further because of the dearth of measurements. There is ample support for the expectation expressed by some experimenter that the coordinates of the critical points are discontinuous as functions of the temperature.

Project 4. Collect all experimental data for one or more of the mixtures that are dealt with above. Make a detailed study of the success or lack of it of the interaction $\alpha\rho_1\rho_2 + \alpha'\sqrt{\rho_1\rho_2}$. *Then use the most successful model to calculate the speed of sound, using the same interaction.*

Chapter 7

Speed of Sound in Mixtures

In a mixture of two ideal gases, without mutual interaction, two sets of degrees of freedom are excited and two normal modes of propagation will appear. The formula for the value of c^2 (c is the velocity of propagation) is replaced by a **propagation matrix**; the two possible values of c^2 are the eigenvalues of this matrix.

For the propagation of sound in air, approximated by the ideal gas model, it was found (first by Laplace) that a crucial condition for a successful prediction is the assumption that the specific entropy density $S = s/\rho$ is uniform. It is evident that, as we approach the problem of sound propagation in a mixture of gases, the entropy needs to be taken into account, and appropriately.

7.1. Sound Speeds in Gibbs–Dalton Mixtures

Equations of motion for ideal gases

We shall do the calculation under the assumption that the equilibrium state satisfies the strong Gibbs–Dalton hypothesis

$$\mathcal{R}_i\rho_i\left(\ln\frac{\rho_i}{T^{n_i}} - n_i\right) + \rho_i S_i = 0, \quad i = 1, 2, \tag{7.1.1}$$

with S_1, S_2 being constant. First-order perturbations are introduced by substitutions

$$T \to \hat{T} = T + \delta T, \quad \rho_i \to \hat{\rho}_i = \rho_i + \delta\rho_i, \quad i = 1, 2. \tag{7.1.2}$$

The perturbations are subject only to the **adiabatic condition**

$$\sum_i \left(\mathcal{R}_i\hat{\rho}_i\left(\ln\frac{\hat{\rho}_i}{\hat{T}^{n_i}} - n_i\right) + \hat{\rho}_i S_i\right) = 0. \tag{7.1.3}$$

In the expansion of this, according to (7.1.2), the term of order zero vanishes by hypothesis and the first-order term simplifies, with the help of (7.1.1), to

$$\frac{\delta T}{T}\sum n_i \mathcal{R}_i \rho_i = \sum \mathcal{R}_i \delta\rho_i. \tag{7.1.4}$$

I repeat that *the perturbation* does not obey the Gibbs–Dalton hypothesis; if it did, it would reduce the number of degrees of freedom and the result would be very different. The Gibbs–Dalton hypothesis serves only to constrain the entropy. With that, the Lagrangian is fixed and the only dynamical equations used are the Euler–Lagrange equations, including the relation (7.1.3) in the linearized form (7.1.4). That we have such a relation is due to the fact that we have modeled the entropy of the mixture as a linear function of the densities,

$$\hat{s} = \hat{\rho}_1 S_1 + \hat{\rho}_2 S_2.$$

The other linearized equations of motion are the **equations of continuity** and the **Bernoulli equation**

$$\frac{\delta\ddot{\rho}_i}{\hat{\rho}_i} = -\dot{v}_i' = \partial_x^2 \delta \frac{\partial W}{\partial \rho_i}\Big|_T, \quad i = 1, 2, \ldots \tag{7.1.5}$$

with

$$W = \sum_i \mathcal{R}_i \rho_i T \ln \frac{\rho_i}{T^{n_i}} + sT, \tag{7.1.6}$$

and

$$\frac{\partial W}{\partial \rho_i}\Big|_T = R_i T\left(\ln \frac{\rho_i}{T^{n_i}} + 1\right) + S_i T, \quad i = 1, 2, \ldots$$

To evaluate (7.1.5), we use the **Gibbs-Dalton condition** (7.1.4) on the equilibrium state (on the coefficients),

$$\delta \frac{\partial W}{\partial \hat{\rho}_i}\Big|_{\hat{T}} = \mathcal{R}_i \delta T + \mathcal{R}_i T \frac{\delta \rho_i}{\rho_i}, \quad i = 1, 2,$$

but only the linearized adiabatic condition (7.1.4) on the variations $(\delta T, \delta\rho_i)$, namely,

$$\frac{\delta T''}{T} = \frac{R_1 \delta\rho_1'' + R_2 \delta\rho_2''}{n_1 \mathcal{R}_1 \rho_1 + n_2 \mathcal{R}_2 \rho_2}. \tag{7.1.7}$$

Let

$$\kappa = \frac{\mathcal{R}_2}{\mathcal{R}_1} = \frac{\mu_1}{\mu_2}; \quad \tau = \frac{\rho_1}{\rho_2},$$

$$A = n_1 \tau + n_2 \kappa.$$

Then

$$\frac{\delta T''}{T} = \frac{\delta\rho_1''}{\rho_1}\frac{\tau}{A} + \frac{\delta\rho_2''}{\rho_2}\frac{\kappa}{A}. \tag{7.1.8}$$

Next,

$$\frac{\delta\ddot{\rho}_1}{\rho_1} = \frac{\mathcal{R}T}{\mu_1}\frac{\tau}{A}\frac{\delta\rho_1''}{\rho_1} + \frac{\mathcal{R}T}{\mu_1}\frac{\kappa}{A}\frac{\delta\rho_2''}{\rho_2} + \frac{\mathcal{R}T}{\mu_1}\frac{\delta\rho_1''}{\rho_1}, \tag{7.1.9}$$

$$\frac{\delta\ddot{\rho}_2}{\rho_2} = \frac{\mathcal{R}T}{\mu_2}\frac{\kappa}{A}\frac{\delta\rho_2''}{\rho_1} + \frac{\mathcal{R}T}{\mu_2}\frac{\tau}{A}\frac{\delta\rho_1''}{\rho_1} + \frac{\mathcal{R}T}{\mu_2}\frac{\delta\rho_2''}{\rho_2}. \tag{7.1.10}$$

The propagation matrix

The matrix of squared velocity is the matrix of coefficients in these equations,

$$\frac{\mathcal{R}T}{\mu_1}\begin{pmatrix} 1+\frac{\tau}{A} & \frac{\kappa}{A} \\ \frac{\kappa\tau}{A} & \kappa+\frac{\kappa^2}{A} \end{pmatrix}, \qquad A = n_1\tau + n_2\kappa.$$

For mixtures, we define the reduced speed c in terms of the Newtonian speed of the first component, $v = c\sqrt{\mathcal{R}_1 T}$. This reduced speed is a zero of the determinant of the matrix

$$M = \begin{pmatrix} 1+\frac{\tau}{A} - c^2 & \frac{\kappa}{A} \\ \frac{\kappa\tau}{A} & \kappa+\frac{\kappa^2}{A} - c^2 \end{pmatrix}. \tag{7.1.11}$$

The density fluctuations satisfy

$$M\begin{pmatrix} \delta\rho_1/\rho_1 \\ \delta\rho_2/\rho_2 \end{pmatrix} = 0. \tag{7.1.12}$$

The theory predicts two modes with speeds that vary with the concentrations. Experimenters report a single mode.

We shall see that the disagreement with experiment can be overcome, but it is of interest to begin by describing the predictions of the naive Gibbs–Dalton model in some detail. There is a surprise: two different types of mixtures must be distinguished.

Two types of mixtures

There will be reference to the figures. In all the figures, the abscissa is the natural logarithm of the ratio of densities, $\tau = \rho_1/\rho_2$. Gas number 1 dominates on the right, where $\tau \to \infty$; gas number 2 dominates on the left, where $\tau \to 0$. The vertical coordinate is c/γ_1, the unit is thus the Laplacian speed of sound in gas number 1, approached asymptotically in the region marked C in Fig. 7.1.1.

In the limit $\tau = 0$, the first solution is

$$c^2 = 1, \quad M = \begin{pmatrix} 0 & 1/n_2 \\ 0 & \kappa\gamma_2 - 1 \end{pmatrix}, \quad \gamma_i = 1 + \frac{1}{n_i}, \quad i = 1,2. \tag{A}$$

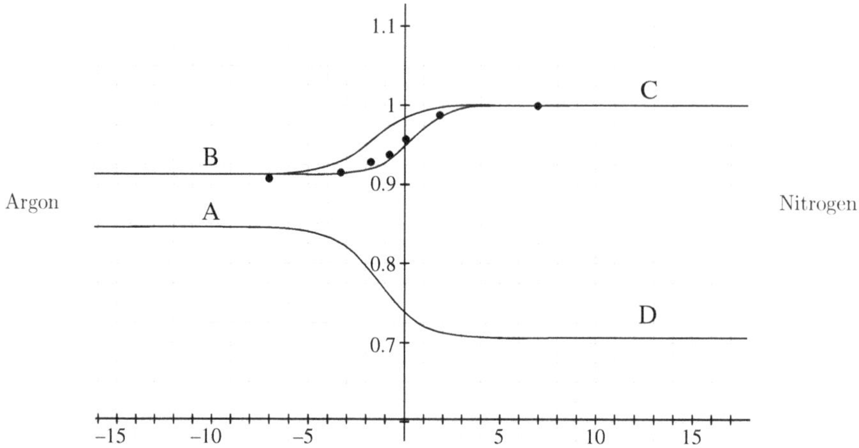

Fig. 7.1.1. The upper and lower curves show the calculated speeds (actually $c/\sqrt{\gamma_1}$) of the two modes of sound in N_2/Ar, a type 1 mixture, from the Gibbs–Dalton model, plotted against $\ln(\rho_1/\rho_2)$. The curve below the data points is from Eq. (7.1.14). The abscissa is $\ln(\rho_N/\rho_{Ar})$, pure nitrogen on the right, pure argon on the left. Data are from by Löfqvist and Delsing (2003). The unit of speed that is used as vertical coordinate is the Laplacian speed of sound in nitrogen.

The speed is the Newtonian speed of gas number 1, $d\rho_2 = 0$ and the energy is carried by $d\rho_1$. The fluctuations are in equal or in opposite phase. To determine which, we have the matrix equation

$$Md\rho = 0.$$

In particular,

$$\frac{\kappa\tau}{A}d\rho_1 + \left(\kappa + \frac{\kappa^2}{A} - c^2\right)d\rho_2 = 0. \tag{7.1.13}$$

When $\tau \approx 0$, the second coefficient is $\kappa\gamma_2 - 1$. The two coefficients have the same sign if $\kappa\gamma_2 > 1$, so in that case the phases are opposed. This may be important for it may make it difficult to excite this mode. The relative phase does not change as the concentration is varied. (Note that M_{22} has the same sign for all values of the concentration.)

The second solution is

$$c^2 = \kappa\gamma_2, \quad M = \begin{pmatrix} 1-\kappa\gamma_2 & 1/n_2 \\ 0 & 0 \end{pmatrix}. \tag{B}$$

The speed is the Laplacian speed of gas number 2 and the energy is shared. The phases are equal if $\kappa\gamma_2 > 1$, so the two solutions are different in this respect.

In the limit $\tau = \infty$, the first solution is

$$c^2 = \gamma_1, \quad M = \begin{pmatrix} 0 & 0 \\ \kappa/n_1 & \kappa - \gamma_1 \end{pmatrix}. \tag{C}$$

The speed is the Laplacian speed of gas number 1 and the energy is shared. The fluctuations are in opposite phase if $\gamma_1/\kappa < 1$. The second solution is

$$c^2 = \kappa, \quad M = \begin{pmatrix} \gamma_1 - \kappa & 0 \\ \kappa/n_1 & 0 \end{pmatrix}. \tag{D}$$

The speed is the Newtonian speed of gas number 2, $d\rho_1 = 0$ and the energy is carried by $d\rho_2$. The fluctuations are in opposite phase if $\gamma_1/\kappa > 1$.

It may be expected that the excitation with opposite phases will be very difficult to achieve.

The solution that is Newtonian at one end will connect to whatever solution has the same relative phase at the other, thus if

$$\kappa\gamma_2 - 1, \quad \gamma_1/\kappa - 1$$

have the same sign, then the speed of one mode will be Newtonian and that of the other will be Laplacian. While if the signs are opposite, each solution will be Newtonian at one end and Laplacian at the other.

Summary

The letters A, B, C, D refer to the four regions of Fig. 7.1.1. The phases of the two modes are opposed in these cases

A, the Newtonian mode, if $\kappa\gamma_2 = (\mu_1/\mu_2)\gamma_2 > 1$.
B, the Laplacian mode, if $\kappa\gamma_2 < 1$.
C, the Laplacian mode, if $\gamma_1/\kappa < 1$.
D, the Newtonian mode, if $\gamma_1/\kappa = (\mu_2/\mu_1)\gamma_1 > 1$.

These excitations are difficult to excite and they may not have been observed.

Type 1 mixture, "similar" gases

This is the case when $\kappa\gamma_2 - 1$ and $\gamma_1/\kappa - 1$ have the same sign, for example,

$$\text{Gas 1/Gas 2 = Nitrogen/Argon: } \kappa = 28/40, \ n_1 = 5/2, n_2 = 3/2,$$

with $\kappa\gamma_2 = 7/6$ and $\gamma_1/\kappa = 2$. There is a "Newtonian mode" that interpolates between case (A) and case (D) with a speed that interpolates between the Newtonian (isothermal) speeds of the two pure gases. See the lower curve in Fig. 7.1.1. The two amplitudes are in opposite phase and the amplitude of the dominant gas tends to zero in the limit when this gas is alone. This mode is almost irrelevant since it is very difficult to excite it. A "Laplacian mode" mode interpolates between case (B) and case (C) (upper curve) with a speed that interpolates between the Laplacian (adiabatic) speeds of the two pure gases. The two amplitudes are in opposite phase. The dominant gas carries most of the energy but the amplitudes are comparable. A simple oscillator will excite this mode only. The success of the Laplacian assumption of fixed entropy is assured in this case.

All this tends to be confirmed by experiment. The experiment, by Löfqvist and Delsing (2003), was done under approximately **normal** conditions, where both gases can be treated as ideal gases with small error. The first mode cannot be excited mechanically[64] and there are other reasons why it may be expected to be unobservable, or at least to justify the fact that it has escaped detection. The prediction of the Gibbs–Dalton model needs only a minor correction to agree with experiment (see Fig. 7.1.1).

Type 2 mixture, "disparate" gases

It is the case when $\kappa\gamma_2 < 1$ and $\gamma_1/\kappa > 1$, for example, helium/xenon, $\kappa = 4/131.3$, $n_1 = n_2 = 3/2$.

There is a mode with speed that approaches the adiabatic speed of sound in the lighter gas in the limit when this gas is alone (C), but in the opposite limit, it approaches the Newtonian value of the same gas (A). The amplitudes are in equal phase.

A second mode has a speed that approaches the Newtonian speed of sound of the heavier gas in the limit when this gas is alone (B) and the Laplacian speed of the same gas in the opposite gas when that is alone (D). The amplitudes are in the same phase.

Experiment yields a single mode with a speed that varies with concentration and that approaches the expected Laplacian values in the limits of either pure gas (see Fig. 7.1.2). It appears that a crossover takes place; in fact, there is evidence of a "critical" concentration. There are frequent hints that there are other, less prominent modes, about which little information is given.

The reason for this disagreement was at first attributed to the fact that no account had been taken of any kind of damping, and for high frequencies, the importance of damping has been confirmed. But at low frequencies, the

[64] Compare the fruitless attempts to excite second sound in superfluid Helium (Peshkov, 1946).

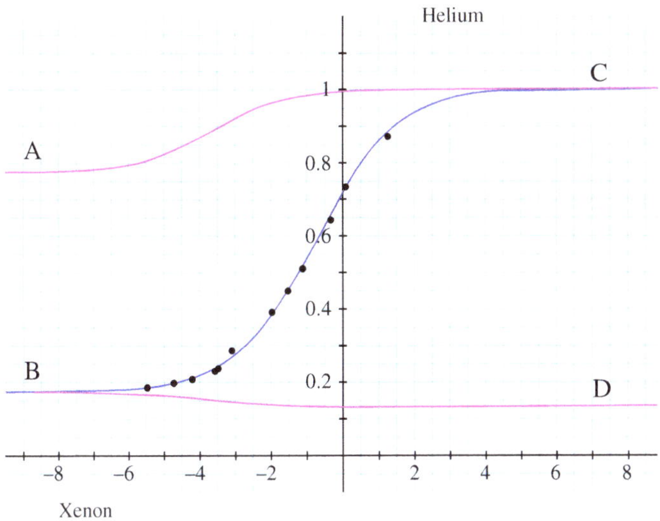

Fig. 7.1.2. The upper and lower curves show the two calculated speeds of sound in He/Xe, a type 2 (disparate) mixture. Coordinates as in Fig. 7.1.1. The interpolating curve is from Eq. (7.1.14). Data points are from Fernández de la Mora and Puri (1986). Pure Helium at right, pure xenon at left. At A, C, the asymptotes are the Newtonian, respectively the Laplacian, speeds in Helium.

loss of energy to absorption is not thought to be important. Indeed, the measured rates of absorption are extremely low (Holmes, 1960).

An alternative explanation is much more appealing. The calculations were made on the assumption that there is no interaction between the two gases. This makes it possible for a signal to be transmitted by either component without involving the other. In the solution represented by the upper line in Fig. 7.1.2, the signal is carried by helium even in the limit when the mixture consists entirely of xenon. In the other case, the signal is carried by xenon even in the limit when this gas is absent. This is clearly unreasonable and we expect that even a weak interaction must result in a crossover.

Plotting these functions, we want to compare with experiments, but besides the data, we shall use a well-known analytic fit. The formula that fits the observations, and much better than could be expected, is (Löfqvist and Delsing, 2003)

$$c^2 = \frac{\tau+\kappa}{\tau+1}\left(1+\frac{\tau+\kappa}{n_1\tau+n_2\kappa}\right). \tag{7.1.14}$$

The origin of this prediction is not the thermodynamic field theory but calculations of particle dynamics and the Boltzmann scattering equations. For a brief account of the calculation, see Fernández de la Mora and Puri (1986). It is predicted to hold in the limit when the effect of diffusion is so strong as to force the two velocities to be nearly equal. That is the key feature of the formula, and that must be the basic fact that is responsible for its success: apparently, something forces the two velocities to be nearly equal. This must be explained in terms of an interaction between the two components, but the conclusion that it is due to strong diffusive damping is not compelling.

The explanations that have been offered to explain the success of (7.1.14) are equations for the two velocities and, sometimes, two temperatures. We have noted that the introduction of two temperatures may be interpreted in terms of a deviation from the Gibbs–Dalton equilibrium conditions. But we are strongly constrained by the need to retain the standard equations, continuity and Bernoulli. Indeed, any modification of the kinetic part of the Lagrangian tends to cause a loss of mass conservation, that is, of particle number conservation. It is therefore of interest to explore alternatives.

In the special case that $\kappa = 1$, the value given by (7.1.14) is one of the two eigenvalues of the matrix M; the other one is $c^2 = 1$. As the van der Waals expression for the free energy of a gas was intended as a phenomenological expression of the inter particle interactions, the interaction term $\alpha\sqrt{\rho_1\rho_2}$ seems to be a means of introducing the interaction between two dissimilar atoms into the thermodynamical formulation.

The data quoted are from experiments at room temperature at frequencies around 1 MHz. More data are needed over a wide range of frequency and temperature.

7.2. Simple Model of Interactions

Let us add the following "**`interaction term`**" to the potential, that is, to the Hamiltonian,

$$\alpha\sqrt{\rho_1\rho_2}, \tag{7.2.1}$$

with α constant. It turns out that α has to be positive, which means that the interaction is repulsive. The reason for this choice will become clear almost immediately. Such a term will affect the formula for the internal energy, but it is a correction to the free energy density and it does not change the formula for the entropy. The hypothesis of Gibbs and Dalton can be maintained and we shall do so. The formula for the pressure is not affected.

The equation of motion

$$\frac{D\vec{v}_j}{Dt} + \text{grad}\left(W + \frac{\alpha}{2}\sqrt{\frac{\rho_i}{\rho_j}}\right) = 0, \quad i, j = 1, 2 \text{ or } 2, 1$$

no longer suggests a useful definition of individual pressures. The acceleration of the molecules of each gas is strongly dependent on the configuration of the other.

Again, we consider first-order perturbations of the equilibrium configuration. The linearized equations (7.1.5) are modified:

$$\frac{d\ddot{\rho}_i}{\rho_i} = -\dot{v}_i' = \partial_x^2 \left(\frac{\partial W}{\partial \rho_i}\Big|_T + \frac{\alpha}{2}\sqrt{\frac{\rho_j}{\rho_i}} \right), \quad j \neq i.$$

Equations (7.1.5) and (7.1.6) gain additional terms,

$$\frac{\alpha}{4}\sqrt{\frac{\rho_2}{\rho_1}}\left(\frac{d\rho_2''}{\rho_2} - \frac{d\rho_1''}{\rho_1}\right) \quad \text{and} \quad \frac{\alpha}{4}\sqrt{\frac{\rho_1}{\rho_2}}\left(\frac{d\rho_1''}{\rho_1} - \frac{d\rho_2''}{\rho_2}\right).$$

What makes the choice (7.2.1) special is the fact that, for a monocromatic wave, these additions are proportional to the difference $v_1 - v_2$ of the velocities. When the interaction is strong, it will favor equalization of the velocities. Our brief analysis of the standard theory shows that this is the feature that accounts for its success.[65]

The reduced speed of propagation is now obtained by setting to zero the determinant of

$$\begin{pmatrix} 1 + \frac{\tau}{A} - \beta\tau^{-\frac{1}{2}} - c^2 & \frac{\kappa}{A} + \beta\tau^{-\frac{1}{2}} \\ \frac{\kappa\tau}{A} + \beta\tau^{\frac{1}{2}} & \kappa + \frac{\kappa^2}{A} - \beta\tau^{\frac{1}{2}} - c^2 \end{pmatrix}, \quad \beta := \frac{\alpha}{4\mathcal{R}_1 T}. \tag{7.2.2}$$

In the limit of large β, the only eigenvector has $d\rho_1/\rho_1 = d\rho_2/\rho_2$, or $\vec{v}_1 = \vec{v}_2$, and the eigenvalue c^2 is precisely as in (7.1.14). This very curious mathematical result is overshadowed by the fact that, as already noted, when $\kappa = 1$ and $\beta = 0$, there is perfect coincidence between the two formulas, except that the vanishing of the determinant of (7.2.2) has the additional solution $c^2 = 1$.

Numerical results for a type 1 mixture are in Fig. 7.2.1. It may be seen that a value $\beta = 0.3$ is sufficient to bring the theory into substantial agreement with the experiment. Higher values of β improve the agreement for low frequencies but tend to spoil it when compared with the few data that are available for the highest frequencies. Negative values of α do not give reasonable results.

---

[65] In the presence of the interaction, we are not justified to use the Gibbs–Dalton hypothesis. It would be interesting to use the modification proposed in Section 6.1. We have not done the calculation.

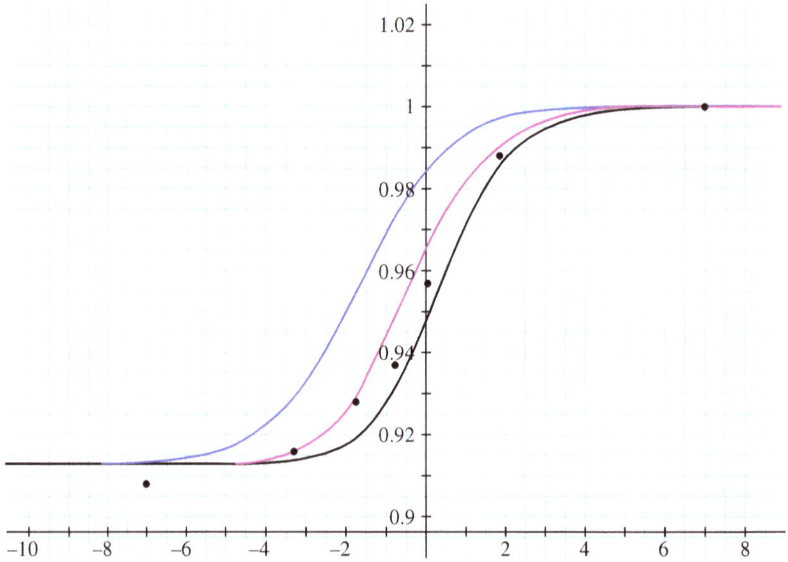

Fig. 7.2.1. The effect of the interaction (7.2.1) on sound speed in N_2/Ar. The blue curve on top (purple curve middle) is obtained with $\beta = 0$ ($\beta = 0.3$). Higher values of β give even better agreement with experiment. The unphysical mode has disappeared. The lower (black) curve is the prediction of Eq. (7.1.14).

The case of type 2 mixtures (Fig. 7.2.2) is more dramatic. Very small values of β are enough to eliminate the isothermal modes in the SE and NW corners of the figure. The former disappears and the latter drops down to join the Laplace branch at the lower left. Agreement with experiment requires a value of β at least equal to 1, near perfect agreement is obtained with $\beta = 5$, and larger values of β only improve the fit, except for the data at the highest frequencies.

Variants

(1) In Chapter 6, we began by following the custom in using an interaction of the form $\alpha\rho_1\rho_2$ ($k = 1$ instead of $k = 1/2$). Since we are bound to use the same interaction Lagrangian, we must use the same value of k in both cases. But this is not a critical test. The quadratic interaction energy ($k = 1$) used to study critical mixing is numerically insignificant at the temperatures and densities at which we have examined sound speeds. When an interaction of the form $\alpha\rho_1\rho_2$ is included, the linearized equations take the form

$$\frac{d\ddot{\rho}_i}{\rho_i} = -\dot{v}_i' = \partial_x^2\left(\frac{\partial W}{\partial \rho_i}\bigg|_T + \alpha\rho_j\right), \quad j \neq i.$$

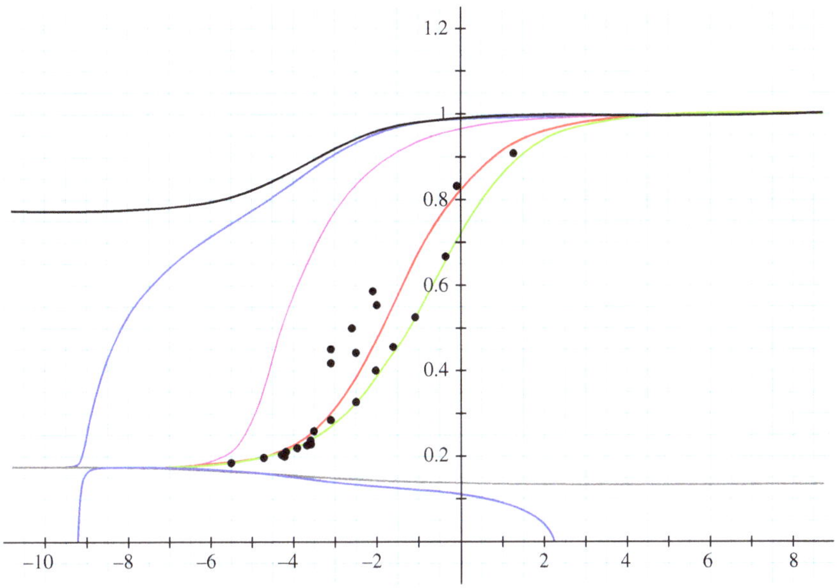

Fig. 7.2.2. Effect of the simple interaction term on the speed of sound in He/Xe. From top to bottom: $\beta = 0$ (black), 0.01 (blue), 0.1 (purple) and 1 (red). The green line is from (7.1.14). Additional data points are from Bowler and Johnson (1985), at higher frequencies.

This, as an attempt to improve the agreement with experiment, is a total failure. A small amount of this type of interaction can be allowed, but it does nothing to improve the agreement with experiment.

We have tried to include a coupling of the form $\alpha\sqrt{\rho_1\rho_2}$ in the calculation of critical mixtures and find that the result is encouraging. See Chapter 6.

(2) The variation of the speed of sound with the frequency can be modeled with interactions of the squared gradient type (van der Waals)

$$\alpha \int d^3x \left(\sum \partial_i\rho_1\partial_i\rho_2\right)^k .$$

It is necessary to collect much more experimental information, at a wide range of temperatures, at appropriate frequencies.

7.3. Two van der Waals Gases: Sound Speed

This section and the Chapter 8 study the gaseous helium–xenon mixture. The treatment is new in the sense that it is a test of a new model of the entropy of mixtures. This does not contradict the usual treatment, it just aims to reduce the number of phenomenological parameters.

The Lagrangian for a mixture of two non-interacting van der Waals gases is

$$\sum_i \left(\rho_i(\dot\Phi_i - \vec v_i^2/2) + \mathcal{R}_i\rho_i T \ln\frac{\rho_i}{(1-b_i\rho_i)T^{n_i}} - a_i\rho_i^2 + \rho_i S_i T\right). \tag{7.3.1}$$

This formula presents our model for the entropy of a mixture of two van der Waals gases as a linear function of the two densities. The usual treatment avoids introducing the entropy but the formula used for the free energy density contains phenomenological terms, linear in the densities, to make up for it.

To this, we shall have to add an interaction term.

As usual, we shall consider, in the first-order perturbation theory, a traveling wave on a uniform background at rest,

$$\rho_i \to \rho_i + \delta\rho_i, \quad T \to T + \delta T, \quad \vec v_i \to \delta\vec v_i.$$

The direction of the velocity is fixed and taken to coincide with that of the x-axis.

The equations of continuity are

$$\delta\dot\rho_i = -\rho_i\frac{d}{dx}\delta v_i,$$

$$\frac{\delta\ddot\rho_i}{\rho_i} = -\frac{d}{dx}\delta\dot v = \frac{d^2}{dx^2}\delta\dot\Phi_i.$$

The Bernoulli equations are, to first order,

$$-\dot\Phi_i = \mathcal{R}_i T\left(\ln\frac{\rho_i}{(1-b_i\rho_i)T^{n_i}} + \frac{1}{1-b_i\rho_i}\right) - 2a_i\rho_i + S_i T$$
$$=: \mu_i(\rho_1,\rho_2,T), \quad i = 1,2. \tag{7.3.2}$$

We need the coefficients in

$$-\delta\dot\Phi_i = \sum_j \frac{\partial\mu_i}{\partial\rho_j}\delta\rho_j + \frac{\partial\mu_i}{\partial T}\delta T.$$

To lowest order, the partial derivatives are

$$\frac{\partial\mu_i}{\partial\rho_j} = \left(\frac{\mathcal{R}_i T}{\rho_i(1-b_i\rho_i)^2} - 2a_i\right)\delta_{ij}, \quad \frac{\partial\mu_i}{\partial T} = \frac{\mathcal{R}_i}{1-b_i\rho_i}. \tag{7.3.3}$$

To derive this simple expression for $\partial\mu/\partial T$, we have used Gibbs–Dalton condition for the equilibrium state. The perturbed state is subject to the adiabatic condition only. This is because we are committed to using the Euler–Lagrange equations of our action principle and nothing else. It gives us one relation between the three variables and allows us to express δT in terms of $\delta\rho_1$ and $\delta\rho_2$. To lowest order,

$$\sum \frac{\mathcal{R}_i \delta\rho_i}{1 - b_i\rho_i} = \sum R_i \rho_i n_i \frac{\delta T}{T}. \tag{7.3.4}$$

Finally, the squared-speed matrix is given by

$$\begin{aligned}\frac{\delta\ddot{\rho}_i}{\rho_i} &= \left(\frac{\mathcal{R}_i T}{(1 - b_i\rho_i)^2} - 2a_i\rho_i\right)\frac{\delta\rho_i''}{\rho_i} + \frac{\mathcal{R}_i T}{1 - b_i\rho_i}\frac{1}{\sum_k n_k \mathcal{R}_k \rho_k}\sum_j \frac{R_j \delta\rho_j''}{1 - b_j\rho_j} \\ &= \sum_j (c^2)_{ij} \frac{d^2}{dx^2}\frac{\delta\rho_j}{\rho_j}. \end{aligned} \tag{7.3.5}$$

The van der Waals formula is not enough to calculate the speed of sound, even in a unary system. More information is needed and one way to introduce it is to define the van der Waals fluid by specifying the formula for the internal energy density. When instead we specify the Lagrangian density, there still remains to choose the values of the two entropy parameters. One way to do that is to apply the Gibbs–Dalton hypothesis to the unperturbed configuration. This is what leads to the simple result (7.3.3).

So far, the formulas had involved only the ratio $\tau = \rho_1/\rho_2$ of the two densities. To draw curves of sound speed versus concentration, we need to hold something fixed. In most experiments, the temperature and the pressure of the unperturbed states are kept fixed. The temperature, where it appears explicitly in our calculation, was fixed at 293 K. The pressure may be taken to be 1 atmosphere and

$$p = \sum \frac{R_i \rho_i T}{1 - b_i\rho_i} - a_i\rho_i^2.$$

To show the sound speeds as functions of the concentration, we shall draw curves for fixed values of the pressure. Note that the fixed pressure refers to the sequence of equilibrium states that serve as zeroth order of the perturbations; the perturbed configurations do not have fixed pressure.

We reduce (7.3.5) to

$$\begin{aligned}\frac{d\ddot{\rho}_i}{\rho_i} &= \left(\frac{R_i T}{(1 - b_i\rho_i)^2} - 2a_i\rho_i\right)\frac{\delta\rho_i''}{\rho_i} + \frac{\mathcal{R}_i T}{1 - b_i\rho_i}\frac{\sum_j R_j\delta\rho_j/(1 - b_j\rho_j)}{\sum_k n_k \mathcal{R}_k \rho_k} \\ &= \sum_j (c^2)_{ij}\frac{d^2}{dx^2} d\rho_j \end{aligned}$$

or

$$c^2 = R_1 T \begin{pmatrix} B_1^2 - \frac{2a_1\rho_1}{\mathcal{R}_1 T} + B_1^2 \frac{R_1 \tau}{\mathcal{R}_1 A} & B_1 B_2 \frac{\mathcal{R}_2}{R_1 A} \\ \frac{\mathcal{R}_2}{R_!} B_1 B_2 \frac{\mathcal{R}_1 \times \tau}{R_1 A} & \frac{R_2}{R_1}\left(B_2^2 - \frac{2a_2\rho_2}{\mathcal{R}_2 T}\right) + \frac{\mathcal{R}_2}{\mathcal{R}_1} B_2^2 \frac{R_2}{\mathcal{R}_1 A} \end{pmatrix}.$$

Here,

$$A = n_1\kappa_2 + n_2\tau_1, \quad B_i = \frac{1}{b_i\rho_i}.$$

Treating helium at normal temperature as an ideal gas, we simplify this,

$$c^2 = R_1T\begin{pmatrix} 1+\frac{\tau}{A} & B_2\frac{\kappa}{A} \\ B_2\frac{\kappa\tau}{A} & \kappa\left(B_2^2 - \frac{2a_2\rho_2}{\mathcal{R}_2T}\right) + B_2^2\frac{\kappa^2}{A}\end{pmatrix}.$$

We wish to compare these curves with experimental values; alternatively, we may compare with the phenomenological formula from Section 7.1,

$$c^2 = \frac{\tau+\kappa}{\tau+1}\left(1+\frac{\tau+\kappa}{n_1\tau+n_2\kappa}\right). \tag{7.1.14}$$

Our formula for the sound speed does not involve the temperature except in the definition of the coupling strength β, and as an overall factor that we have factored out. But it does contain the densities through ρ and τ

$$\rho_1 = \frac{\rho}{1+1/\tau}, \quad \rho_2 = \frac{\rho}{1+\tau}.$$

In order to pursue this study, we must have more data, especially near condensation.

Chapter 8

Superfluid Helium

8.1. Experiments and Interpretation

Here is a brief summary of the first important experiments on **superfluid Helium**, the development that culminated with Landau's two-fluid theory. It is natural that we should take a special interest in this system, for it stands out as the most important example of two-fluid thermodynamics. This chapter is written under the perspective of potential hydrodynamics, consequently, there is no reference to two kinds of flows, to phonons and rotons. The development of conservative hydrodynamics, in Chapter 10, will allow us to deal with those aspects.

The **fountain experiment**

The story of superfluid Helium started with revolutionary experiments by Kapitza (1938) and by Allen and Misener (1939). Liquid Helium is cooled in vessels A and B; initially, the level is higher in A. A capillary that connects vessel A to the vessel B is very fine and no significant amount of liquid gets through, until the temperature descends below 2.19 K. Both vessels are then thermally isolated. Subsequently, if a temperature difference is established, with $T_A > T_B$, then the level in vessel B will rise due to a flow through the tube (Fig. 8.1.1), until the pressure due to gravity provides a sufficient counter-force and an equilibrium is established. The mass flows toward the lower temperature.

The effect is quite dramatic: The flow is independent of the pressure and, what is more, it is nearly independent of the diameter of the tube though this be varied as much as by a factor of 50!

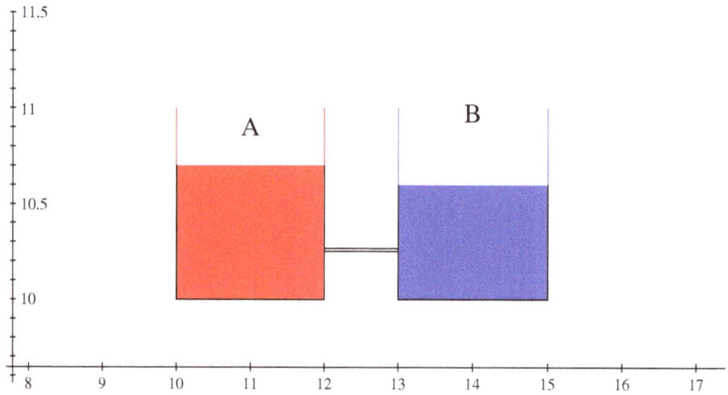

Fig. 8.1.1. The fountain effect. The central portion (light blue) is filled with a fine powder that is impermeable to Helium at temperatures above the λ point.

This highly unusual behavior led F. London (1938) to suggest that the fluid passing through carries no entropy. At about the same time, Tisza (1938) suggested that liquid Helium below 2.19 K is a mixture of two components or phases, a normal component and a **superfluid component** with two highly unusual properties. It has no viscosity, which helps it get through the tube, and it carries no entropy with it, so that no heat transfer takes place. This is known as the two-fluid model.

In 1945, Landau, in a very imaginative analysis of this experiment, suggested an alternative interpretation; instead of two fluids, he introduced two kinds of flows. Let us call this the two-flow theory. For our purpose, it is natural to look at the problem as a competition between two theories: two fluids or two flows. Historically what emerged was a compromise that adopted elements from both. The idea of two fluids is still thought to be essential, although Landau's warning against postulating that there are two kinds of Helium atoms suggests that it is not to be taken literally (see e.g., Tilley and Tilley, 1986).

Further variations of this type of experiment are described and interpreted in lucid detail in a review by Brooks and Donnelly (1974). The arrangement of two vessels connected by a tube so thin that only the superfluid gets through is the same, but instead of heating one of the vessels it is raised to a higher level. The difference in gravitational potential gives rise to a flow from the higher to the lower vessel, with the same unusual characteristics as in the original experiments. What is more, this process is completely reversible. On the basis of this experiment, and others to be described later, it is established that, for a fixed temperature and pressure, only one equilibrium concentration is possible, only one equilibrium configuration exists for given total density and temperature.

Consequently, the result is not a separation of the two components: With the establishment of a new **equilibrium, the concentration** in either vessel is determined uniquely by the local temperature and pressure. It is impossible to produce another mixture, at the same temperature and pressure, with a different concentration. The equilibrium concentration is not an independent variable.

This conclusion is highly significant and difficult to reconcile with the interpretation of other experiments. We shall not describe these other processes here, but we are referring to statements of the type "the superfield seems to flow through the normal fluid without resistance". If only one concentration is possible for any fixed value of temperature and pressure, then it is difficult to understand what it means that the superfluid flows through the normal fluid. Actually, this difficulty arises when we regard an equilibrium configuration as a snapshot of a quasi-static sequence that involves flow, as would seem to be justified in the case of flows at very low velocity. We seem to be in a situation that resembles the equilibrium in the dissociating mixture of atomic and molecular gases that we have examined in Section 5.8.

At this stage, we do not yet have an understanding of hydrodynamics that allows us to examine Landau's two-flow theory; for this reason, we shall not discuss experiments that involve flow. Instead, we shall pursue the idea that Helium below the λ point is a mixture of two fluids, a normal component and a superfluid component, with an interaction that transforms a quantity of the one into the same quantity of the other, preserving mass

$$\textit{Normal} \quad \leftrightarrows \quad \textit{Super} + \epsilon,$$

where ϵ is the difference in energy per unit of volume, assumed constant.

That the **entropy of the superfluid** component is near zero (supposing that it is) is thought to be a reflection of the excitation spectrum of the Helium atom; this information can be used but not justified within our context. A reasonable inference is that the temperature of the liquid in equilibrium cannot be lowered without converting some of the liquid from normal fluid to superfluid. And this is what is observed.

The underlying, atomic description of what may be going on will not be discussed in this book. The (by now) standard explanation is that the individual Helium atom has a ground state of very low energy. When the temperature is lowered, the atoms, being bosons, all seek the lowest level, and since there is only one isolated level at this lowest energy, the effective number of degrees of freedom is zero and there are, effectively, no populations of the higher energy levels (Bose–Einstein condensation). According to what is now generally accepted, the fluid divides into two parts, with the atoms of one, the "superfluid", condensing into the lowest level. London speaks of a phase separation in momentum space and Landau later advanced the same concept.

But there is another view. Landau, in his first paper on this subject, says that it would be wrong to think of two different types of atoms; instead, he writes, there are two different types of flows, referred to as **phonons and rotons**. Phonons are whorls of **locally irrotational flow**, rotons are said to carry vorticity; they are also described as flows of the solid body type. This statement by Landau is repeated in the last edition of *Fluid Mechanics* by Landau and Lifshitz (1959). The unusual behavior does not by itself imply that liquid Helium is governed by a different kind of thermodynamics; we shall see, in Section 8.2, that London's intuition about the entropy is sufficient to explain some of the unusual behavior of this liquid, so long as we limit our attention to quasi-static phenomena.

The theory of flows to be developed later will make this question of interpretation much more acute and much more complicated.

The entropy

We have proposed the representation

$$s = \rho_1 S_1 + \rho_2 S_2, \tag{8.1.1}$$

of the **entropy functional**; (density) of a mixture. The specific entropies S_1 and S_2 are adiabatic invariants. In the analogous case of saturation, heating results in an increase of the amount of gas while the specific entropies S_1 and S_2 remain fixed or almost so.

This formula appears to help us understand the fountain experiment. As was discovered in Chapter 5, if each component is treated as if it has no interaction with the other, the equations of motion for a system at rest, namely,

$$\dot{\vec{v}}_i = \vec{\nabla}\frac{\partial(f + sT)}{\partial \rho_i}, \quad i = 1, 2,$$

can be expressed in the form (when $S_2 = 0$)

$$\dot{\vec{v}}_1 = -\frac{1}{\rho_1}\vec{\nabla} p_1, \quad \dot{\vec{v}}_2 = -\frac{1}{\rho_2}\vec{\nabla} p_2 - S_1\vec{\nabla} T. \tag{8.1.2}$$

The last term represents a force that drives the superfluid component into the connecting tube, just as in Landau's theory. The difficulty with this argument is, as we have said, that the "partial pressures" are not defined.

Table 8.1.1. Relation between measured entropy density and the equilibrium values of densities and temperature, covering the whole interval from 0.1 K to 2.1 K.

$p = 00.0$	$s = 1.59\rho_1 + 0.0058\, T^3 \rho_2$
$p = 05.0$	$s = 1.45\rho_1 + 0.0042\, T^3 \rho_2$
$p = 10.0$	$s = 1.36\rho_1 + 0.0025\, T^3 \rho_2$
$p = 15.0$	$s = 1.289\rho_1 + 0.0025\, T^3 \rho_2$
$p = 20.0$	$s = 1.18\rho_1 + 0.0020\, T^3 \rho_2$
$p = 25.0$	$s = 1.129\rho_1 + 0.0017\, T^3 \rho_2$

Note: That entropies given by Brooks and Donnelly (1977) are specific densities. Our "s" is density in J/cm^3; the entropy quoted by Brooks and Donnelly multiplied by mass density. In the table, the pressure is given in atmosphere.

Very precise measurements of the entropy density s are available. In the upper half of the temperature interval, where the normal gas dominates, the first term is expected to make the larger contribution and on this assumption it is seen that the specific entropy S_1 varies only slowly with the temperature, which confirms the view that the lowering of entropy takes place, mainly, through conversion of the normal fluid to superfluid. But the measurement of partial densities is indirect and depends on the interpretation.

There is no significant dependence of the specific entropy S_1 on the temperature when $1 < T < 2$. Towards still lower temperatures, when the normal gas is absent, the entropy of the superfluid becomes dominant and there is no way to go except to reduce the specific entropy S_2. We see from Table 8.1.1 that it is proportional to the third power of the temperature.

Note that, experimentally, the entropy of the superfluid is not zero, though small, but that its dependence on the temperature is very different from S_1.

The equilibrium configuration

A relation between the basic variables ρ_1, ρ_2 and T defines the equilibrium configurations of the system. Using data found in Putterman's book, we plot all available physical points in the interval $1.3 < T < 2.1$ in Fig. 8.1.2 and find that the following equation describes it well,[66]

$$Q(\rho_1, \rho_2, T) := (\rho_1 + a\rho_2 + b)(\rho_1 + c\rho_2 + d) + 10^{-4} = 0,$$

[66]See however recent experimental work on superfluid Helium at negative pressure.

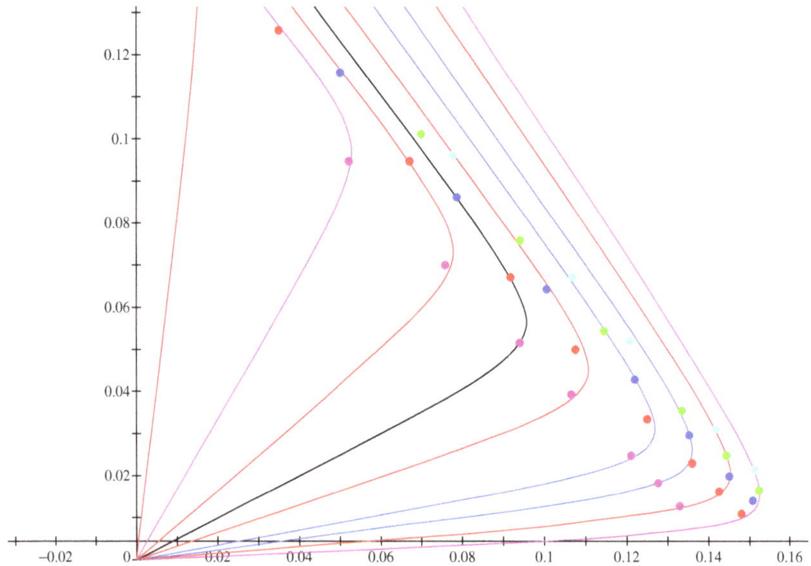

Fig. 8.1.2. The set of equilibrium configurations as given by Putterman (1974). The ordinates are the two densities. The color goes with the pressure; $p = 0$ for the leftmost purple dots and $p = 25\,\text{atm}$ for the (also purple) dots at the right end. The lines connect points with the same temperature, 1.3 K for the outer line and 2 K for the inner line. The extrapolation from the purple dots to the origin (negative pressure) is not justified by the experimental data.

with

$$a = -0.15, \quad b = 0.014 - 0.21(T - 1.3)^2, \quad c = 1.86 - 0.31T,$$
$$d = -0.398 + 0.113T.$$

Using data collected by Brooks and Donnelly (1977), we get a slightly different result, see Fig. 8.1.3. A more convenient expression for the equilibrium relation between ρ_1, ρ_2 and T will be found later.

The data points used are tabulated at the end of this chapter.

In Table 8.1.2, the values of the two densities can be represented with high precision, for each of six values of the pressure, the densities in mg/cm^3 and the pressure in atmospheres. The variation with pressure, though relatively weak, makes this representation inconvenient (Fig. 8.1.4).

Second sound

Landau's theory is distinguished by the fact that it treats the two components of a mixture as independent degrees of freedom: two densities and two velocities. It predicted the existence of two kinds of sound, a normal

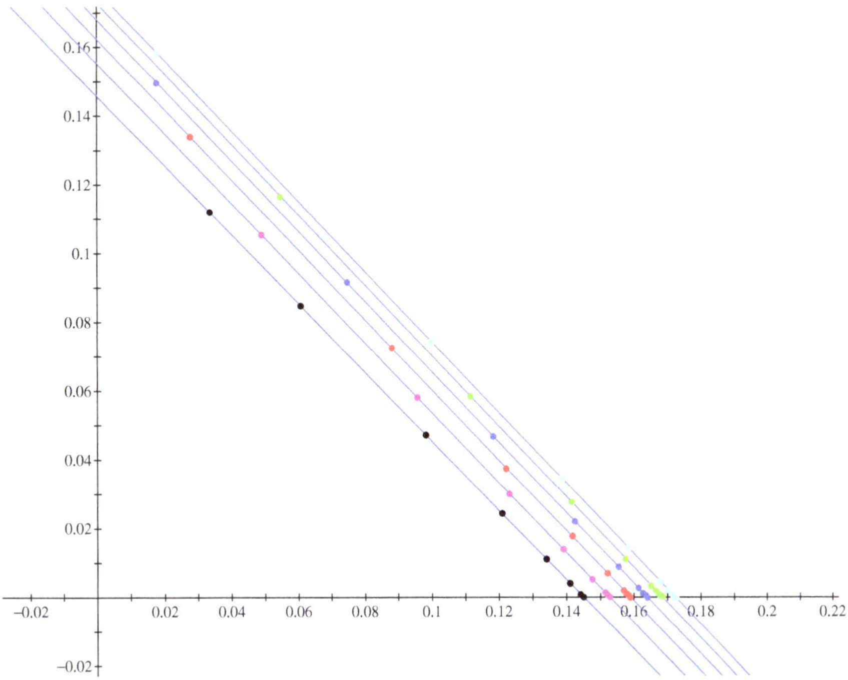

Fig. 8.1.3. The set of equilibrium configurations according to Brooks and Donnelly (1977). Blue lines are loci of equal pressure.

Table 8.1.2. This representation of the concentration is valid over the whole temperature interval, from 0.1 K to 2.1 K. Later, we shall see how it was measured. From Brooks and Donnelly (1977).

$p = 00$	$\rho_1 = 1.90\ T^{5.5}$	$\rho_2 = 145.0–1.74\ T^{5.6}$
$p = 05$	$\rho_1 = 2.17\ T^{5.6}$	$\rho_2 = 152.5–1.86\ T^{5.8}$
$p = 10$	$\rho_1 = 2.56\ T^{5.7}$	$\rho_2 = 159.0–2.20\ T^{5.9}$
$p = 15$	$\rho_1 = 3.03\ T^{5.8}$	$\rho_2 = 164.0–2.65\ T^{6.0}$
$p = 20$	$\rho_1 = 3.70\ T^{5.9}$	$\rho_2 = 168.5–3.20\ T^{6.1}$
$p = 25$	$\rho_1 = 4.40\ T^{6.0}$	$\rho_2 = 172.5–4.00\ T^{6.2}$

mode (first sound) in which both oscillations are in phase and another, second sound, in which they are in opposite phase. This was a real prediction made before any experiments confirmed the existence of two sounds in liquid Helium. This feature of Landau's theory appears to contradict the conclusion that we had reached earlier: that there is only one equilibrium mixture at each temperature. A possible resolution of this difficulty may be that

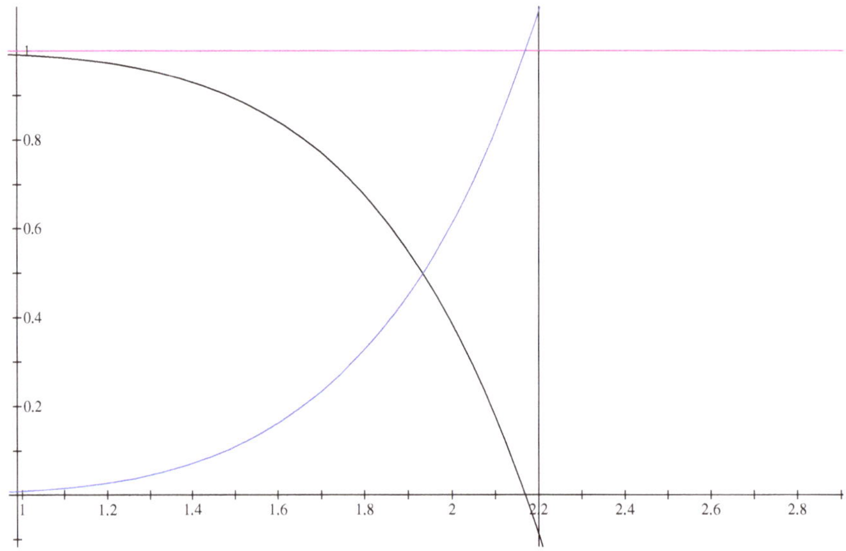

Fig. 8.1.4. Sketch of the concentrations of the normal fluid (increasing) and the superfluid component (decreasing) between $T = 1\,\mathrm{K}$ and $T = \lambda$, for fixed pressure. The actual measurements are indirect; one method relies on an interpretation of the cross-section for **`neutron scattering`**.

both flows are preserved in the bulk; transformations take place only with a transfer of heat, as in collisions with the walls. A true equilibrium is reached only after this process has had time to adjust the concentration. The value of the equilibrium concentration would be established by the process of collisions with the wall. It would be reasonable to ask for a more convincing explanation.

Having postulated the existence of two sound modes, we can affirm with some assurance that there must be two densities and **`two velocities`**. With the present limitation to irrotational vector fields, the kinetic part of the Lagrangian density is

$$\rho_1(\dot\Phi_1 - \vec v_1^2/2 - \phi) + \rho_2(\dot\Phi_2 - \vec v_2^2/2 - \phi). \tag{8.1.3}$$

The gravitational potential ϕ will be neglected. The appearance of an additional degree of freedom naturally implies that two sound modes are likely to exist; the surprise is not the existence of two sounds in liquid Helium but the fact that the observation of a similar phenomenon is rare in other mixtures. The reason, as was seen, is that most mixtures are characterized by a strong interaction between the components, interactions that effectively quash the mode in which the two fluctuations are in opposite phase. Other materials do not exhibit these special features because at comparable temperatures they are in a solid state.

Unless strong interactions between the two components prevent it, two sounds come naturally.

An implication of this is that we shall have two equations of continuity,

$$\dot{\rho}_i + \text{div}(\rho_i \vec{v}_i) = 0, \quad i = 1, 2. \tag{8.1.4}$$

The observation that only the total mass is conserved suggests that there can be only one equation of continuity. But we remember that conservation of mass is not an automatic consequence of the equation of continuity; it only follows when, in addition, one imposes the condition that the **boundary velocity** be tangential to the boundary. Here, we follow Landau in that we shall impose this boundary condition on the total flow

$$\vec{j} = \sum_i \rho_i \vec{v}_i,$$

but not on the individual velocities. Both mass densities are conserved in the bulk but only the total mass is preserved in the interaction with the boundary.

In the bulk, away from the boundary, and in the absence of external manipulation of the system, such as irradiation, the two liquids flow past each other, conserving their respective identities. This is the usual interpretation expressed by comments in the literature; see, for example, Khalatnikov (1965, p. 93).

The kinetic terms in our model, including the terms $\rho_i \dot{\Phi}_i$, are chosen so as to make possible a Lagrangian formulation. Recall, from Section 3.11, that the inclusion of the velocity potential helps to make the Lagrangian invariant under **Galilei transformations**. Khalatnikov, in a book that has become a classic reference (Khalatnikov, 1965), also aimed at a Lagrangian formulation, but without the help of the velocity potentials he was constrained to reduce the kinetic energy to the invariant expression $\rho(\vec{v}_1 - \vec{v}_2)^2/2$. According to Putterman (1974), there is an analogous contribution to the pressure.

Thermodynamic pressure

The thermodynamic pressure is defined as

$$p = \rho_1 \frac{\partial f}{\partial \rho_1} + \rho_2 \frac{\partial f}{\partial \rho_2} - f. \tag{8.1.5}$$

The free energy is a function of the three independent variables ρ_1, ρ_2 and T. It is common usage in thermodynamics of mixtures since Gibbs, and especially so in Landau's treatments of liquid Helium, to reduce the concentration to the role of a parameter, denying it the respect that it deserves as a thermodynamic variable. Such treatments are dynamically incomplete, being partially on shell.

The pressure on the wall has an additional contribution whenever the wall is maintained at a temperature that is different from the fluid. In the case of a heated wall, heat is transmitted to the liquid by the conversion of superfluid to normal fluid. This upsets the balance and, to compensate, normal fluid flows towards the wall and superfluid flows away from the wall. The additional pressure has been measured (Hall, 1954). The standard interpretation of this effect gives results that are in qualitative agreement with experiment.

The on-shell pressure can be expressed in terms of T, ρ_1 and ρ_2, this relation is found by measurements at equilibrium to take the following form: for example,

$$T = 0.1, \quad p = 25 - 1360(0.17246 - \rho) + 14300(0.17246 - \rho)^2. \qquad (8.1.6)$$

See Fig. 8.1.3 where the nearly exact dependence of the pressure on the total density $\rho = \rho_n + \rho_s$ alone is manifest. The derivative,

$$\frac{dp}{d\rho} = 28600\rho - 3.6384 = c^2,$$

varies from 526.7 at $p = 0$ to 1308.8 at $p = 25$. The unit of pressure is one atmosphere. To convert to cgs units, multiply by 10^5 to obtain the square of isothermal velocity of normal sound in cm/sec. At $T = 0.1$, $c = 229$ m/sec at $p = 0$ and $c = 362$ m/sec at $p = 25$. At higher temperatures, much the same. The quasi-independence of the speed on the temperature is significant.

Critical behavior

According to Arp (1989), all singular features of the thermodynamic functions can be fitted with a specific heat that is singular at $T = \lambda$. A term like

$$(T - \lambda) \ln\left(1 - \frac{T}{\lambda}\right)$$

in the free energy will produce just such a term in the heat capacity. Although we shall use a similar phenomenological function, we are not going to take the point of view of Arp, who emphasized this singular behavior in a trial function with 41 adjustable coefficients. Arp's very thorough work suffers from an important omission: it deals with a single fluid. There is no account of the separate densities of the two components, consequently no mention of second sound. All thermodynamic relations ignore the concentration.

Let us remember that van der Waals' highly successful modeling of a gas near and above the critical gas/liquid point did not predict the "correct critical behavior". This refers to a very small region that involves altogether different concepts. We do not attempt to achieve agreement with experiment in this small neighborhood.

8.2. First Analytic Models

We shall construct idealized Lagrangian models, with features that resemble those of He II, but simplified and, in one respect, exaggerated.

Thermodynamic variables

We postulate that the correct choice of thermodynamic variables includes T, the densities ρ_1, ρ_2 and the individual, specific entropies S_1, S_2. It is by far the most convenient choice though others are of course possible. In what follows, we are concerned almost exclusively with equilibrium configurations; with no flow ($\vec{v}_1 = \vec{v}_2 = 0$) and all fields uniform.

Structure of the Lagrangian

Landau's equations for the two-fluid model was constructed as a set of equations, without a Lagrangian variational structure, with a Galilei invariant expression for "the energy density" (Putterman, 1974; Khalatnikov, 1965). As we know, it is the Lagrangian that must be Galilei invariant, not the energy.

It is expected that the Lagrangian density, in a first approximation, is a sum of three kinds of contributions, the kinetic terms (8.1.3), the free energy, and the entropy term $-sT$ with the entropy density in the form (8.1.1): $s = \rho_1 S_1 + \rho_2 S_2$. The principal terms in the free energy are associated with each component separately,

$$f = f_1(T, \rho_1) + f_2(T, \rho_2) + \cdots ,$$

but additional terms will be needed to take interactions into account.

An adiabatic system is defined by fixing the specific entropies S_1, S_2 and the pressure (or the volume). How the total entropy is apportioned between the two species can be understood in terms of an underlying atomic theory. In the event that a successful model is found, it should be reconciled with atomic theory.

The pressure

The term attributed to f so far, the kinetic part in Eq. (8.1.3), being linear in the density, does not contribute to the pressure (8.1.5). In He II, a good approximation is (8.1.6), with variations of c^2 of up to 5 percent over the range of temperatures. Our first model, before necessary improvements, has pressure

$$p = c^2(\rho - 0.1455), \tag{8.2.1}$$

with c^2 constant. This pressure implies another term to be included in the free energy, for example,

$$c^2(\rho_1 \ln \rho_1 + \rho_2 \ln \rho_2 + 0.1455) \quad \text{or} \quad c^2(\rho \ln \rho + 0.1455). \tag{8.2.2}$$

The first expression is appropriate for two non-interacting components, the other for a single component with the density $\rho = \rho_n + \rho_s$. Both expressions result in the pressure (8.2.1). The absence of any dependence on the temperature in this expression ensures that the velocity of propagation of ordinary (first) sound in the model has no direct dependence on the temperature, although it may depend on the concentration. In He II at equilibrium it varies from 227 to 366 m/sec. We are attempting to construct a simple model with properties that are similar to those of Helium, but not one to be identified with it.

Entropy

The distribution of the **entropy** — between the two species, as a function of pressure and temperature, must be determined from experiments. For liquid Helium, what is known is summarized in Table 8.1.1. Any model must make some assumption, to be justified either by experiment or by atomic theory, preferably by both. Our experience with phase transitions, dissociation and immiscible fluids is not very helpful in the present instance. The only reliable guide is observation.

We take as a principal feature of our idealized model the axiom that the measured entropy is carried by the normal component; that is,

$$s = \rho_1 S_1. \tag{8.2.3}$$

This defines a family of adiabatic systems, parameterized by a single parameter S_1. The value of S_1 determines the adiabatic system and its equilibrium configuration, just as is the case for unary systems. This feature emulates one of the most dramatic aspects of He II, in an extreme form, by setting $S_2 = 0$, while experimentally S_2 is non-zero but insignificant, except at the very lowest temperatures.

For this reason, it is natural to take the temperature-dependent terms in the model free energy to be independent of ρ_2. We make this another postulated property of our model. The simplest possibility is that the T dependence of f is confined to a term

$$a(T)\rho_1. \tag{8.2.4}$$

Linearity in ρ_1 implies that this term makes no contribution to the pressure. The adiabatic condition is

$$S_1 = -\partial a/\partial T. \tag{8.2.5}$$

It is a feature of most familiar systems that the internal energy is first order in T. To achieve this, $a(T)$ must include terms of the type $T \ln T$. Since a singularity is expected at $T = \lambda$, we may be tempted to follow Arp, including a term $b(\lambda - T)\ln(\lambda - T)$, with b and λ constant. We leave the function $a(T)$ undetermined for the present, but notice that the adiabatic condition makes T uniform for all adiabatic processes.

Sound speeds

So far, our partial expression for f contains no interaction term, no explicit coupling between the two fluid components. At first, this seems to imply that there are two sound modes, one in which the superfluid remains at rest, another where the normal component is quiet. This is, of course, very different from what is observed. The mistake is to overlook the imposition of a common temperature. Indeed, we have seen, in Section 7.1, that the two modes are coupled, even in a mixture of two ideal gases, without an explicit interaction term in the free energy density.

In our treatment of mixtures of ideal gases or van der Waals fluids, to get a unique prediction of the two speeds, we asked that the unperturbed state satisfy the Gibbs–Dalton hypothesis; this has the effect of relating the two entropy parameters to each other. In superfluid Helium, this is not a problem if $S_2 = 0$ for the unperturbed state. The predicted speed matrix is thus unique, and so are the speeds as functions of the temperature and the density. The result is that there are two modes, one in which the two densities are approximately in phase; another in which the phases are 180 degrees off.

As far as we understand, this is not the way that the speeds are calculated in Landau's theory. Instead, one assumes (more or less correctly) that the phases are the same in one case and opposite in the other. It is perhaps only fair to ask for an estimate of the errors. A complete calculation by Tilley and Tilley leads to a "small" internal contradiction. See Tilley and Tilley (1986, p. 110). The finding that the first sound is a density variation and that the second sound is a temperature fluctuation is based on writing the equations of motion in the form (8.1.3) and the assumption that the last term is negligible for the first sound and dominant for the second sound. In this respect, there is little difference between our model and Landau's theory. This observation tells us something interesting: The existence of two sounds in normal fluid (one of which is usually unobservable) and in the superfluid does not constitute a qualitative difference between superfluids and normal fluids, the distinction is only quantitative.

Guided by our previous experience with sound propagation in mixtures, we would add a term $\alpha\sqrt{\rho_1\rho_2}$ or $\alpha\rho_1\rho_2$ to the free energy density, but let us make the more general assumption that the interaction takes the form

$$f_{\rm int} = \alpha(\rho_1, \rho_2). \tag{8.2.6}$$

For a small deviation from an equilibrium mode, we have the following equations:

$$\frac{\ddot{\rho}_i}{\rho_i} = -\dot{\vec{v}}_i = \frac{\partial^2}{\partial x^2}\frac{\partial}{\partial \rho_i}(f + sT),$$

where f is the sum of the three terms (8.2.2), (8.2.4) and (8.2.6), $\partial/\partial\rho_i$ are adiabatic derivatives — at constant S_1, S_2. Ignoring an additive constant,

$$\frac{\partial}{\partial \rho_1}(f + sT) = a(T) + c^2 \ln \rho_1 + S_1 T + \alpha_1, \tag{8.2.7}$$

$$\frac{\partial}{\partial \rho_2}(f + sT) = c^2 \ln \rho_2 + \alpha_2, \tag{8.2.8}$$

where suffixes on α stand for derivatives. The adiabatic condition (8.2.5) makes T constant and the T-dependent terms do not contribute. Hence,

$$\frac{\ddot{\rho}_1}{\rho_1} = c^2\frac{\rho_1''}{\rho_1} + \rho_1\alpha_{11}\frac{\rho_1''}{\rho_1} + \rho_2\alpha_{12}\frac{\rho_2''}{\rho_2},$$

$$\frac{\ddot{\rho}_2}{\rho_1} = c^2\frac{\rho_2''}{\rho_2} + \rho_2\alpha_{22}\frac{\rho_2''}{\rho_2} + \rho_1\alpha_{21}\frac{\rho_1''}{\rho_1}.$$

The squared-speed matrix is

$$c^2 + \begin{pmatrix} \rho_1\alpha_{11} & \rho_2\alpha_{12} \\ \rho_1\alpha_{21} & \rho_2\alpha_{22} \end{pmatrix}.$$

The determinant of the matrix (without the c^2 term) is

$$\rho_1\rho_2 \det(\alpha_{ij}).$$

The case that the function α is everywhere singular; that is, when this determinant vanishes identically, is especially easy to analyze, for in that case the eigenvalues are

$$c_1^2 = c^2, \quad c_2^2 = c^2 + \rho_1\alpha_{11} + \rho_2\alpha_{22}.$$

If α is zero, either component moves independently with the same speed. If instead we had replaced the pressure potential $c^2(\rho_1 \ln \rho_1 + \rho_2 \ln \rho_2)$ by the expression favored by Arp and others, namely, $c^2\rho \ln \rho$, then only the total density would oscillate; that is, one of the speeds would be zero.

For a simple example, consider

$$\alpha(\rho_1, \rho_2) = \beta\rho \ln \rho.$$

Then the speed-squared matrix is

$$c^2 + \frac{\beta}{\rho}\begin{pmatrix} \rho_1 & \rho_2 \\ \rho_1 & \rho_2 \end{pmatrix}.$$

The eigenvectors are: $(1, 1)$ with eigenvalue $c_1^2 = c^2 + \beta$ and $(-\rho_2, \rho_1)$ with eigenvalue $c_2^2 = c^2$. The first is the ordinary sound, both components oscillating in phase, the other is the second sound with the components in opposite phase.

With a slight change of notation, the complete expression for f takes the form

$$f = A + B, \quad A = a(T)\rho_1, \tag{8.2.9}$$
$$B := (c_1^2 - c_2^2)\rho \ln \rho + c_2^2(\rho_1 \ln \rho_1 + \rho_2 \ln \rho_2) + C(\rho_1 - \rho_2),$$

where c_1 (c_2) are the constant speeds of the two kinds of sound in the model. Other forms of interaction will give us speeds that vary with the concentration. The pressure is

$$p = c_1^2\rho.$$

The last term in B makes no contribution to the pressure or to the speed of sound.

The result does not validate the approach, for the interaction turns out to overwhelm the partial free energies, contradicting the evidence that the interaction between the two fluids is weak.

Equilibrium

As was seen, the existence of a unique equilibrium concentration for a given value of the temperature is complicated.

This is a state of rest, with both velocities equal to zero. It is determined by equalizing the chemical potentials

$$\mu_i = \frac{\partial}{\partial \rho_i}(f + sT)\Big|_{\rho,T}. \tag{8.2.10}$$

The relation that determines the concentration at equilibrium is $\mu_1 = \mu_2$ or

$$\tilde{a}(T) + c_2^2 \ln \frac{\rho_1}{\rho_2} + 2C = 0, \tag{8.2.11}$$

where the function $\tilde{a}$ is

$$\tilde{a}(T) = T\frac{da}{dT} - a. \tag{8.2.12}$$

The model is thus characterized by a fixed relationship between concentration and temperature, independent of ρ and p.

Can this be justified? In order to explain other experiments, we have admitted that there are two independent densities and two independent vector fields, and that equilibrium is established solely by scattering off the wall. But Eq. (8.2.10) is the criterion that was used to determine the equilibrium of a mixture of hydrogen atoms and hydrogen molecules, two gases that transform freely from one to the other. If we admit that the superfluid "mixture" has this property, then it becomes difficult to understand the experiments that were discussed at the beginning of this chapter.

In liquid Helium below the λ point, the concentration is often quoted as a fixed function of the temperature. In the model, this relationship between concentration and temperature is determined by the choice that we make of the function $a(T)$, and we may choose this function so as to further enhance the similarity of the model to Helium II.

The experimental temperature profile for Helium shown in Table 8.1.2 is approximately

$$\frac{\rho_1}{\rho} = \left(\frac{T}{2.19}\right)^6, \quad T < 2.19. \tag{8.2.13}$$

Thus, for our model to have similar properties, we should take, for $0 < T < 2.19$,

$$\tilde{a}(T) = -2C - c_2^2 \ln\left(\left(\frac{T}{2.19}\right)^{-6} - 1\right). \tag{8.2.14}$$

The heat capacity

In the case of a heterogeneous system without chemical reactions or phase changes, what is measurable and usually quoted as C_V is

$$C_{\rho_1,\rho_2} = C_{\rho,x} = \frac{\partial}{\partial T} U(S,T)\Big|_{\rho_1,\rho_2},$$

where U is the specific internal energy density. It is measurable and well defined because it is possible to raise the temperature without affecting the values of the two densities.

But in the case of Helium II and other systems with two mixed phases, it is not possible to keep the concentration fixed. Heating takes the system through a sequence of equilibrium states in which the concentration is determined by ρ and T. What can be defined, and measured, is

$$C_{\rho_1,\rho_2} = C_{\rho,x} = \frac{\partial}{\partial T} U(S,T)\Big|_{\rho}.$$

The full expression for the free energy density of the model is (8.2.9). The internal energy density, expressed in terms of the variables ρ_1, ρ_2 and T,

$$u(s,T) = \tilde{a}(T)\rho_1 - B$$

and the specific internal energy density, is

$$U\left(\rho, \frac{\rho_1}{\rho_2}, T\right) = \tilde{a}(T)\frac{\rho_1}{\rho} - \frac{B}{\rho}.$$

The specific heat at constant volume is

$$C_V = \frac{d}{dT} U(\rho,\rho_1,T)\Big|_{\rho}, \quad \frac{\rho_1}{\rho} = \left(\frac{T}{2.19}\right)^6.$$

The relevant total derivative is

$$\frac{d}{dT} = \frac{\partial \rho_1}{\partial T}\frac{\partial}{\partial \rho_1} + \frac{\partial}{\partial T}$$

and

$$C_V = \frac{\partial \tilde{a}}{\partial T}\frac{\rho_1}{\rho} + \frac{\tilde{a}}{\rho}\frac{\partial \rho_1}{\partial T} - \frac{1}{\rho}\frac{\partial B}{\partial \rho_1}\frac{\partial \rho_1}{\partial T}.$$

At equilibrium, the second and third terms cancel and

$$C_V = \frac{6c_2^2}{t}\frac{1}{t^{-6}-1}, \quad t := \frac{T}{2.19}.$$

This is in close agreement with the observed specific heat for Helium II for temperatures below the λ point, if $c_2^2 = 1/10$ (in Joules/mol).

8.3. Dynamical Equations of the Model

Since Landau's theory can be presented as a collection of formulas, let us list the equations of our model. All are derived from the Lagrangian density

$$\mathcal{L} = \sum_{i=1,2} (\rho_i(\dot{\Phi}_2 - \vec{\nabla}\Phi^2/2)) - f - sT,$$

with

$$f = a(T)\rho_1 + B(\rho_1\rho_2).$$

- Two equations of continuity, derived from variation of the action with respect to the velocity potentials.
- An adiabatic condition, from variation of T, one of the fundamental relations of thermodynamics. In the simplest model, it fixes the temperature in terms of the specific entropies.
- Variation of the densities gives the Bernoulli equations

$$-\frac{D\vec{v}_1}{Dt} = \text{grad}\left(\frac{\partial f}{\partial \rho_1} + S_1 T\right) = 0, \quad -\frac{D\vec{v}_2}{Dt} = \text{grad}\frac{\partial f}{\partial \rho_2}.$$

Finally, at equilibrium,

$$\frac{\partial f}{\partial \rho_1} + S_1 T = \frac{\partial f}{\partial \rho_2} + S_2 T;$$

this is from variation of the densities holding ρ fixed. That is all.

The adiabatic index

We have not referred to the interpretation of this system as a manifestation of the **Bose-Einstein** condensation, for it would take us far from the field in which we are comfortable. We are tempted, however, to speculate about what it is that characterizes, from our perspective, a system where a phase transformation makes part of the fluid move to a state in which the entropy is very near to zero. In the Bose–Einstein condensation, the lowest level in the atomic spectrum gets a high population and this appears to have an effect that resembles that of an energy gap.

This concept, that relates the properties of a liquid to interactions between the atoms, as well as interactions between the liquid and the walls, is a basic but difficult mainstay of kinetic theory. One of the important examples is the attempt to predict the value of the adiabatic index. The underlying idea is a principle of equipartition of energies. It suggests that $2n$, twice the adiabatic index, is the number of degrees of freedom of each atom. This idea needs some development and phenomenological adjustments; for example, vibrational degrees of freedom should not be counted. The idea encountered much resistance before being finally accepted.

To explain the fountain experiment, it was suggested (London, 1938) that the liquid that gets through the narrow channels carries no entropy. Entropy is related to the number of degrees of freedom but the adiabatic index is more directly so. While the entropy is a variable, not exactly zero and not constant, the adiabatic index of an ideal gas is a fixed parameter. For this reason, we suggest that what characterizes the condensate, from the point of view of thermodynamics, is the adiabatic index, and that it is close to zero. This implies a low viscosity and it has an interesting effect on the propagation of sound in the mixture.

As we have seen, the existence of two modes of sound propagation is not unique to Helium, but a basic property of mixtures, although interactions between the components tend to damp one of the two. But in the case of this condensate model for superconducting Helium, the interaction would be weak and it makes sense to calculate the two velocities under the assumption that it is nearly negligible.

The speed of sound at 1 atmosphere and 293 K is 985 m/s, at the λ point, it is 133.3 K. It is well approximated in the interval by the formula of Laplace; a slightly better fit is achieved with the van der Waals equation of state, with $n = 3/2$.

We have done all necessary preparation for this in Section 7.2. If liquid Helium is a mixture of two ideal gases, then the speeds v_1, v_2 are obtained as

$$v_i = c_i\sqrt{\mathcal{R}_1 T}, \quad i = 1, 2,$$

where $c = c_1$, c_2 are the zeros of the determinant of the matrix

$$M(c) = \begin{pmatrix} 1 + \frac{\tau}{A} - c^2 & \frac{\kappa}{A} \\ \frac{\kappa\tau}{A} & \kappa + \frac{\kappa^2}{A} - c^2 \end{pmatrix},$$

with

$$A = n_1\tau + n_2\kappa,$$

$$\kappa = \frac{\mathcal{R}_2}{\mathcal{R}_1}, \quad \tau = \frac{\rho_1}{\rho_2}.$$

Below the critical point, we use these formulas with $m_1 = m_2$, $\kappa = 1$. The adiabatic index of the normal component is $n_1 = 3/2$, that of the superfluid component is denoted n, a free parameter. In this case, the zeros of the matrix (7.1.11) are at

$$c^2 = 1 + \frac{\tau + 1}{n_1\tau + n} \quad \text{and at} \quad c^2 = 1.$$

The first value corresponds to the eigenvector $\delta\rho_1 = \delta\rho_2$; the two densities oscillate in phase. In the other case, the eigenvector satisfies $\rho_1\delta\rho_1 = -\rho_2\delta\rho_2$, the densities are in opposite phases. For an ideal gas this means that the pressure does not participate, as is characteristic of second sound in Helium II; it is an oscillation of the temperature only.

At equilibrium, the ratio

$$x = \rho_1/(\rho_1 + \rho_2) = \frac{1}{1 + 1/\tau}$$

is related to the temperature,

$$x \approx \left(\frac{T}{2.19}\right)^6.$$

This is an experimental fact, and not applicable to an ideal gas. Nevertheless, let us make the naive assumption that the model can be tweaked to satisfy this relation. This is not completely unreasonable, since the formula for the total entropy should yield the value zero at $T = 0$. On this assumption, we plot, in Fig. 8.3.1, both sounds below the λ-point, namely,

$$v_1 = \sqrt{5/3\mathcal{R}T}\sqrt{1 + \frac{\tau + 1}{3\tau/2 + n}}, \quad v_2 = \sqrt{5/3\mathcal{R}T}.$$

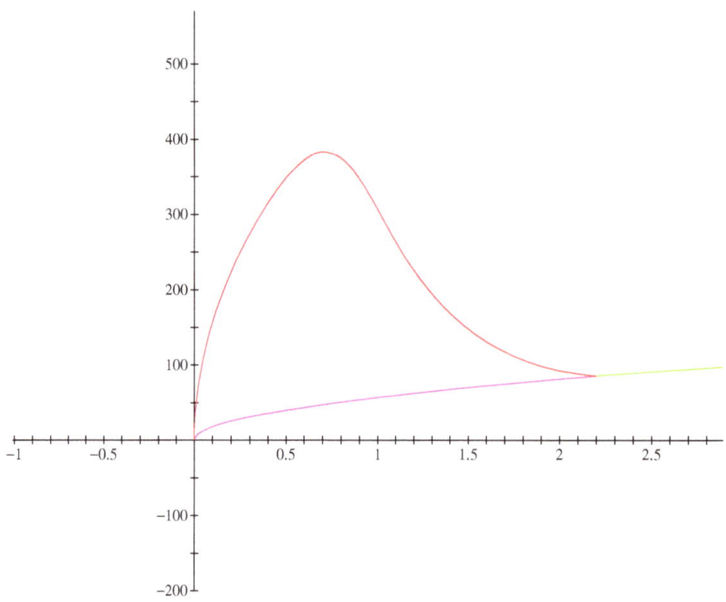

Fig. 8.3.1. The two speeds in Helium II, treated as a mixture of two ideal gases, a normal component with adiabatic index 3/2 and a superfluid component with adiabatic index 1/100. The agreement with experiment is significant for $T > 1$.

With diminishing value of the adiabatic parameter n of the superfluid, the sound profiles develop a greater similarity with those of superfluid Helium. A best fit in the interval $1\,\mathrm{K} < T < 2.19\,\mathrm{K}$ is obtained for $n = 1/100$. With the usual kinetic interpretation of the adiabatic index, we believe that this can be interpreted as representing a fluid with a very small space of excitations as well as a very weak interaction with the walls, thus providing a tentative explanation of the fountain effect.

Rotational flow

A complete description of superfluids must deal with "rotons", rotational flow, beyond the reach of the Fetter–Walecka theory. We shall see, in Section 8.4, that the required generalization involves two kinds of flows. The extension of this theory that is needed to deal with mixtures, and the correct approach to superfields, are far from clear (Tables 8.4.1–8.4.6).

Project 5. Try to understand flux quantization.

8.4. Superfluid Helium

Two fluids or two flows

Landau, in his first, path-breaking paper, strongly recommended that we should not think of Helium below the λ point as a mixture of two kinds of fluids; "there is only one kind of Helium atom". Instead, there are two kinds of "flow". Nevertheless, Tisza proposed his 2-fluid theory, and this point of view has since become dominant. Yet Tilley and Tilley (1974, 1986) (and others) make it a point to say that the two-fluid hypothesis should not be taken literally.

In the first experiments, Helium below the λ point flowed from vessel A to vessel B through a very narrow channel, apparently without resistance and without transporting heat. This suggested to London that the transported fluid has zero entropy and to Tisza that the flow consists of only one of two components that make up the fluid, a superfluid with zero entropy. But the fluid that accumulates in vessel B is not pure superfluid and in fact pure superfluid exists only at absolute zero of temperature. Other experiments have confirmed that the concentration (or what is interpreted as concentration) is completely determined by temperature and pressure. The "concentration" is not an independent dynamical variable and the idea of a mixture has no sense. Measurements of concentration are indirect and other interpretations are possible.

One set of experiments of this kind, by Andronikashvili (1946) and others, has a stack of parallel discs suspended by a wire, rotating as a torsion pendulum. The interpretation depends on determining the angular momentum and the period. We have seen, in the Appendix of Chapter 10, that the former is defined in a conservative context. The calculation that was done to justify a conclusion about the concentration is a guess that cannot be justified unless the flow is either irrotational or solid-body type.

In a similar experiment, a cylindrical vessel was filled with Helium and rotated at constant angular speed. The temperature was lowered below the λ point and the cylinder was brought to rest. After some time, the fluid appeared to have come to rest as well. Then, after a delay of as much as 2 weeks, the temperature was brought back up above the λ point and the fluid started to flow, demonstrating that the system remembered some of its original angular momentum. The accepted interpretation is that, at the lower temperature, the normal fluid was at rest while the superfluid moved through it without encountering any resistance. A more attractive hypothesis, possible in a fluid subject to Conservative Hydrodynamics, is that the flow velocity $\vec{v}$ is zero while the angular momentum is not. This would make the angular momentum invisible.

Perhaps it may be permitted to ask whether conservative hydrodynamics may offer some hope of an alternate understanding of the odd behavior of superfluids. It is commonly held that what makes this fluid different is that it remains liquid at very low temperatures. That is to say that, except for this accident, other fluids would behave the same way. That might have made it possible, on the basis of a more advanced form of hydrodynamics, to predict the odd behavior of liquid Helium.

Table 8.4.1. Pressure = 0.

T	ρ_2	ρ_1	ρ	C_V	ρ_2/ρ_1	s_{BC}	$s_{BC}\rho$	C_V/ρ_1
0.1	0.14513	0.17497E–8	0.14513	0.20597E–4	829E+5	0.68889E–5	0.1E–5	1.2056E–8
0.5	0.14513	0.11607E–5	0.14513	0.23749E–2	125030	0.80709E–3	0.117E–3	0.8E–5
0.8	0.14513	0.11607E–5	0.14513	0.21058E–1	1250.4	0.42709E–2	0.6202E–3	0.8E–5
0.9	0.14473	0.37579E–3	0.14511	0.47598E–1	385.145	0.8084E–2		0.2596E–2
1.0	0.14415	0.95892E–3	0.14511	0.10504	150.32	0.15562E–1	0.00226	0.6652E–2
1.2	0.14107	0.40407E–2	0.14511	0.33523	34.912	0.51689E–1	0.00750	0.2864E–1
1.4	0.13403	0.11109E–1	0.14513	0.80923	12.0649	0.13301	0.0193	0.8289E–1
1.6	0.12090	0.24307E–1	0.14520	1.5506	4.97407	0.28431	0.0413	0.20105
1.8	0.098205	0.47147E–1	0.14535	2.9278	2.08299	0.54155	0.0757	0.4801
2.0	0.060925	0.84682E–1	0.14561	5.1900	0.58158	0.95098	0.1385	1.390
2.1	0.033582	0.11222	0.14582	6.6428	0.299258	0.12394E+1	0.1807	0.03358

Note: s_{BC} is from Brooks and Donnelly (1977). My s is ρs_{BC}.

Table 8.4.2. Pressure = 5.

T	ρ_2	ρ_{-}	ρ	C_V	ρ_2/ρ_1	s	s/ρ_1	
0.1	0.15284	0.87476E–9	0.15284	0.12912E–4	∞	0.43073E–5	3937.19	0.5723E–9
0.5	0.15282	0.95820E–6	0.15283	0.16557E–2	159491	0.53774E–3	561.198	0.627E–5
0.8	0.15263	0.18344E–3	0.15283	0.22425E–1	832.043	0.37520E–2	20.4536	0.1202E–2
0.9	0.15226	0.55459E–3	0.15283	0.53154E–1	274.543	0.79541E–2	14.3423	0.3642E–2
1.0	0.15148	0.13510E–2	0.15284	0.11564	112.12	0.16276E–1	12.0474	0.8919E–2
1.2	0.14765	0.52329E–2	0.15287	0.35856	34.912	0.55231E–1	10.5546	0.7E–2
1.4	0.13902	0.13998E–1	0.15296	0.86374	9.931420	0.14263	10.1893	0.1007
1.6	0.12315	0.30126E–1	0.15313	1.6423	4.08776	0.30331	10.068	0.2446
1.8	0.095685	0.57957E–1	0.15345	3.1132	1.65273	0.57448	9.91218	0.60588
2.0	0.049022	0.10553	0.15405	5.4286	0.464536	1.0145	9.61338	2.1527

Table 8.4.3. Pressure = 10.

T	ρ_2	ρ_1	ρ	C_V	ρ_2/ρ_1	s	s/ρ_1
0.1	0.15892	0.53584E–9	0.15892	0.92471E–5	∞	0.30835E–5	5757.27
0.5	0.15891	0.12284E–5	0.15892	0.12825E–2	12935.6	0.39716E–3	323.315
0.8	0.15862	0.29087E–3	0.15892	0.26776E–1	545.33	0.38804E–2	14.3347
0.9	0.15809	0.8369E–3	0.15892	0.6383E–1	188.9	89332E−2	10.6742
1.0	0.15701	0.19438E–2	0.15893	0.13393	80.7728	0.18815E–1	9.67949
1.2	0.15220	0.68628E–2	0.15899	0.41524	22.1766	0.62507E–1	9.10809
1.4	0.14167	0.17631E–1	0.15911	0.99358	8.035	0.15848	8.98871
1.6	0.12221	0.37530E–1	0.15938	0.1.9024	3.25641	0.33491	8.9303
1.8	0.088137	0.72376E–1	0.15991	3.5295	1.21774	0.63384	8.75833
1.95	0.046985	0.11446E–1	0.16059	5.246	0.481813	0.97517	8.51974
2.0	0.027802	0.13405	0.16091	5.9711	0.207394	1.1252	8.39388

Table 8.4.4. Pressure = 15.

T	ρ_2	ρ_1	ρ	C_V	ρ_2/ρ_1	s	s/ρ_1
0.1	0.16403	0.36709E–9	0.16403	0.71329E–5	∞	0.23779E–5	64777.69
0.5	0.16402	0.18453E–5	0.16403	0.10838E–2	88887.9	0.31462E–3	170.498
0.8	0.1636	0.424E–3	0.16403	0.32024E–1	0.99742	0.42917E–2	170.498
0.9	0.16289	0.11714E–2	0.16404	0.75851E–1	139.056	0.10319E–1	8.80912
1.0	0.16147	0.26347E–2	0.16405	0.10950	61.2859	0.21955E–1	8.333024
1.2	0.15540	0.88647E–2	0.16412	0.44072	17.5302	0.71957E–1	8.11725
1.4	0.14242	0.22080E–1	0.16429	0.86039	6.45712	0.17897	8.10553
1.6	0.11839	0.46810E–1	0.16465	2.0192	2.5292	0.37664	8.04614
1.8	0.74742E–1	0.91507E–1	0.16536	3.7607	0.816794	0.71670	7.83219
1.95	0.17795E–1	0.14971	0.16641	6.2825	0.118869		

Table 8.4.5. Pressure = 20.

T	ρ_2	ρ_1	ρ	C_V	ρ_2/ρ_1	s	s/ρ_1
0.1	0.16849	0.26997E–9	0.16849	0.57664E–5	∞	0.19223E–5	7120.42
0.5	0.16848	0.29171E–3	0.16849	0.99665E–3	57755.7	0.26359E–3	90.3603
0.8	0.16791	0.59442E–3	0.16849	0.38572E–1	88887.9	0.49635E–3	170.498
0.9	0.16695	0.15873E–2	0.16850	0.89913E–1	88887.9	0.1216E–1	170.498
1.0	0.16512	0.34758E–2	0.16852	0.18135	47.506	0.25837E–1	7.4334
1.2	0.15750	0.11308E–1	0.1861	0.50271	13.9287	0.83426E–1	7.37761
1.4	0.14159	0.27633E–1	0.16883	1.1844	5.12395	0.20455	7.40238
1.6	0.11154	0.58433E–1	0.16931	2.2859	1.90884	0.42826	7.32908
1.8	0.54659E–1	0.11660	0.17028	4.4834	0.816794	0.82223	7.05172

Table 8.4.6. Pressure = 25.

T	ρ_2	ρ_1	ρ	C_V	ρ_2/ρ_1	s	s/ρ_1
0.1	0.17246	0.22085E–9	0.17246	0.4815E–5	0.78E–9	0.16053E–5	7268.73
0.5	0.17245	0.18453E–5	0.17246	0.10069E–2	88885	0.23324E–3	126.40
0.8	0.17425	0.83354E–3	0.17247	0.47252E–1	209	0.59974E–3	48.32898
0.9	0.17166	0.21514E–2	0.17248	0.10767	88887.9	0.14726E–2	170.498
1.0	0.16803	0.45751E–2	0.17251	0.21193	36.7273	0.30896E–1	6.75308
1.2	0.15847	0.14396E–1	0.17262	0.58045	11.0077	0.97621E–1	6.78112
1.4	0.13868	0.34682E–1	0.17291	1.3577	8.994	0.23620	6.81045
1.6	0.99891E–1	0.74342E–1	0.17356	2.635	1.34368	0.49733	6.668976
1.8	0.17527E–1	0.15821	0.17512	5.4955	0.11079	0.99186	6.26926

Chapter 9

Atmospheres

9.1. Introducing Gravitation

Gravitation enters as a component of most, if not all applications of physics. There is a set and tested procedure for introducing the gravitational field into a dynamical context. In classical, non-relativistic physics, it consists of adding the **gravitational potential energy** to the Hamiltonian. For an isolated, ideal gas, one adds

$$\int_\Sigma d^3x\, \rho\, \phi,$$

where ρ is the mass density and ϕ is the **Newtonian potential**.

The adiabatic Lagrangian becomes

$$\int \mathcal{L} = \int_\Sigma d^3x(\rho(\dot\Phi - \vec{v}^{\,2}/2 - \phi) - f - sT).$$

In the case of an ideal gas,

$$\int \mathcal{L} = \int_\Sigma d^3x \left(\rho(\dot\Phi - \vec{v}^{\,2}/2 - \phi) - \mathcal{R}T\rho \ln \frac{k}{k_0}\right), \quad k := \frac{\rho}{T^n}. \tag{9.1.1}$$

This prescription for including the effect of gravity has its ultimate justification in **General Relativity**. The most important effect of gravity is contained in the time component of the **metric** and it usually appears in connection with time derivatives, thus as an addition to the Hamiltonian.

In this section, the constant k_0 stands for the specific entropy,

$$S/\mathcal{R} = n - \ln k_0, \tag{9.1.2}$$

and $k := \rho/T^n$ is sometimes referred to as the adiabatic temperature. The gravitational field appears in the equation of motion,

$$\dot{\Phi} - \vec{v}^2/2 - \phi = \mathcal{R}T\left(\ln\frac{k}{k_0} + n + 1\right).$$

When the effect of gravity is taken into account in the dynamics, we get a theory of **atmospheres**. In the case of **terrestrial gravity**, $\phi = gz$, where g is a constant and z is the elevation. The adiabatic condition makes $k = k_0$, constant, and the Euler–Lagrange equation that comes from variation of the density

$$\dot{\Phi} - gz = \mathcal{R}T(\ln x + n + 1), \quad x := \frac{k}{k_0} = 1. \tag{9.1.3}$$

predicts a constant **temperature gradient** (lapse rate) for an isolated atmosphere that consists of an ideal gas, at rest. This result was first obtained by Herapath in a paper submitted to the Royal society in (1820); according to Brush (1976), the paper was not only rejected, but the society also refused to return the manuscript to the author. The effect has not, to our knowledge, been conclusively tested in the laboratory, (see Liao, 2009; Graeff, 2007), but the constant lapse rate is a feature that is observed in the earth's atmosphere, in the oceans and in the internal structure of stars (Lane, 1870; Ritter, 1938; Emden, 1907; Eddington, 1926; Levy, 2012). For the **earthly atmosphere**, this formula actually gives a value for the temperature gradient that is close to observation, at low atmospheric humidity. This successful application of a theory originally constructed to account for laboratory experiments where gravitational effects are insignificant, modified in standard fashion to include gravity, does *not* attribute the observed temperature gradient to the radiation from the Sun or to any external source other than gravity (Fronsdal, 2011). The actual atmosphere is close to isentropic, not because of the effect of convection and radiation but in spite of it. The effect of humidity on the lapse rate will be examined below (Good intentions, at least). This was the attitude adopted by Emden (1907) and all others in astrophysics, and currently, in the context of planetary atmospheres.

It is evident that, in the absence of the Sun, the present state of the earthly atmosphere could not endure. The extinction of the Sun would lead to a general cooling of the atmosphere. This cooling is not described by any process encompassed by adiabatic Lagrangian dynamics, but by the spontaneous emission of **infrared radiation**. As this is a slow effect, we can regard the progressive cooling as a sequence of equilibria of adiabatic dynamics, a slow loss of entropy and an increase in the value of k_0 (Section 2.3).

The reverse effect is also possible and if the terrestrial atmosphere is found to be stable over long times, then we must conclude that spontaneous cooling by emission is balanced by the heating that is indirectly provided by the Sun (the **greenhouse effect**).

The problem of the gravitation-induced temperature gradient can be debated within the framework laid out in the appendix to this chapter. Important aspects of historical atmospheric science, including the popular polytropic atmospheres and Emden's stability theorem, are also relegated to that Appendix.

9.2. Mixed, Isentropic Atmospheres

At one time, it was believed that, in an atmosphere consisting of several components with different molecular weights, the lighter gas would float on top, contrary to the popular idea that all gases are miscible. This was surely based on observation since cooking gas, entered at ground level, tends to remain there; however, it does so only for a short time. Dalton made the radical proposal that each gas behaves as if the other were absent. This was an overstatement and led to much misunderstanding and debate, some of it recorded in Dalton's book (Dalton, 1808). In the case of gases that have had time to settle, Dalton's prediction had some initial success, but the most recent observations indicate that the concentrations tend to be nearly independent of elevation.

A complete investigation of the atmosphere must recognize that it involves a dissipative process, but if the problem that concerns us is the density and temperature profile of a final, stationary state of the mixed atmosphere, then it is not the issue. If an equilibrium is ultimately reached, then we are mainly interested in the end result, and less in describing the process that leads to it. The process includes **dissipation** but the net entropy of the atmosphere is maintained by radiation. We shall assume that it can be treated as a constant in a study of the final, stationary configuration and, following Laplace, in the adiabatic perturbations of it.

We have embraced the Gibbs–Dalton hypothesis and we have used it for mixtures that are not under the influence of gravity. But for a mixture of ideal gases in the gravitational field, it leads to a difficulty. Namely, since the hydrostatic conditions for the individual gases also remain valid in the mixture (assuming no mutual interaction), we would deduce that, at equilibrium,

$$\dot{\Phi}_i - \phi = R_i(n_i + 1)T, \quad i = 1, 2,$$

from which two expressions for the lapse rate T' are:

$$g = R_1(n_1+1)T' = R_2(n_2+1)T',$$

which is contradictory whenever $(n_1+1)/m_1 \neq (n_2+1)/m_2$.

This is where it is important to have a generalization of the **Gibbs-Dalton** hypothesis; one was introduced in Chapter 5.

In this section, the mass densities are in g/cm^3, not molar, and the constants $\mathcal{R}_1 = \mathcal{R}/m_1$, $\mathcal{R}_2 = \mathcal{R}/m_2$.

In what follows, we shall take the Lagrangian for the mixture to be the sum of the Lagrangians for the two pure gases, eventually to be corrected by an interaction term.

9.3. Nitrogen and Oxygen

We begin by examining the atmospheric concentration of **nitrogen and oxygen**, ignoring the rarer components. According to a number of sources, the relative concentration does not vary with altitude, but no authoritative scientific information was found.

Consider a vertical column of air consisting of nitrogen and oxygen and confined to the range $0 < z < z_1$ of elevation. The equations of motion that determine the equilibrium configurations are, if entropic forces and other interactions are ignored,

$$\dot{\Phi}_1 - gz = \mathcal{R}_1 T(\ln x + 1 + n_1), \quad x := \frac{\rho_1}{T^{n_1}k_{01}} = \frac{k_1}{k_{01}}, \tag{9.3.1}$$

$$\dot{\Phi}_2 - gz = \mathcal{R}_2 T(\ln y + 1 + n_2), \quad y := \frac{\rho_2}{T^{n_2}k_{02}} = \frac{k_2}{k_{02}}, \tag{9.3.2}$$

and

$$\mathcal{R}_1\rho_1 \ln x + \mathcal{R}_2\rho_2 \ln y = 0. \tag{9.3.3}$$

The last equation is the **adiabatic condition**. The adiabatic parameter k_0 and the "adiabatic temperature" k were defined in Eqs. (9.1.1) and (9.1.2). In the case of pure gases without gravity, $x = y = 1$; the Gibbs–Dalton hypothesis would assign the same values in the mixture. The concentration would be constant and related to the specific entropies. This would be possible if nothing linked the two systems, but not here, for the adiabatic condition amounts to indirect forms of interaction.

The fixed parameters are

$$\mathcal{R}_1 = \mathcal{R}/28, \quad \mathcal{R}_2 = \mathcal{R}/32, \quad \mathcal{R} = 0.8314 \times 10^8, \quad n_1 = n_2 = 5/2.$$

The two $\dot{\Phi}$'s are constants, determined by the **boundary conditions**. The most interesting question concerns the adiabatic parameters k_{01} and k_{02}; that is, the specific entropies S_1 and S_2. We harbor a strong prejudice for taking them to be uniform, independent of z; the first question is whether this allows for a solution. We shall show that, when $n_1 = n_2$, the answer is affirmative. The data at the sea level ($z = 0$) are, at $T(0) = 293.159$ (normal conditions),

$$\rho(0) = 0.0012,\ \rho_1(0) = 0.9184\times 10^{-3},\ \rho_2(0) = 0.2816\times 10^{-3},\ \frac{\rho_1(0)}{\rho_2(0)} = 3.2614,$$

and at $T(0) = 273.159$ (standard conditions)

$$\rho(0) = 0.0016346,\ \ \rho_1(0) = 1.251 \times 10^{-3},$$
$$\rho_2(0) = 0.3836 \times 10^{-3},\ \ \frac{\rho_1(0)}{\rho_2(0)} = 3.2612.$$

This gives us the predicted pressure at the sea level ($z = 0$),

$$p(0) = [\mathcal{R}_1\rho_1(0) + \mathcal{R}_2\rho_2(0)]T = 1.0139 \times 10^6,$$

against the experimental value that ranges from 1 to 1.05 million ergs.

We shall look for a solution with x, y both uniform. The specific entropies S_1 and S_2 are also uniform though not with the same values as for the separate gases. In this case, the concentration is also constant, by (9.3.3).

Combining the three equations of motion to eliminate the logarithms, we find, since the concentration is uniform,

$$\left.\frac{T(0) - T}{z}\right|_{z=0} = \frac{g}{\sum(n_i + 1)\mathcal{R}_i\rho_i/\rho} \tag{9.3.4}$$

which gives a constant temperature lapse rate of 9.715 degrees Kelvin per kilometer at normal conditions. Consequently, Eqs. (9.3.1) and (9.3.2) are equivalent to

$$\frac{-g}{\mathcal{R}T'} = \frac{\ln x + 3.5}{28} = \frac{\ln y + 3.5}{32},\quad T' := \frac{\delta T}{dz}. \tag{9.3.5}$$

Combining this with Eq. (9.3.3), we find that, if x, y are constant, the unique solution is

$$x = 0.902428,\quad y = 1.4662,\quad \frac{x}{y} = 0.61240,\quad \frac{\rho_1}{\rho_2} = 3.2614,\quad \frac{k_{01}}{k_{02}} = 5.2988.$$

The densities for the pure gases at standard conditions are: nitrogen 0.0001251 and oxygen 0.0001429.

For any set of values of k_{01}, these figures fix the concentration in terms of T, which is in turn determined by the elevation z. Thus, k_{01} is unrestricted and can be taken to be uniform. Using the data at the ground level, we obtain

$$k_{01} = \frac{0.9184 \times 10^{-3}}{0.902428}(293.159\mathrm{e})^{-2.5} = 5.6771 \times 10^{-11}, \quad k_{02} = 1.0714 \times 10^{-11}$$

and

$$S_1 = -\mathcal{R}\ln(xk_{01}) = 23.6947\mathcal{R}/\mathrm{mol}, \quad S_2 = 24.8768\mathcal{R}/\mathrm{mol}.$$

Note that, because x and y are not equal to 1, these differ from the entropy of the corresponding pure ideal gas with the same density and temperature. The deviation of x and y from unity is a measure of the effect of mixing; nitrogen effectively passes some of its entropy to oxygen. The adiabatic parameter for pure nitrogen under **`normal conditions`**[67] is 6.58×10^{-11}; for oxygen, it is 7.52×10^{-11}; that is,

$$S_N = -\mathcal{R}\ln(5.6771 \times 10^{-11}) = 23.44\mathcal{R}/\mathrm{mol}, \quad S_O = 23.31\mathcal{R}/\mathrm{mol}.$$

These values are not normalized to agree with the third law, but the difference $S_N - S_O = 1.12$ J/mol compares well with the measured value reported as 1 J/mol.

The difference from the values that apply to pure gases is of course greater for the minor component of the mixture.

Discussion. We justify taking x, y uniform as follows. Consider the mixture without the presence of gravity. Then, surely all variables are uniform. The effect of mixing is merely to change the values of the entropy parameters k_{01}, k_{02}. These values may conform to the Gibbs–Dalton hypothesis or they may not. The variables $\dot{\Phi}_1$, $\dot{\Phi}_2$ play their limited role as constants of integration. What happens as we turn on gravity? Only one thing is changed: we are to add the term $-gz\rho$ to the Lagrangian density. Nothing suggests a direct effect of gravitation on the entropy distribution, so the assigned specific entropy densities should not change, so long as the equations of motion have a solution. But the equations of motion do demand that the entropy parameters allow for the existence of solutions, and this dynamical requirement is inconsistent with the Gibbs–Dalton hypothesis. Although turning on gravitation does not involve a change in entropy, the entropies need to have values that allow for the existence of solutions of the equations of motion, and that is the requirement that we are using to determine these parameters, pending more data.

[67] "Standard conditions": $T = 0°\mathrm{C} = 273.16\,\mathrm{K}$, the density of N is 0.001251 and $0.001251/(300\mathrm{e})^{5/2} = 6.587 \times 10^{-11}$. This was my original number. "Normal conditions": $T = 293.15\,\mathrm{K}$. I should have used $0.00125/(273.16e)^{5/2} = 8.2368 \times 10^{-11}$.

Alternatively, the Gibbs–Dalton hypothesis can be saved by admitting a small interaction between the two gases. But for this to be convincing, this interaction must be confirmed by other means.

9.4. Air and Argon

The densities for the pure gases at standard conditions are: air 0.00012, Argon 0.0001784. The adiabatic parameters for pure are air: 6.32E–11, Argon: 7.66E–8; that is,

$$S_{\rm Air} = 23.48\mathcal{R}/{\rm mol}, \quad S_{\rm Ar} = 16.38\mathcal{R}/{\rm mol}.$$

This case is different since the two gases have different adiabatic index; see (9.3.1) and (9.3.2). By "air", we mean an ideal gas with $n = 5/2$ and atomic weight 29. Equations (10.3.1)–(10.3.3) apply, but the data are different.

The fixed parameters are

$$\mathcal{R}_1 = \mathcal{R}/29, \quad \mathcal{R}_2 = \mathcal{R}/40, \quad \mathcal{R} = 0.8314 \times 10^8, \quad n_1 = 5/2, \quad n_2 = 3/2.$$

The data at ground level, $z = 0$,

$$T = 293.159, \quad \rho_1(0) = 1.2 \times 10^{-3}, \quad \rho_2(0) = 1.116 \times 10^{-5}, \quad \frac{\rho_1}{\rho_2} = 107.527.$$

This case is more difficult. Observations seem to indicate that the ratio ρ_1/ρ_2 varies slowly, but there is no solution that makes it exactly independent of altitude. Because air is very dominant, we expect it to be only a little perturbed by the presence of Argon; that is, the specific entropy is likely to have a value close to what would be observed in the absence of Argon. In other words, the variable x is likely to be close to 1 under normal conditions at the ground level. Taking $x = 0.99$, we get from (9.3.3) the value $y = 4.68249$.

Equations (9.3.1) and (9.3.2) tell us that

$$c := T\left(\frac{\ln x + 3.5}{29} - \frac{\ln y + 2.5}{40}\right)$$

must be uniform. Setting $x = 0.99$, $y = 4.44$ at the sea level, we get the value $c = 7.0462$. Also, at the sea level,

$$\frac{k_{01}}{k_{02}} = \frac{\rho_1}{\rho_2}\frac{y}{eTx} = 0.68249.$$

Table 9.4.1. The air–Argon atmosphere.

x	0.99	0.991	0.992	0.993	0.994	0.995
y	3.72	3.285	2.931	2.628	2.350	2.094
T	300	266	241	222	205	189
ρ_1/ρ_2	100	100.7	102.4	105	109	113
z	0	3.4	5.9	7.9	9.6	11.2

Having thus obtained good trial values of the parameters, we calculate x, y and T by

$$T^{-1} = \frac{1}{c}\left(\frac{3.5 + \ln x}{29} - \frac{2.5 + \ln y}{40}\right) = -e\frac{k_{01}}{k_{02}}\frac{x \ln x}{y \ln y}\frac{40}{29}$$

and finally the altitude by

$$g(z - z_0) = \mathcal{R}_1 T(\ln x + 3.5)$$

to obtain values from zero to +11 km. The results are illustrated in Table 9.4.1. The lapse rate is steady at 10 K/km.

The theory predicts a small reduction in the concentration of Argon at high altitudes.

Since the values of x and y are determined, we can calculate the adiabatic parameters; at the ground level

$$k_{01} = \rho_1\frac{(\mathrm{e}T)^{n_1}}{x} = 8.485 \times 10^{-14}, \quad k_{02} = \rho_1\frac{(\mathrm{e}T)^{n_2}}{y} = 1.6735 \times 10^{-13}.$$

Then we are free to fix these parameters independent of the elevation and use them to evaluate the individual densities. Calculate entropies. Apply the same method to the case that $n_1 = n_2$.

Table 9.4.1 shows the temperature profile and the line $T = 300 - 10z$; the lapse rate is very close to uniform.

The significance of a **`constant lapse rate`**

So far, we have made scant reference to Fourier's heat equation, namely,

$$\dot{T} \propto \Delta T.$$

If this venerable relation is to be respected, then any stationary atmosphere must have $\Delta T = 0$. This property is met by the **`polytropic atmosphere`** and that is indeed a principal reason why it is of interest, as emphasized by Emden in 1908. Here, as in the following example, this property is satisfied to a very good approximation, but not exactly.

The presence of water vapor in our atmosphere has a pronounced effect on the measured lapse rate; we shall try to calculate this effect later.

Van der Waals atmospheres

As far as the atmosphere of the earth is concerned, the van der Waals modification of the ideal gas may seem to be of little interest, since the excluded volume factor $(1 - b\rho)$ differs from unity by about one part in 1000. Nevertheless, we shall find a useful application of the van der Waals model to study the humidity of our atmosphere, as discussed in the following section. Applications to stellar atmospheres are interesting but they must be done within the framework of General Relativity.

9.5. Air and Water: Introduction

This is the mixture that controls life on earth. We shall take up the study with respect for generations that have studied it before us and conscious of the special difficulties that characterize H_2O.

Water presents a problem in that we know of no simple, analytic equation of state. The classical van der Waals equation was developed to account for the domain near the critical temperature of $T = 647\,\mathrm{K}$. To agree with the measured values of the critical temperature and the critical pressure, the density predicted is off by nearly a factor of two, giving 10.95 mol/Lit instead of the measured value 17.87. Conversely, the density of water in normal conditions is 998.2 g/Lit or 54 mol/Lit, which would imply that $b\rho = 1.6 > 1$. In the van der Waals model, this parameter varies from 0 to 1.

Data for water. The measured values are, molecular weight 18.015,

$$T_{\mathrm{cr}} = 647\,\mathrm{K}, \quad p_{cr} = 220\,\mathrm{atm}, \quad \rho_{cr} = 17.87\,\mathrm{mol/Lit}.$$

Standard van der Waals parameters

$$a = 5.536\,\mathrm{atm}(\mathrm{Lit/mol})^2, \quad b = 0.03049\,\mathrm{Lit/mol}.$$

Check, the function

$$p(\rho) = \frac{0.08314 \times 647\rho}{1 - 0.03049\rho} - 5.536\rho^2$$

has the value 220.45 at the inflection point $\rho = 11.069$.

Our study will be directed at near-normal conditions. Much work has been done to create an accurate equation of state for water, over a very wide range of temperatures, but they lack the attractive feature of the van der Waals model where the coexistence of liquid and water is built in. The solution is to use a van der Waals model specifically adapted to near-normal conditions.

At $T = 293\,\mathrm{K}$, the experimental data are as follows: Vapor pressure 2.3 kPa, densities

$$\rho_{\rm liq} = 998.2\,\mathrm{g/Lit} = 55.4\,\mathrm{mol/Lit}, \quad \rho_{\rm vap} = 17.3\,\mathrm{g/m^3} = 0.00096\,\mathrm{mol/Lit} \tag{9.5.1}$$

and the ratio of densities is

$$\frac{\rho_{\rm vap}}{\rho_{\rm liq}} = 1.73285 \times 10^{-5} = \frac{y^u}{y^v}. \tag{9.5.2}$$

We shall construct a van der Waals model on these data.

The van der Waals formula can be written as

$$\frac{p}{p_{\rm cr}} = 8\frac{T}{T_{cr}}\frac{y}{1-y} - 27y^2$$

and the free energy density can be expressed as

$$\frac{f}{p_{\rm cr}} = \frac{8T}{T_{cr}} y \ln\frac{y}{1-y} - 27y^2 + \cdots, \tag{9.5.3}$$

where $y = b\rho$ and $+\cdots$ stands for a term that is linear in y and that does not affect our calculations of the equilibrium. Because both expressions involve only one parameter, the `common tangent` method allows to calculate $T_{\rm cr}$ from (9.5.2). We find that $T_{\rm cr}$ must have a value near 1178.55.

The two normalized densities and the associated pressure, calculated with the common tangent method using (9.5.2), are

$$\begin{aligned} y_{\rm liq} &= b'\rho_{\rm liq} = 0.9199258777 = 55.4b'\,\mathrm{mol/Lit}, \\ y_{\rm vap} &= 1.59401 \times 10^{-5} = 0.00096b'\,\mathrm{mol/Lit}, \end{aligned} \tag{9.5.4}$$

where a', b' are the new van der Waals parameters required. Both data give the same value for b'. To sum up, our modified van der Waals model, chosen to represent the data near normal conditions, is

$$T'_{\rm cr} = 1178.6, \quad b' = .016605, \quad a' = 5.5382. \tag{9.5.5}$$

The value of a' was obtained from $647(a'/b'^2) = 1187.55(a/b^2)$. We drop the primes from now on. The units are molar densities in g/Lit, and atmospheres.

The many significant figures in some of these numbers bespeaks the low compressibility of the liquid, $xdp/dx \approx 6.8 \times 10^{-3}$; the compressibility is the inverse.

Table 9.5.1. Vapor pressure of water without interaction.

T	$p(9.5.2)$	p exp.	y^u (9.5.2)	y^u exp.	y^v
10	1.5040E–3	1.2281E−3	1.0243E−5	0.866E−5	0.922908668
20	2.3389E–3	2.3388E−3	1.5940E−5	1.594E−5	0.919925878
30	3.7790E–3	4.2455E−3	2.4056E−5	2.801E−5	0.91692
50	8.4718E–3	1.2344E−2	5.0605E−5	7.372E−5	0.9108489
70	1.7280E–2	3.1176E−2	9.7250E−5	1.86E−4	0.904682723
90	3.2519E–2	7.0117E−2	1.7304E−4	3.85E−4	0.898422971
110	9.24309E–2	8.4277E–2	2.8904E−4	5.25E−4	0.892034

Finally, the following formula was used for the pressure of the liquid–vapor equilibrium at near-normal conditions

$$p = \frac{0.08314 \times T\rho}{1 - 0.0166\rho} - 5.5382\rho^2. \tag{9.5.6}$$

The model of water determined, we easily calculate, using the common tangent method, the equilibrium configuration for a modest range of temperature.

Table 9.5.1 shows the pressure in atm, temperature in degrees celsius, $y = b\rho$ dimensionless.

Standard method

It is common to invoke the **ideal gas law**. If we replace the parameters a and b by zero, our formula for the vapor pressure takes the form

$$p = \mathcal{R}T\rho = \frac{\mathcal{R}T}{0.0166}y^u = 5Ty^u. \tag{9.5.7}$$

This does not differ significantly from (9.5.6).

9.6. Phases in a Mixture of Air and Water

Data, for "air", I used nitrogen, molecular weight 29,

$$T_{\rm cr} = 128\,\mathrm{K}, \quad p_c = 34\,\mathrm{atm}, \quad \rho_c = 0.044\,\mathrm{mol/Lit} = 0.0012754\,\mathrm{gr/cm^3},$$
$$a = 1.408, \quad b = 0.03913.$$

For H_2O, we use the van der Waals model constructed above.

We shall test the merit of regarding the atmosphere as a two-phase mixture of two fluids.

$$f = f_1 + f_2 + \alpha \rho_1 \rho_2.$$

To discover the spinodal line, we use the formula from Section 6.3 with a slightly different notation.

$$\left(\frac{T/T_1}{x(1-x)^2} - \frac{27}{4}\right)\left(\frac{T/T_2}{y(1-y)^2} - \frac{27}{4}\right) = \beta := \frac{\alpha^2}{b_1 b_2 T_1 T_2 \mathcal{R}^2}. \tag{9.6.1}$$

Here, T_1, T_2 are the two critical temperatures. When $\beta = 0$, there are only the spinodals of each separate fluid. For small, positive β, there appear ovals, below 210 K for $\beta = 1$, below 1149 K when $\beta = 10$. These ovals very quickly grow to the limits of the domain $0 < x, y < 1$ for lower temperatures.

We know the saturated state of water at normal conditions. We now increase the pressure by admitting air, and calculate the coexistent two-phase configurations for a range of values of the pressure. This is easy, for the water densities do not change.

Having established a pair of coexistent modes of water,

$$y^u = 1.59405, \quad y^v = 0.919926,$$

it is easy to include air into the mixture, at least so long as the interaction strength α is zero. We just choose a value of p and adjust the parameters p and p. The result is in Table 9.6.1.

Table 9.6.1 shows the saturation densities for the system of air and water. The density ratios are $c = 0.01$ (humidity in the air) and $c' = 0.0017$ (air dissolved in water).

At the lowest pressure, we have the ideal-gas approximation

$$p \approx 0.02323 + 622.5x = 0.02323 + 21.5\rho_1.$$

Here, ρ_1 is the molar density of water in mol/Lit as usual and the pressure is in bars. The first term is the contribution of water vapor, the other is

Table 9.6.1. Air and water.

p	ρ^u	c	ρ^v	c'
0.02323	(0.00000, 1.5942E−5)	0	(0.000000, 0.919926)	0.0000
0.2	(0.00028, 1.594E−5)	0.057	(0.000283, 0.919926)	0.00031
0.4	(0.00060, 1.594E−5)	0.027	(0.000605, 0.919926)	0.00066
0.6	(0.00093, 1.594E−5)	0.017	(0.000925, 0.919926)	0.00010
0.8	(0.00125, 1.594E−5)	0.0127	(0.001250, 0.919926)	0.00140
1.0	(0.00161, 1.594E−5)	0.0102	(0.001606, 0.919926)	0.00170

$\mathcal{R}\rho T$. In fact, this approximation is valid over the whole range, the partial pressure of water being fixed. The result concerning the saturation density of water in air is somewhat model independent, yet it is off by several orders of magnitude.

At normal conditions, the mass ratio water/air in the atmosphere is

$$\frac{\rho_2}{\rho_1}_{\text{molar}} = c\frac{b_1}{b_2} = 0.0236 \quad \text{or} \quad \frac{\rho_2}{\rho_1}_{\text{mass}} = c\frac{29b_1}{18b_2} = 0.0380.$$

This is within a factor of two of measured values for saturated air. The ratio of air to water in the ocean is

$$\frac{\rho_1}{\rho_2}_{\text{molar}} = c'\frac{b_2}{b_1} = 0.000721 \quad \text{or} \quad \frac{\rho_1}{\rho_2}_{\text{mass}} = c'\frac{18b_2}{29b_1} = 0.001166.$$

This result is a prediction for the solubility of water in air that is very different from what is observed. We take this to mean that a form of interaction is needed.

Interaction

We try $k = 10$ and $k = 25$. The last row in Table 9.6.2 shows the effect of interaction when $p = 1$.

The humidity in air does not budge, the ratio c does not change, but we observe a reduction of c', and the specific proportion of air in water that, for $k = 25$, is a factor of $1/100$, and this brings the prediction in line with observation.

We end our investigation of the earth's atmosphere at this point for others to finish.

We have chosen to use the confrontation of theory with experiment to discover useful formulas for the internal energies of mixtures, formulas from which all properties of the mixtures can be calculated. This allows for very little freedom and, consequently, strongly enhanced predictive power. Our first attempt at constructing a model for water does not aim at a detailed agreement with experiment, but only, by experience, to see if a simple and partly justified expression for the free energy density reproduces general features, as was the case with the van der Waals formula.

Table 9.6.2. Air and water: Effect of interaction when $p = 1$.

k	ρ^u	c	ρ^v	c'
10	(0.00156992, 1.5925E−5)	0.01014	(0.00023688, 0.919929)	0.0002575
25	(0.0015699, 1.58971E−5)	0.010126	(0.1.388E−5, 0.919931)	1.509E−5

Project 6. Build a model of our atmosphere with interaction, then use it to calculate the speed of propagation in air and compare with measurements at various elevations and atmospheric conditions. Compare the value of the interaction strength with that found in other applications. Try to find a single, simple model that fits all the data.

Appendix: Polytropic Atmospheres

Emden's book of 1907 laid the foundations for 100 years of astrophysics and it had a profound effect on atmospheric science as well, as can be verified by a modern web survey. It is concerned with the stability of a polytropic atmospheres.

A polytropic atmosphere is one in which the pressure and the density are related this way:

$$p = a\rho^\gamma, \quad \gamma > 1,$$

with a and γ constant, independent of the altitude. An isentropic atmosphere that consists of an ideal gas is polytropic with $\gamma = 1 + 1/n$. We shall consider an atmosphere that consists of an ideal gas, polytropic but not isentropic, with the constant γ unrelated to the adiabatic index n.

Emden's theorem

The polytropic atmosphere is stable to convection if and only if

$$\gamma < 1 + 1/n.$$

The most stable among polytropic atmospheres is the isothermal atmosphere with $\gamma = 1$.

More precisely, the expectation is that if the polytropic relation is satisfied as initial conditions at $t = 0$, then the inequality determines what will happen later: a small convective perturbation will grow unless it is satisfied. (It is assumed that there is no flow at $t = 0$.)

The proof offered by Emden in 1907 and repeated in today's textbooks is the same. Suppose that a small volume of gas is displaced upwards adiabatically, that is, without any heat transfer. The pressure in the displaced gas will adjust, adiabatically, to the new environment, almost immediately. Suppose the inequality is satisfied, then the displaced volume will have a larger density than the gas nearby and it will therefore revert downwards. If, instead, the small volume is displaced downwards, then it will become thinner and again it will rebound.

This argument is an account of the balance of forces: pressure against gravity. The unspoken assumption is that there is a dynamical theory behind the argument. The number n is the adiabatic index, hence the fluid is an ideal gas. Ideal gases have been extensively studied and the only known dynamical theory is the thermodynamics developed over many years and reported in this book.

As we have seen in Section 3.1, the forces are the pressure, gravitation and, unless the specific entropy is uniform, the entropic gradient. The latter is zero in the isentropic case when $\gamma = 1 + 1/n$. The "proof" has no weight for other values of γ because the entropic forces are not taken into account.

The effect of entropic forces

The paradigm of the type of stability argument offered in the "proof" of Emden's theorem is this. A ball resting on top of a cylinder is subject to no forces because the contact force is balanced by gravitation; this is true if the position of the ball is precisely at the highest point, where the energy is stationary to first order of perturbations. This is first-order equilibrium. To decide whether the equilibrium is stable or not, one carries the calculations to second order. The initial conditions envisaged in the proof do not define an equilibrium except in the case that $\gamma = 1+1/n$, since the effect of entropic gradient has not been included; the forces are not balanced and the argument cannot proceed.

When the gravitational force is included, and $s = \rho S$, Eq. (3.1.6), the hydrostatic condition becomes

$$\vec{\bigtriangledown} p = -\vec{\bigtriangledown}(gz) - \rho T \vec{\bigtriangledown} S.$$

In accordance with the Gibbs variational principle, the entropy is a given, fixed function of position; suppose we have solved this differential equation for the density (after elimination of the temperature by the adiabatic relation).

To study the stability of this point of equilibrium, we must study second-order perturbations. Turning once more to Gibbs (and our own experience), we do not vary S but keep it fixed. The required calculation is the same as for sound propagation and the answer is the same, real number, $c = \sqrt{\gamma \mathcal{R} T}$. The equilibrium is stable.

The isentropic atmosphere

A famous controversy concerns the temperature profile of an isolated atmosphere in a constant gravitational field. A pillar of statistical mechanics proclaims that, at equilibrium, the temperature of any extended system is uniform, and more, that this statement remains true in the presence of a gravitational field. One of the first to voice disagreement with this dictum was Loschmidt, who was crushed by the united authority of Boltzmann and Maxwell. A steady flow of papers have been published over a period of more than 100 years, all proving that there can be no temperature gradient. A few papers have managed to get past editors and referees with proofs of the contrary. Taking up the question again at this time seems almost redundant, and yet....

It must be said that the belief in the isothermal atmosphere is, in many cases, just that, a belief or conviction, unsupported or contradicted by the evidence. That said, what would qualify a scientist to issue an expert opinion?

What physics is involved?

1. *Gravitation.* What is absolutely clear, yet frequently overlooked, is that it is futile to try to make a decision without a deep knowledge of gravitation. That the controversy involves gravitation is in fact fortunate, for no force is better understood. At the deepest level, we have General Relativity with its Equivalence Principle. It tells us that gravitation is a force that resides in the space-time metric and that the interaction is strongly restricted by invariance under general coordinate transformations. An intimate knowledge of General Relativity is probably not required, but one should know that, in non-relativistic physics, the only component of the metric that is needed is the time–time component g_{00} and that it takes the form

$$g_{00} = c^2 - 2\phi.$$

Here, c is the velocity of light and ϕ is the Newtonian gravitational potential. In field theories, this component of the metric appears in conjunction with a time derivative, and that is the origin of the following rule:

> *To introduce the effect of gravitation in any dynamical framework, simply add the Newtonian potential to the energy density.*

In some less developed contexts, this prescription fails to be meaningful, or insufficiently precise, but then there is no alternative. I think that it may be agreed that, in any context in which it cannot be applied, we lack enough knowledge about gravitation to have a meaningful discussion about the influence of gravitation on an atmosphere.

2. *Thermodynamics.* This word is frequently taken to refer to the study of equilibria exclusively. But that is not enough. There are several cogent

reasons why a dynamical theory is needed. In the first place, any discussion of equilibria *per se*, without a description of the systems of which they are the equilibria, is likely to be inconclusive. The concept of equilibrium is, in my opinion, meaningful only in the context of a dynamical system. And again, without a dynamical theory, we shall not know how to introduce the gravitational field or to determine its effects.

Any discussion that restricts the input of thermodynamics to statements about abstract equilibria is not likely to be useful. That leads us to:

3. *Non-equilibrium thermodynamics.* Under this banner, there has been a great deal of activity in the last half-century. This point of view is probably not going to be of much help for the following reason. The term non-equilibrium strongly suggests that any attempt to get away from the equilibrium is an exploration of the unknown. For example, it is widely held that there is not, until now, a reliable account of entropy. But without an understanding of the role of entropy, we are lost, just as we should be lost without an understanding of gravitation.

4. *Adiabatic thermodynamics.* Actually, thermodynamics is advanced well beyond the study of equilibria. Already, in 1825, Laplace formulated a successful theory of sound propagation based in part on the postulate that this phenomenon is isentropic; that is, the entropy per particle does not change when the density and the temperature are executing the motion that is sound. Furthermore, the calculation is done within a dynamical framework involving the equation of continuity and the Bernoulli equation. Finally, this is an apt example, for it is very close to the issue at hand. So, let us examine the calculation in a little more detail.

There is nothing controversial about the equation of continuity; nothing needs to be said about it. When Laplace postulates that sound propagation is isentropic, he refers to the fact that the entropy of an ideal gas is related to the temperature and the density by the polytropic relation,

$$\rho/T^n = \text{constant},$$

where n is a constant that is characteristic of the gas. That is, this combination of density and temperature is a constant of the motion. The value of the constant is directly related to the entropy per particle; that is, to the specific entropy.

This dynamical calculation of the speed of sound, and many others, demonstrate that something is in fact known about entropy away from equilibrium. Furthermore, it strongly suggest that a reliable dynamical framework is available that may serve to investigate the effect of gravitation on an isolated atmosphere that consists of an ideal gas.

The theory of ideal gases has been around for more than 150 years. Let us use it. Textbooks published in recent years do not mention Emden's theorem and derive the formula for the constant lapse rate exactly as in this book.

Reply to Maxwell

It has to be maintained that the existence of isentropic atmosphere in a gravitational field is in contradiction with the second law of thermodynamics. If that were so, the second law would share the faith of all other laws that have been formulated by scientists over the ages. The existence of the isentropic atmosphere is a matter for experimental science; unfortunately, a decisive experiment would be very difficult. Besides, scientists with enough reputation to be respected seem not to be enthusiastic about investing either their time or their reputation on this problem. The evidence presented by our actual atmosphere has not proved convincing.

The experiment should certainly be done and with the astounding advance in technology, it shall eventually lead to a conclusion. And then what? What should we do if the conclusion is that the isentropic atmosphere exists?

It is unlikely, even in that case, that thermodynamics, or even just the second law, would become obsolete; since the arguments advanced by Maxwell and others are far from conclusive, as will be shown.

Maxwell. Two tubes are filled by two different gases, placed vertically next to each other and isolated from each other and from the environment, except for being connected to a reservoir at the top. At equilibrium, that is, after some time, if no further changes are taking place, the temperatures at the bottom of the two tubes will be different. A heat engine can be installed to exploit this difference and energy can thus be extracted.

Student. But that would not be very shocking, not like constructing a perpetuum mobile, since the energy is supplied by the system that will eventually get cold.

Maxwell. Consider instead the following system.

The *U*

Imagine a tube, in the form of the letter U, upright, divided into two parts in the middle (the low end), containing two different ideal gases, gas A on the left, gas B on the right, the whole in complete isolation except for terrestrial gravity, the two gases in thermal contact at the dividing wall. The tubes are closed at the top; the volume and the geometry are fixed. This is "the open

U". There is no heat bath. For the sake of the argument, assume that the equilibria, in the terrestrial gravitational field, are isentropic.

The tubes are thermally connected only at the bottom. Wait for equilibrium to be established; then the temperatures will be equal there, but different at the top. Now, connect the tops of the U with a heat conducting wire, thus creating "the closed U". We are going to neglect the heat loss in the wire as well as its heat capacity. The temperature will eventually equalize at the top, finding a new equilibrium at which the temperatures at the top are equal. This means that there has been a flow of heat and, because the change is spontaneous, involving no heat transfer from the environment, there has been a net increase of the total entropy. The process that has just taken place is therefore irreversible.

Next, disconnect the wire and let the system return to the original configuration. But this is impossible, for it would require a decrease in the entropy. Therefore, the isentropic equilibrium does not exist.

Student. Let us examine the process in a little more detail, starting with no thermal connection at the top, and different temperatures T_1, T_2 there. Let S_0 be the total entropy of the system. Now, two things will happen in succession.

First, the thermal connection at the top is connected. This constitutes a change of the system by external interference. It leaves the system out of equilibrium. If the entropy of this new configuration can be defined, call it S_1.

Second, the system, left out of equilibrium, will find a new equilibrium, with entropy S_2. Given the properties of each gas, we can calculate both S_2 and S_0. But what is the value of S_1?

Maxwell. Because the system, after the end of the first stage, is out of equilibrium, the entropy that you call S_1 is not defined.

Student. In that case, I may make a suggestion. Is it not possible to assign the value S_2 to S_1? I may perhaps propose this as the definition of S_1. By this definition, the change in entropy is due to the interference with the system (the establishment of the thermal connection) and the subsequent approach to the new equilibrium would be adiabatic since there is no interference from the outside during the second stage.

Maxwell. I do not like that definition of S_1 but I admit that my argument has been dealt a severe blow.

Student. Thank you professor.

The standard approach

There is no standard approach to constructing the thermodynamic potential for mixtures. The general condition for equilibrium is that

$$\mu_1^u = \mu_1^v, \quad \mu_2^u = \mu_2^v. \tag{A.9.1}$$

Remember that this comes from minimizing the energy. The chemical potentials can be expressed as derivatives of the free energy, the enthalpy or the Gibbs energy. Using the enthalpy for illustration, we have

$$(h_1 - S_1T)^u - (h_1 - S_1T)^v = 0, \quad (h_2 - S_2T)^u - (h_2 - S_2T)^v = 0,$$

see Section 4.3. The first equation is often ignored if only the vapor is the subject of interest. The value of $h_2^u - h_2^v$ is taken from tables. Then follows a step that I do not understand. It is averred that the total differential

$$d[(h_2 - S_2T)^u - (h_2 - S_2T)^v] = 0,$$

but this can be justified only when projected on a tangent that points along the manifold of coexistent configurations. The completion of the calculation involves additional reference to tables. Tables are useful for engineers who do not have to review the theory in order to get information. But physicists prefer to use the accumulation of experimental results to gradually pin down a useful formula for the free energy density, one that contains the information assembled in all the tables.

Chapter 10

Conservative Hydrodynamics*

As we have seen in Chapters 1–3, Lagrange (1760) has provided us with an efficient and elegant action principle for hydrodynamics. It applies in the case when the velocity field is irrotational, when it can be expressed as

$$\vec{v} = -\vec{\nabla}\Phi. \tag{A}$$

This is a strong limitation. There have been many attempts to lift it, especially in dealing with superfluids, and some have come very close to formulating an action principle (Hall and Vinen, 1956). In Chapter 3, this theory was integrated with thermodynamics, still under the strong limitation to potential flow. In the intervening chapters, we have studied thermodynamic applications focusing on liquids at rest. It is time to take the crucial step that allows us to confront the most interesting applications of hydrodynamics and thermodynamics.

Conservative hydrodynamics is the result of 10 years of intensive effort to create an action principle for hydrodynamics that includes rotational flows and vorticity. It puts us in a position to study traditional subjects in hydrodynamics, such as the flow in pipes, aerodynamic lift and flight, the internal constitution of stars and General Relativity, and superfluids.

This chapter develops an action principle for hydrodynamics that incorporates the Lagrange–Fetter–Walecka theory, but without the restriction to potential flow. It has four independent, canonical variables and it allows for vorticity. As such, it may be unique, but hydrodynamics includes complicated systems that require still more variables and about them we have nothing to say. The theory has not yet been extended to mixtures.

---

*An important development of this material may be found under the title: "Stability analysis for cylindrical Couette flow." In fronsdal.physics.ucla.edu.

For more than 100 years, theoretical hydrodynamics has been dominated by the Navier–Stokes equation. It is an equation with some merits, but it is not a theory, as can be seen from the fact that almost all advanced applications depend on supplemental input based on the concept of energy. Unfortunately, the Navier–Stokes equation is incompatible with an effective kinetic energy and the related concept of kinetic potential. The first three sections of this chapter are devoted to demonstrating this strong statement by a very detailed examination of the famous special case, cylindrical Couette flow.

10.1. The **Variables of Hydrodynamics**

Traditional hydrodynamics has four **independent variables**, the density and the three components of the velocity. The Action Principle that we know, for irrotational hydrodynamics, has two independent variables, so two more are needed. Here, we shall try to identify the two variables that have been missing. It will not do to seek a generalization of Eq. (A), as we shall try to make clear.

The first step is the realization that

> **the scalar velocity potential is essential to hydrodynamics.**

In the first place, the Euler–Lagrange equation

$$\frac{\delta}{\delta\Phi}A = \dot{\rho} + \vec{\nabla}\cdot(\rho\vec{v}) = 0$$

is the equation of continuity, the very essence of hydrodynamics. A scalar equation, it comes from variation of the action with respect to a scalar field.

Second, even though the equations of hydrodynamics are Galilei invariant, only the action principle, with velocity potential, offers an invariant action and invariant equations of motion. The Hamiltonian is *not* Galilei invariant; the search for a Galilei invariant expression for the energy is misdirected.

As important is the tradition that incorporates gravitation into field theories, especially into hydrodynamics, by including the **Newtonian potential** into the Hamiltonian density, when this is available. We regard hydrodynamics as a non-relativistic approximation to General Relativity, but is there any generally relativistic field theory that has the expected non-relativistic limit? Granted that it is suggested by the correspondence between hydrodynamics and particle mechanics, but our subject is field theories.

Gravitation is represented by the **metric**, in the non-relativistic limit by the time–time component. Consider the generally relativistic field theory with Lagrangian density

$$\mathcal{L} = \int d^4x\sqrt{-g}\big(\rho(g^{\mu\nu}\psi_{,\mu}\psi_{,\nu} - c^2) - W[\rho]\big). \qquad (10.1.1)$$

To ensure the existence of a **non-relativistic limit**, we must set

$$g_{00} = c^2 - 2\phi, \quad g_{11} = g_{22} = g_{33} = -1, \qquad (10.1.2)$$

other components zero, expand in powers of (1/c) and take the limit $c \to \infty$. To ensure the cancellation of the terms of lowest order, we need to set

$$\psi = c^2 t + \Phi, \qquad (10.1.3)$$

then terms of order c^2 cancel and we are left with the Fetter–Walecka Lagrangian density

$$\mathcal{L}_{FW} = \rho(\dot{\Phi} - \vec{\nabla}v^2/2 - \phi) - W[\rho]. \qquad (10.1.4)$$

In the context of field theories, this is the only confirmation that we have of the expectation that Newtonian hydrodynamics, with the inclusion of the gravitational potential, is the non-relativistic limit of General Relativity. As we have already seen, in Section 3.11, Eq. (10.1.2) leads to the correct transformation law for the field Φ under Galilei transformations. We see that, in this limited context, it is indeed legitimate to look at hydrodynamics as a classical limit of General Relativity. But this theory is limited to irrotational flows.

Let us be absolutely clear: In any relativistic field theory in which the non-relativistic limit is more straightforward, there is no interaction left in the limit $c \to \infty$. The Newtonian interaction survives only because of the special cancellation that accompanies the expansions (10.1.2) and (10.1.3).

We conclude that the Fetter–Walecka action principle is not to be regarded as incorrect, but as incomplete. The required generalization of (10.1.4) must include additional fields with two more **degrees of freedom**, with an action that reduces to (10.1.4) by projection. Two more dynamical variables are needed, one pair of canonical variables, either an additional scalar field or else some gauge field with only one propagating component. But the scalar velocity potential must be retained, to preserve the equation of continuity and the connection to General Relativity.

The simple introduction of another scalar velocity potential would not lead to rotational flow, and of gauge theories with just one propagating mode there is just one: the antisymmetric 2-form field ($Y_{\mu\nu}$) of Ogievetskiĭ and Polubarinov (1963a, 1963b, 1964). This is a topic that shall be taken up below.

The connection of potential flow to an action principle is intimately related to the fact that the velocity field that is used is a gradient of a more basic field, a scalar potential. The vector field that is used in an alternative approach to hydrodynamics, usually denoted $\dot{\vec{x}}$, is also a derivative and also a natural candidate for an action principle. We shall call this other potential $\vec{X}$, to distinguish it from the coordinates. Since the equations of motion are second order in the time derivative, this field appears to have six independent degrees of freedom. We already have two degrees of freedom with ρ and Φ, conventional hydrodynamics has four, so this appears to be far too much. It will be seen, however, that it is possible to impose a constraint that reduces the number of degrees of freedom of the new field to 2.

Anticipating that, we add an additional kinetic term $\rho\dot{\vec{X}}^2/2$ to our Lagrangian,

$$\mathcal{L} = \rho(\dot{\Phi} + \dot{\vec{X}}^2/2 + \kappa\rho\dot{\vec{X}} \cdot \vec{\bigtriangledown}\Phi - \vec{\Phi}^2/2 - \phi) - f - sT. \tag{10.1.5}$$

The aptness of the second term will be confirmed in Section 10.2, by comparison with Navier–Stokes theory. The mixing term (with constant coefficient κ) is included in order to allow for vorticity. It is gauge invariant and in a sense unique, as will be seen in Chapter 12.

The genesis of this proposal; that is, the idea that (10.1.4) is a proper generalization of the Fetter–Walecka Lagrangian needed to include rotational flows, will be presented below in the context of an important application: cylindrical Couette flow. The **relativistic generalization** is needed to discover the gauge theory constraint of this non-relativistic gauge theory; it will be developed in Section 12.1. In Section 10.2, we shall anticipate by introducing the constraint without explanation.

10.2. Constraint and Euler–Lagrange Equations

The **constraint** comes from variation of the **gauge field** (Section 12.1),

$$\vec{\bigtriangledown} \wedge \vec{m} = 0, \quad \vec{m} := \rho\vec{w}, \quad \vec{w} := \dot{\vec{X}} + \kappa\vec{\bigtriangledown}\Phi. \tag{10.2.1}$$

This is the constraint that reduces the number of degrees of freedom carried by the field $\vec{X}$ to just 1 canonical pair. The Euler–Lagrange equations associated with the Lagrangian (10.1.5) are as follows.

Variation of Φ gives the **equation of continuity**,

$$\dot{\rho} + \vec{\bigtriangledown} \cdot \rho\vec{v} = 0, \quad \vec{v} := \kappa\dot{\vec{X}} - \vec{\bigtriangledown}\Phi. \tag{10.2.2}$$

The flow vector is thus $\rho\vec{v}$. Variation of $\vec{X}$ leads to

$$\frac{d}{dt}\vec{m} = 0, \quad \vec{m} := \rho(\dot{\vec{X}} + \kappa\vec{\bigtriangledown}\Phi). \tag{10.2.3}$$

Finally, variation of the density gives

$$\dot{\Phi} + \dot{\vec{X}}^2/2 + \kappa\rho\dot{\vec{X}} \cdot \vec{\nabla}\Phi - \vec{\nabla}\Phi^2/2 - \phi = \mu + sT. \tag{10.2.4}$$

The vector field $\vec{v}$ is the visible flow field; the vector field $\dot{\vec{X}}$ is associated with stress, as the first application in Sections 10.3–10.5 of this chapter will show. The significance of the parameter κ will become clear immediately below. Galilei invariance of this non-relativistic, gauge-fixed Lagrangian will be studied later (Chapter 12).

The Euler–Lagrange equations of the proposed Lagrangian, Eqs. (10.2.2)–(10.2.4), have much in common with modern theories of rotational flow; they all involve potential flow plus additional velocity fields. To initiate a close comparison, and a confrontation with experiment, we note the following. From Eqs. (10.2.1)–(10.2.4) we easily deduce that the vorticity field

$$\vec{\nabla} \wedge \vec{v} = \kappa\vec{\nabla} \wedge \vec{w} = \kappa\vec{\nabla}\frac{1}{\rho} \wedge \vec{m}. \tag{10.2.5}$$

It follows that the gradient of the density is orthogonal to the vorticity:

$$\vec{\nabla}\rho \cdot (\vec{\nabla} \wedge \vec{v}) = 0. \tag{10.2.6}$$

A canonical theory of incompressible fluids can be developed only as the limit of a family of theories of compressible fluids.

10.3. Cylindrical Couette Flow

Couette flow has been studied for 150 years, with the main focus on the spontaneous onset of turbulence. The experiment has a homogeneous fluid contained in the space between two concentric cylinders, as in Fig. 10.3.1, both cylinders turning independently on their common axis.

The initial series of experiments, by Couette and Mallock, were followed by a paper by G.I. Taylor (1921). This work included new experiments but also a pioneering interpretation, the main impact of which was the vindication of the **non-slip** boundary condition. Though it is in need of refinement, this boundary condition is still being used for liquids and it will be used here. It is the statement that the limit of the flow velocity, at the material boundary of the containing vessel, is the same as the local velocity of the container at the same point.

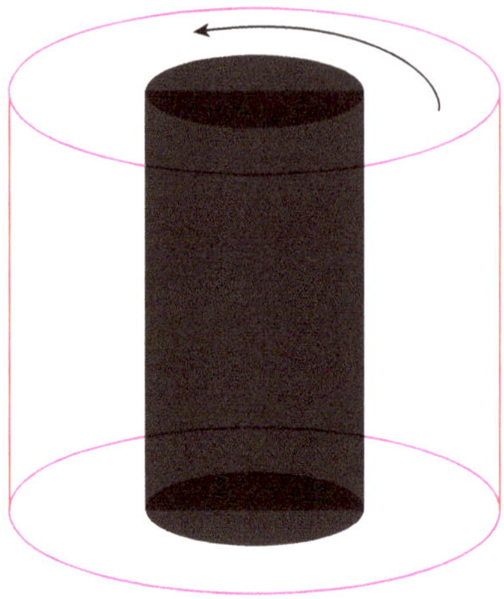

Fig. 10.3.1. **Cylindrical Couette flow**; both cylinders turning independently.

Laminar Couette flow

The simplest type of flow velocity is horizontal, stationary and circular, in cylindrical coordinates,[68]

$$\dot{\vec{X}} = \omega(r)(-y, x, 0), \quad r := \sqrt{x^2 + y^2}, \tag{10.3.1}$$

with angular velocity ω depending on r only. The boundary conditions are "no-slip", meaning that the velocity of the flow, at the cylindrical boundaries, is the same as the velocity of the walls (see Fig. 10.3.2). To simplify the analysis, one thinks of the cylinders as being long, and ignores end effects. The particular case

$$\dot{\vec{X}} = b(-y, x, 0), \quad b \text{ constant}, \tag{10.3.2}$$

is the flow of a solid body. It gives rise to a kinetic potential

$$K = -\dot{\vec{X}}^2 \propto -r^2 \tag{10.3.3}$$

(all proportionalities with positive coefficients) and a centrifugal force $-\vec{\nabla} K \propto \vec{r}$ that is balanced by a pressure (constant + r^2) that presents a force: $\vec{\nabla} p \propto -\vec{r}$. It follows, for ordinary fluids with a positive adiabatic derivative $dp/d\rho$, if there are no other forces acting, that $\vec{\nabla} K + (\vec{\nabla} p)/\rho = 0$ and that the density must increase outwards. For simplicity, we have taken the fluid to be incompressible, setting $\rho = 1$.[69]

---

[68] A review of the strategy pursued below is at the end of the section.

[69] When the dynamical field $\dot{\vec{X}}$ in (10.2.4) is replaced by the fixed field (10.3.2), and $\kappa = 0$, we get a Lagrangian proposed by Hall and Vinen (1956) in a much quoted paper on superfluids. We shall see that this is appropriate in the limit of incompressible fluids. For a modern version of this theory, see Fetter (2009).

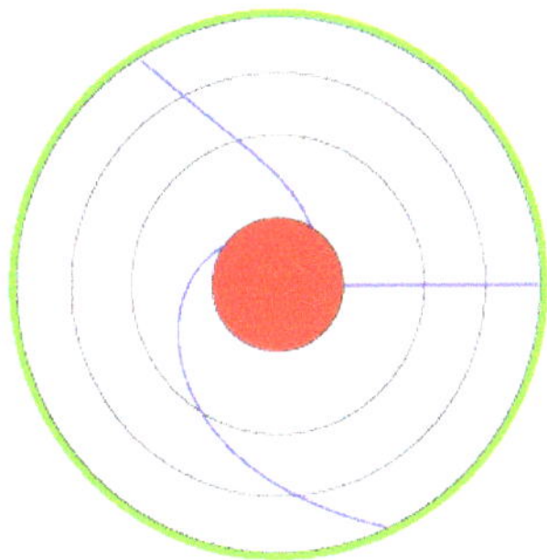

Fig. 10.3.2. Illustration of cylindrical Couette flow. The blue lines are successive snapshots of a set of particles originally on a radial configuration.

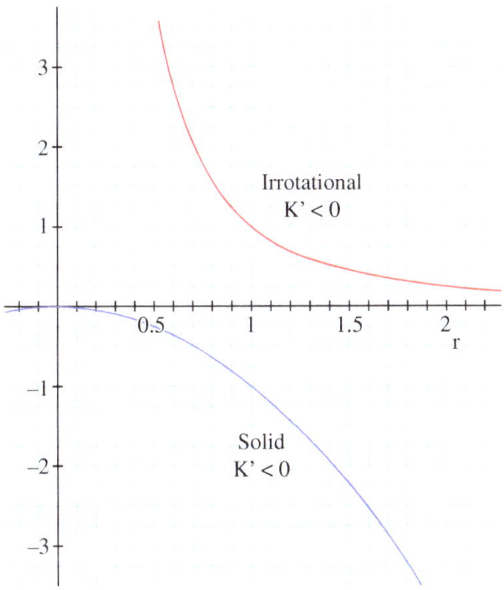

Fig. 10.3.3. The kinetic potentials, $K \propto \vec{v}^2 \propto 1/r^2$ (irrotational flow) and $K \propto -\vec{v}^2 \propto -r^2$ (solid-body flow).

Figure 10.3.3 contains a sketch of this kinetic potential, as well as the analogous function associated with irrotational flow, to be discussed below. Disturbing question: what is the expression for the "kinetic potential" when the velocity has two parts, one rotational and one solid-body type?

The basis for every traditional analysis of Couette flow is the **Navier--Stokes** equation

$$\frac{D}{Dt}\vec{u} = -\frac{1}{\rho}\vec{\nabla}p + \bar{\mu}\rho\Delta\vec{u}. \tag{10.3.4}$$

Here $\bar{\mu}$ is the viscosity of the fluid, in the simplest case a constant parameter. This equation agrees with (10.2.4) in the special case that $\vec{u} = -\vec{\bigtriangledown}\Phi$ and $\dot{\vec{X}} = 0$. It also agrees with (10.2.4) in the complimentary case, when $\vec{\bigtriangledown}\Phi = 0$ and $\vec{v} = \dot{\vec{X}}$.

The velocity field in (10.3.4) is denoted $\vec{u}$ to emphasize that it should not be identified with either $-\vec{\bigtriangledown}\Phi$ or $\dot{\vec{X}}$ or $\vec{v}$.

Stationary flow is possible only if the effect of viscosity is negligible; this requires that $\bar{\mu} = 0$ or else that $\Delta\vec{v} = 0$.[70] Moreover

$$\Delta\vec{u} = 0. \tag{10.3.5}$$

When the flow is stationary, as postulated (and as observed at low speeds), this condition allows the general solution

$$\vec{u} = \frac{a}{r^2}(-y, x, 0) + b(-y, x, 0), \quad a, b \text{ constant}, \tag{10.3.6}$$

where the parameters a, b to be determined by the non-slip boundary conditions.

The first is irrotational for $r > 0$,

$$\frac{a}{r^2}(-y, x, 0) = a\vec{\bigtriangledown}\theta, \quad \theta = \arctan\frac{y}{x}.$$

This gives rise to a kinematical potential, cf. Eq. (10.3.3),

$$K \propto \vec{\bigtriangledown}\Phi^2 \propto 1/r^2 \tag{10.3.7}$$

and a centrifugal force

$$-\vec{\bigtriangledown}K \propto \vec{r}/r^4,$$

and in this case too the force points outwards. Both of these types of flow are of great interest, especially so in connection with superfluid Helium. The second term in (10.3.6) is the flow of a solid body, see Eqs. (10.3.1) and (10.3.2). We observe that dK/dr is negative in both cases and that K is proportional to u^2 in one case and to $-u^2$ in the other (all proportionalities with positive coefficients).

This last circumstance already suggests that there are difficulties ahead. What is the kinetic potential when the velocity field is of neither type, but a combination of both, as in (10.3.6)?

[70] The usual justification for this restriction is more involved: With the hypothesis of circular flow, the density can only depend on the coordinate r and the same is true of the pressure; hence the pressure term is a radial force and the projection of the last term along the direction of motion must also be zero; hence, $\bar{\mu}\Delta\vec{u} = 0$.

In experiments, the rotation speeds of the two cylinders are increased slowly from zero, these types of flows are observed at low speeds. We shall describe this regime.

The boundary conditions will be analyzed below.

So far, the application of the Navier–Stokes equation to Couette flow is a brilliant success, being in very good agreement with experiments, at low speeds of rotation. The only difficulty is that there is no clear guide to constructing an expression for the energy density, or a kinetic potential. It would be natural to expect that the expression

$$\rho\vec{u}^2/2, \quad \text{respectively} \quad \vec{u}^2/2,$$

might be an important part of it, serving as a kinetic energy resp. kinetic potential and the latter giving rise to the centrifugal acceleration; we have seen that this is not the case in general, though it is true in the case of irrotational flow, as we shall now show, again.

The question of sign, again

Consider the simplest cases. The total derivative in (10.3.4) is

$$\frac{D}{Dt}\vec{u} := \frac{d}{dt}\vec{u} + (\vec{u}\cdot\vec{\nabla})\vec{u}.$$

The last term is (minus) the centrifugal force. When the flow is of the solid body type, Eq. (10.2.2),

$$(\vec{u}\cdot\vec{\nabla})\vec{u} = b^2(-y\partial_x + xd_y)(-y,x,0) = -b^2(x,y,0) = -\vec{\nabla}\vec{u}^2/2.$$

That is, the acceleration is outwards and the centrifugal potential K is $-\vec{u}^2/2$. In the complimentary case of potential flow,

$$(\vec{u}\cdot\vec{\nabla})\vec{u} = \frac{a^2}{r^2}(-y\partial_x + x\partial_y)\frac{1}{r^2}(-y,x,0) = -\frac{a^2}{r^4}(x,y,0) = \vec{\nabla}(\vec{u}^2/2);$$

it is in the same direction, hence correct, but now the centrifugal potential is $\vec{u}^2/2$. The difference in sign between the two cases means that

> **One cannot, in general, associate the Navier–Stokes equation with an energy density $E[\rho,\vec{u}]$ or with a kinetic potential $K[\vec{u}]$.**

This conclusion is long overdue.

Rayleigh's mistake and what it tells us

To theorists, the principal aim has been to understand the limits on the **stability** of the lowest flow regime that was described above and the first notable attempt to do so was that of Rayleigh (1889, 1916) who concluded that the laminar flow should be stable if and only if

$$\omega_o > \omega_i \quad \text{(approximately)}.$$

Here, ω_o (respectively, ω_i) is the angular velocity of the outer (respectively, inner) cylinder and ω_i is positive by convention. In particular, this implies that the laminar flow would be unstable whenever the two cylinders rotate in opposite directions. The result of observation was different: the condition for stable Couette flow is, approximately,

$$\omega_o^2 > \omega_i^2,$$

which allows for stable flow for either sign of ω_o/ω_i. Raleigh's reaches his conclusion twice.

(1) **Angular momentum** (about the axis of rotation) is conserved, the angular momentum is $L = vr$, "*...so that the centrifugal force acting on a given portion of the fluid is* L^2/r^3". That is, $-L^2/r^2$ is regarded as a kinetic potential. But this is without justification and, as it happens, actually wrong; if there is any kinetic potential K in the context of the Navier–Stokes equation, it must satisfy

$$(\vec{u} \cdot \vec{\nabla})\vec{u} = \vec{\nabla} K. \tag{10.3.8}$$

(2) We quote (Rayleigh, 1916): "*We may also found our argument upon a direct consideration of the* ***kinetic energy*** *of the motion...*". He concludes that stability requires that this function increase outwards. This is the same argument and here we see more clearly what the difficulty is: We have shown that there is no justifiable way to introduce the energy concept, or to choose an expression to serve as energy, except in the context of an action principle.

Since the Navier–Stokes equation is an expression of the balance of forces, one must ask if there is an additional force that must be taken into account. Yes, there is; but what is without justification is the claim that there is an "energy density" proportional to L^2/r^2 and that it functions as a kinetic potential.

We have seen (1) that any valid expression for "kinetic potential" must have opposite signs for the two types of flow and that (2) if there is a suitable expression for a kinetic potential in the context of the Navier–Stokes equation, then it is not related to an energy density.

Rayleigh's choice of "energy" leads to conclusions that are contradicted by experiment. If we merely restrict ourselves to the two special cases examined in the preceding section, we see that there is, in either case, an effective kinetic potential, proportional to $\vec{u}^2$, but with a positive coefficient in one case and a negative coefficient in the other; this disqualifies it from being interpreted as a kinetic potential in the general case and blocks Rayleigh's argument.

During the 100 years that followed the publication of Rayleigh's paper, the calculation has been repeated in numerous textbooks, including these: Chandrasekhar (1939), Landau and Lifshitz (1959), Drazin and Ried (1981), Tilley and Tilley (1974, 1986), Koschmieder (1993) and Wikipedia (2017).

That Rayleigh's prediction was contradicted by experiments must have been known to himself in 1916; that it was known to the other authors mentioned is not in doubt. Yet, there is no suggestion in the later literature that Rayleigh's argument is faulty!

Landau and Lifshitz (1959, p. 100) make the unsupported claim that viscosity invalidates Rayleigh's argument if the cylinders rotate in opposite directions! Chandrasekhar (1939, p. 275) argues that the criterion is inconsequential because the region in which it is violated is small, Drazin and Ried (1981, p. 79) repeat Chandrasekhar's argument. Koschmieder (1993, Chapter 11) offers the most comprehensive discussion of traditional methods, but he too finds no fault with Rayleigh's argument (*op. cit.* Chapter 10). Finally, Wikipedia repeats Rayleigh's argument but offers no clue to why the criterion fails when the cylinders are rotating in opposite directions.

The Navier–Stokes equation by itself does not imply, and in general it does not allow, the existence of an expression with the attributes of energy. But a kinetic potential can be constructed in the special case of stationary, laminar Couette flow, when the velocity field is of the form (10.3.6); it is (Fronsdal, 2014)

$$-K = \frac{b^2}{2}r^2 + ab\ln r^2 - \frac{a^2}{2r^2}, \quad (\vec{u}\cdot\vec{\bigtriangledown})\vec{u} = \vec{\bigtriangledown}K. \tag{10.3.9}$$

By the Navier–Stokes equation, this makes a contribution $-\vec{\bigtriangledown}K$ to the acceleration. Therefore, this function is the only kinetic potential that is consistent with the Navier–Stokes equation in the simplest case when the velocity is of the form (10.3.6). It is not ($1/\rho$ times) the kinetic energy.[71]

[71] In some highly regarded books, we find definitions of normalized gravitational, internal and total energies. It is taken for granted that the name given to the expression for the "total energy" endows it with physical properties. In this manner are reproduced results

10.4. Conservative Hydrodynamics has a Kinetic Potential

The Hamiltonian density is

$$h = \rho(\dot{\vec{X}}^2/2 + (\vec{\nabla}\Phi)^2/2 + \varphi) + f + sT. \tag{10.4.1}$$

The kinetic part, instead of the expected square of a "total velocity", is the sum of two squared velocities. But the equation of motion, obtained by taking the gradient of (10.2.4), is

$$\vec{\nabla}\dot{\Phi} - \vec{\nabla}(K + \varphi) = \frac{1}{\rho}\vec{\nabla}p,$$

where p is the thermodynamic pressure and

$$K = -\dot{\vec{X}}^2/2 - \kappa\dot{\vec{X}}\cdot\vec{\nabla}\Phi + (\vec{\nabla}\Phi)^2/2 \tag{10.4.2}$$

is the kinetic potential. Equations (10.4.1) and (10.4.2) both have the correct signs, the first gives the Hamiltonian density with the correct, positive sign for both terms, the second equation agrees with the Navier–Stokes equation (when applicable) with the two different signs. The expression denoted K is a kinetic potential, but it is not simply related to an energy density.

More precisely, the Navier–Stokes equation deals correctly with each type of velocity separately. But combining the two is like adding apples and oranges; the sum is meaningless.

The Navier–Stokes equation combines the two types of flow in a single velocity field and deals correctly with each part, but that is as far as one can go with a single velocity field, for the irrotational flow is Eulerian and the solid-body flow is Lagrangian.

The action principle is not in conflict with the traditional treatment of this type of flow, but it completes it by giving us the equation of continuity, as well as an energy density; the Hamiltonian is a first integral of the motion and the Lagrangian, not the Hamiltonian, contains the kinetic potential.

With a Hamiltonian and a kinetic potential in hand, we can apply standard methods. The first result is, of course, that a stable configuration must satisfy the Euler–Lagrange equations, making the energy

---

previously proved by Chandrasekhar (1939). Other examples of such shortcuts are legion, See, e.g., Eddington (1926, p. 142). Here is a different type of example. Arnold and Khesin in their book (1998, pp. 2, 19, 37, 75, 119) lay down axioms that, for them, define hydrodynamics. They always use the standard definition of "energy", $E = v^2/2$.

stationary with respect to all perturbations. A deeper analysis studies harmonic perturbations to first and second order of perturbation theory, as in Chandrasekhar (1939), Landau and Lifshitz (1959), Drazin and Ried (1981) and Koschmieder (1993).

In the presence of **viscosity**, there can be no energy conservation and no action principle. But there is a standard and natural way to modify the Euler–Lagrange equations. Taking our clue from the Navier–Stokes equation and the standard approach, we postulate that, in the presence of viscosity, Eq. (10.2.3) is to be replaced by

$$\frac{d}{dt}\vec{m} = \bar{\mu}\rho\Delta\vec{v}. \tag{10.4.3}$$

In our treatment of cylindrical Couette flow, we have limited our attention to stationary flows, with $\Delta\vec{v} = 0$, once again taking our inspiration from the traditional point of view.

10.5. Stability of Laminar Couette Flow

According to (10.4.2), the kinetic potential is the function

$$K = -\dot{\vec{X}}^2/2 - \kappa\rho\dot{\vec{X}} \cdot \vec{\nabla}\Phi + \vec{\nabla}\Phi^2/2. \tag{10.5.1}$$

By the gauge constraint, for horizontal, circular flow, there are constants a, b such that

$$-\vec{\nabla}\Phi = \frac{a}{r^2}(-y, x, 0), \quad \vec{m} = -\vec{\nabla}\tau = \frac{b\kappa}{r^2}(-y, x, 0). \tag{10.5.2}$$

The second formula solves the gauge constraint (10.2.1). The velocity of mass transport is given by (10.2.2) or

$$\begin{aligned}\vec{v} = \kappa\frac{\vec{m}}{\rho} - (1+\kappa^2)\vec{\nabla}\Phi &= \left(\frac{\kappa^2 b}{r^2\rho} + (1+\kappa^2)\frac{a}{r^2}\right)(-y, x, 0)\\ &= \frac{1+\kappa^2}{r^2}\left(\frac{c^2 b}{\rho} + a\right)(-y, x, 0),\end{aligned} \tag{10.5.3}$$

with $c^2 := \kappa^2/(\kappa^2+1)$. The only constraint on the field $\vec{v}$ is that it must be harmonic — see Eq. (10.3.5); thus

$$\frac{1}{\rho} = 1 + \alpha(1 - r^2), \quad \alpha \text{ constant}. \tag{10.5.4}$$

This density profile implies that the **vorticity** (Eq. (10.2.5)) is uniform, parallel to the axis, with magnitude $2\alpha\kappa b$.

We have referred to the classical intuitive feeling that the pressure must increase outwards. For any normal fluid, for which the adiabatic derivative

$dp/d\rho$ is positive, this implies that the density must increase outwards, so the constant α must be positive, but this conclusion will have to be reviewed. We have normalized the density at the outer boundary to unity, $\rho_o = 1$. This relates α to ρ_i,

$$\alpha = 4.54\left(\frac{1}{\rho_i} - 1\right). \tag{10.5.5}$$

In terms of these vector fields,

$$-K = \frac{\vec{m}^2}{2\rho^2} - (\kappa^2+1)\vec{\nabla}\Phi^2/2 = \frac{1+\kappa^2}{2r^2}\left(\frac{b^2c^2}{\rho^2} - a^2\right), \quad c^2 := \frac{\kappa^2}{1+\kappa^2}. \tag{10.5.6}$$

By (10.5.5),

$$-K = \frac{1+\kappa^2}{2}\left(b^2c^2\left(\frac{(1+\alpha)^2}{r^2} - 2\alpha(\alpha+1) + \alpha^2r^2\right) - \frac{a^2}{2r^2}\right) \tag{10.5.7}$$

and

$$-K' = \frac{1+\kappa^2}{r}\left(\frac{a^2}{r^2} + b^2c^2\left(\alpha^2r^2 - \frac{(1+\alpha)^2}{r^2}\right)\right). \tag{10.5.8}$$

Now that we have a kinetic potential, we can try Rayleigh's criterion, $K' < 0$.

Boundary conditions

The walls of the two cylinders move with angular velocities

$$\omega_i\hat{\theta} = \frac{\omega_i}{r}(-y,x,0), \quad \omega_o\hat{\theta} = \frac{\omega_o}{r}(-y,x,0) \quad \omega_i,\omega_o \text{ constant}. \tag{10.5.9}$$

The velocity of mass transport is $\vec{v}$, Eq. (10.5.3), this is the velocity that must satisfy the no-slip boundary conditions, whence

$$\frac{1+\kappa^2}{r_i^2}\left(a + \frac{bc^2}{\rho_i}\right) = \omega_i, \quad \frac{1+\kappa^2}{r_0^2}\left(a + \frac{bc^2}{\rho_o}\right) = \omega_o, \tag{10.5.10}$$

and

$$a = \frac{r_o^2\rho_o\omega_o - r_i^2\rho_i\omega_i}{(\rho_o-\rho_i)(\kappa^2+1)}, \quad b = \rho_o\rho_i\frac{r_i^2\omega_i - r_o^2\omega_o}{\kappa^2(\rho_o-\rho_i)}. \tag{10.5.11}$$

Limits on Couette flow

Couette (1887; 1888a,b; 1890a,b), Mallock (1888), Taylor (1923), Andereck *et al.* (1983) and others have investigated the behavior of the fluid at low angular speeds. The most complete results, by Andereck *et al.* (1986), is summarized in Fig. 10.5.1, where the lowest line — similar to a hyperbola

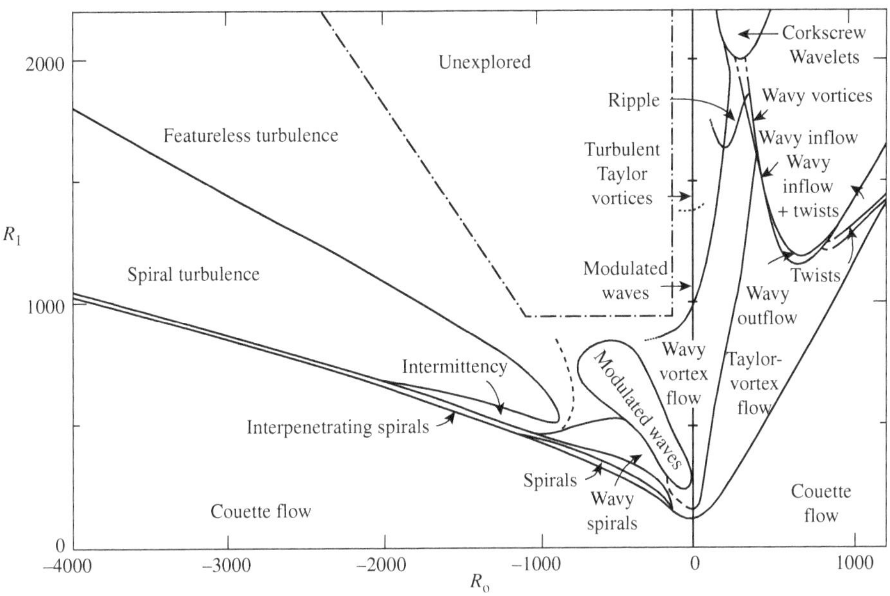

Fig. 10.5.1. Experimental results of Andereck *et al.* (1983, 1986). The abscissa (respectively, ordinate) is the angular velocity of the outer (respectively, inner) boundary. The lower "hyperbola" is the upper limit of stability of laminar Couette flow.

open upwards — is the upper limit of a flow of the type (10.3.6) or, in another presentation, Eq. (10.5.3).

Rayleigh (1880) based his pioneering investigation on the belief that the gradient of a kinetic potential must be negative. Now that we have a legitimate kinetic potential — in (10.5.1) and (10.5.7) — we can test the merits of this idea.

When K' is expressed in terms of the angular velocities, we see that the locus $K' = 0$ consists of two straight lines through the origin of the angular velocity diagram. The best result that can be hoped for with the assumption that K' must be negative, is for the straight lines $K' = 0$ to coincide with the "asymptotes" of the experimental curve, as in Fig. 10.5.2. That would leave a substantial region that is stable, although K' is positive. But even that qualitative success is not granted.

Extensive search has shown that the locus $K' = 0$ never imitates the shape of the experimental limit in Fig. 10.5.1.

A proposal

It is known that some types of instability are accompanied by bubble formation (as in the case of propellers and sonoluminescence). We suggest that this

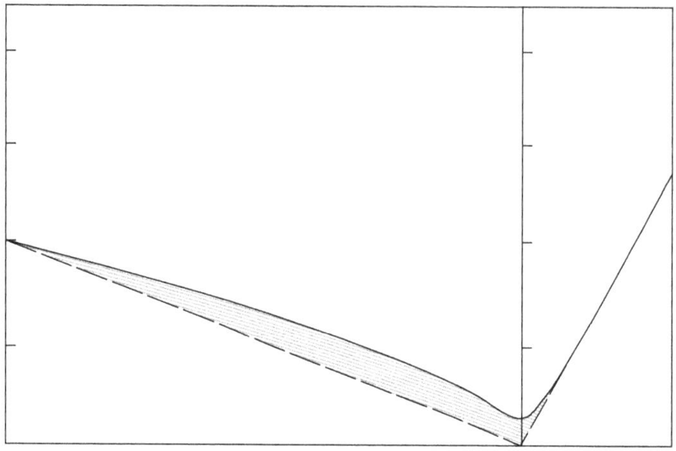

Fig. 10.5.2. A pair of straight through the origin is the best possible fit of a locus $K' = 0$ to the "hyperbolic" curve that could be hoped for, but it is not realized.

may be related to local evacuation (or more generally to a physical breakdown of the laminar nature of the flow) and that it happens at a particular value of the chemical potential, the density and of the corresponding value the kinetic potential K.

We shall test this hypothesis. We are not rejecting the idea that K' must be negative; it turns out that it is negative almost everywhere.

Let us subtract $K(0)$ from $K(r)$ in (10.5.7). As we see from that equation, the locus $K(r) - K(0)$ consists of two straight lines. At the inner boundary,

$$\frac{\omega_i}{\omega_0} = \left(\frac{r_o}{r_i}\right)^2 \frac{1 \pm c}{1 \pm c(\rho_i/\rho_o)} \tag{10.5.12}$$

and these must have opposite signs, hence $1 > c > \frac{\rho_o}{\rho_i}$. At the outer boundary,

$$\frac{\omega_i}{\omega_o} = \left(\frac{r_o}{r_i}\right)^2 \frac{1 \pm c(\rho_o/\rho_i)}{(1 \pm c)}$$

and $1 > c > \rho_i/\rho_o$. This suggests that, as the velocity is increased, the instability first manifests itself at the boundary with the higher density.

The value of K varies with the radius r. If, as the velocity is increased, the instability makes a first appearance at the boundary "B", the inner or the outer cylinder, then we expect that the parameters can be tweaked to make a locus $K(r_i) = A$ or $K(r_o) = A$, in the plane of angular momenta, an adequate approximation to the experimental limit curve and that the family of loci, $K(r) = A, r_i < r < r_o$, displace upwards as we move away from the boundary B.

We first examined the possibility that the instability first manifests itself at the outer boundary, which would imply that the density is higher there, as had been expected. This resulted in a failure; we shall say no more about it. Then we examined the alternative, that the instability first manifests itself at the inner boundary, in which case the density is higher there, which may be unexpected.

Main results

Assume that $\rho_i > \rho_o$. Figure 10.5.1, taken from Andereck *et al.* (1983, 1986), is a summary of their results; the lowest line, resembling a hyperbola, is the upper limit of observed, basic Couette flow.

In Fig. 10.5.3, the dotted curve is the experimental limit in Fig. 10.5.1. The other curves are K loci; the lower one is the one that best fits the experimental curve, with $\rho_o/\rho_i = 0.9$, $\kappa = 5.3$ and $r = r_i$. Identical curves are obtained with $\rho_o/\rho_i = 0.99$, $\kappa = 17.5$, $\rho_o/\rho_i = 0.999$, $\kappa = 55$ and so on, calculated up to $\rho_o/\rho_i = 1 - 10^{-6}$.

The slope of the right asymptote is predicted without ambiguity to be very close to $(r_o/r_i)^2$, — a bullseye for the theory. See Eq. (10.5.12). The slope on the left is very sensitive to the value of κ; this too can be seen by inspection of Eq. (10.5.12).

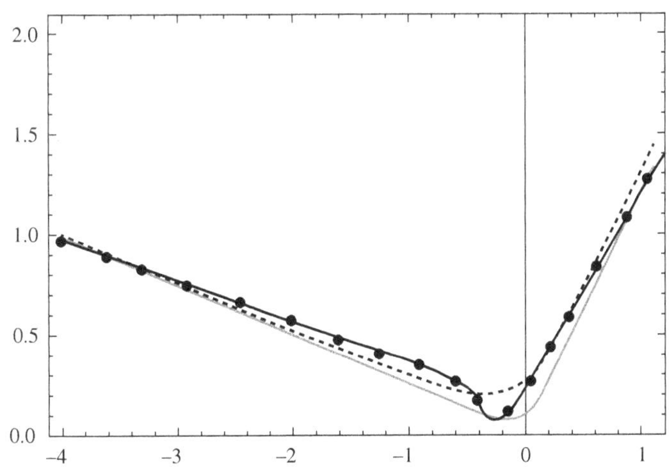

Fig. 10.5.3. The experimental limit curve, dotted, and two loci $K(r_i) = A > 0 = \text{constant}$.

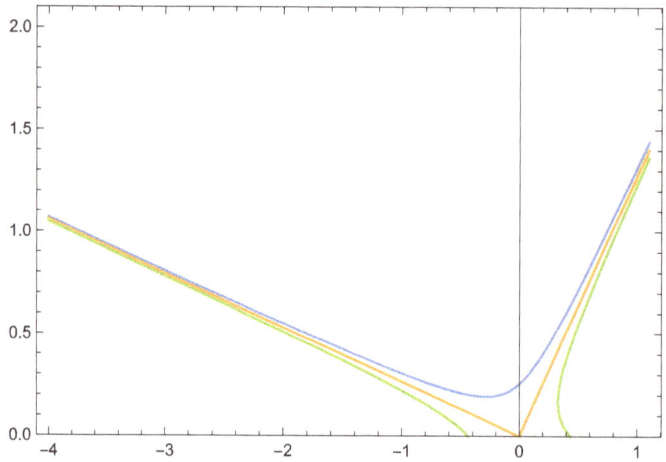

Fig. 10.5.4. Loci $K = A_- < 0$, 0, $A_+ > 0$ at the inner boundary, $r = r_i = 0.883$. The three curves are drawn for the same values of K in all the diagrams that follow.

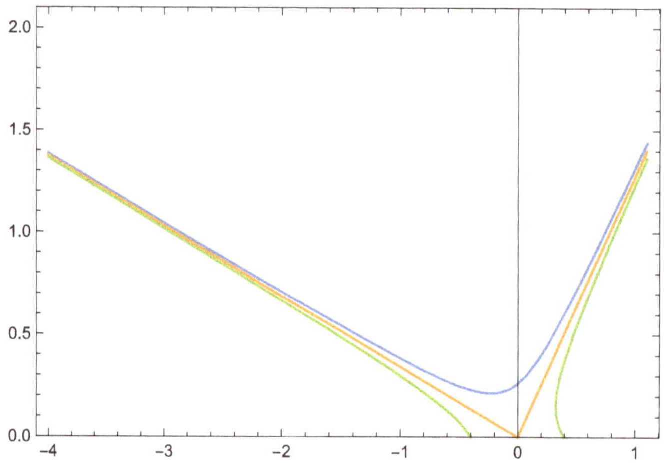

Fig. 10.5.5. The locus $K = A$ with the same values of A at $r = 0.889$.

Figures 10.5.4–10.5.8 show the result of repeating the calculation, with the same values of the parameters and the same value of A, at several points in the interval $r_i < r < r_o$.

In each diagram, two straight lines through the origin make up the $K = 0$ locus. The hyperbola that lies between them is a locus $K = A_+ > 0$. The two lines that lie outside form a locus $K = A_- < 0$. All these figures have the same parameters and the same values of A_+ and A_-. The series starts at the inner boundary and proceeds to the outer boundary.

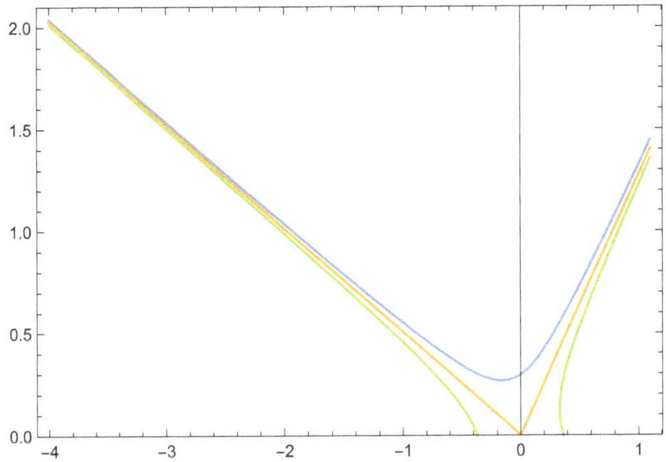

Fig. 10.5.6. At $r = 0.898$. the locus is moving upwards, almost everywhere.

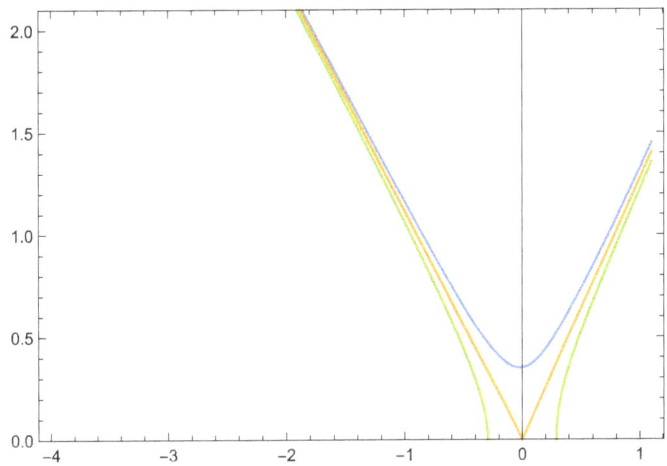

Fig. 10.5.7. At $r = 0.92$.

Discussion

This may mark the first time that the stability of Couette flow has been treated in the case of a compressible flow. That the instability first manifests itself at the inner boundary (in the case that the cylinders are rotating in opposite directions) has not yet been confirmed by observation, as far as we know. When the sense of rotation is the same, a uniform breakdown at all radii is expected.

The next two illustrations are flow lines, in this sense. If at some instant we drop a line of saw dust on the horizontal straight line that connects the

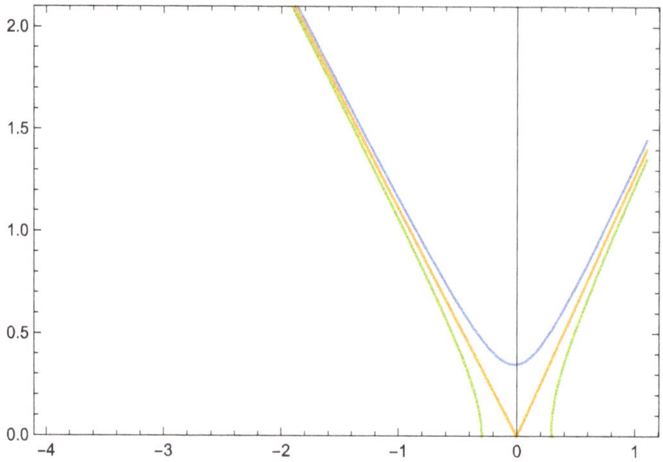

Fig. 10.5.8. At $r = 0.98$.

two cylinders on the right side (on the positive "x"-axis), then this is what we shall see at some future instant. The curved lines in Figs. 10.5.9 and 10.5.10 are flow lines. One gets the impression that the volume elements are being dragged along and stretched by the ones closest to the walls, and this does indeed suggest an increased pressure near the inner cylinder.

This is the first indication that the new vector potential represents liquid stresses.

The fact that the density increases towards the inner boundary is an almost unavoidable consequence of including a potential flow component. The illustrations confirm that, when the cylinders are rotating in opposite directions, the model predicts that the instability is likely to begin at the inner boundary.

Internal tension

The success of the no-slip boundary condition is evidence of a considerable surface tension. The no-slip boundary condition implies a strong adhesion of the fluid to the walls. Directly, this affects only the first layer of molecules near the walls, but water has a great tensile strength, as is known by the ease with which it withstands negative pressure. See Fig. 10.5.11.

The cylinder is filled with water, at atmospheric pressure. The piston is pulled outwards, increasing x and the volume, decreasing the pressure, until the bubble point is reached and the water begins to boil. But if care is exercised, the system may instead go into a metastable state and the pressure eventually becomes negative, evidence of the tensile strength of

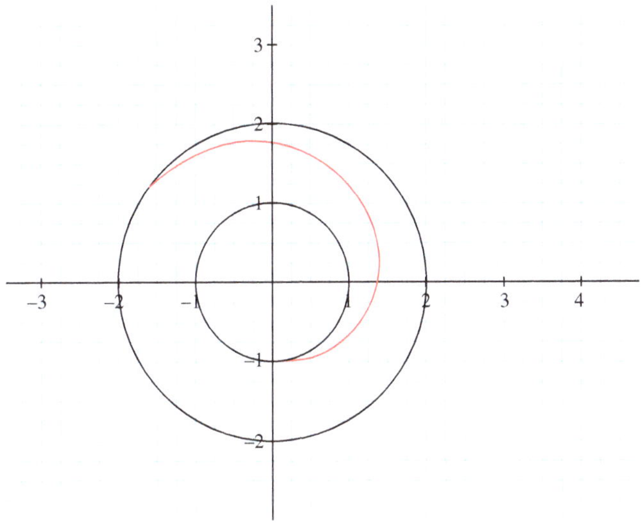

Fig. 10.5.9. Snapshot: Contra-rotation.

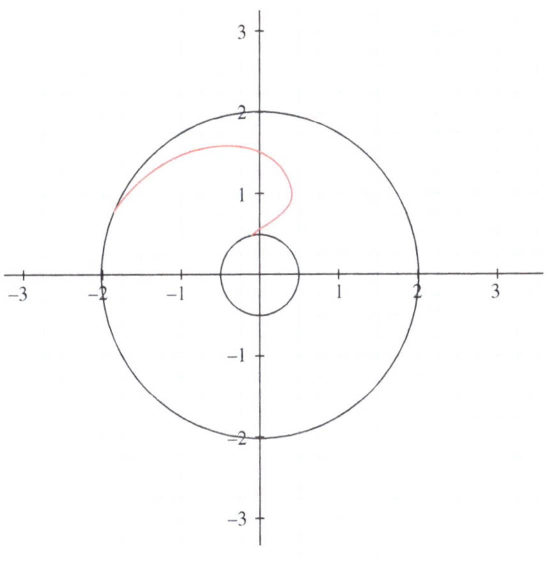

Fig. 10.5.10. Co-rotation.

water. Responding to any perturbation, if the laws of adiabatic thermodynamics are obeyed, the water will begin to boil, creating steam with a higher specific entropy density. (It appears as if water stretches like a rubber band, with an increase in entropy.) When stretched to the breaking point, it evaporates explosively since it already has the energy needed to produce

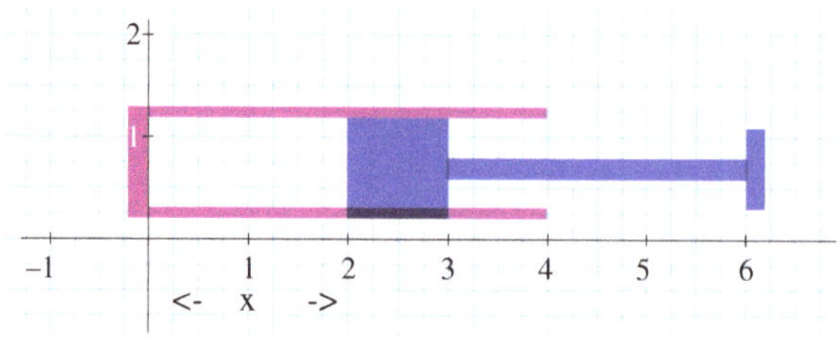

Fig. 10.5.11. Negative pressure.

some steam without further energy transfer. (The event is described as "rupture".)

A very special case, with application to superfluid He

Consider two volume elements, one on the inner cylinder and the other at the outer cylinder, and a thin smooth tube connecting the two points. Make this more precise by assuming that the tensile strength of the fluid is transmitted along the tube. Then there is one case when this force is constant, when the instantaneous, linear velocity of the two end-points are equal. In that case, $\omega_i/\omega_o = r_o/r_i$. At that point, the centrifugal force and the force produced by the gradient of the pressure, both vanish.

Going back to Eqs. (10.5.10) and (10.5.11), we call attention to a special case. If the angular velocities are decreased, then both velocity fields tend to zero, except in the very special case when

$$a = -bc^2, \quad \rho_o = \rho_i, \quad \omega_o = \omega_i.$$

In this case, both the kinetic potential K and the kinetic energy remain finite, with limits

$$K = -\frac{a^2}{2c^2r^2} = -\frac{1}{2c^2}\vec{\nabla}\Phi^2$$

and

$$h = \frac{1}{2c^2}\vec{\nabla}\Phi^2.$$

The mass flow velocity $\vec{v}$ tends to zero, but some tension remains. The angular momentum density also takes a non-zero limiting value, $-a\rho = bc^2$, directed along the axis. This brings to mind an experiment with liquid Helium at temperatures below the λ point. A cylinder filled with the liquid was cooled to a very low temperature while rotating; then the cylinder was brought to rest and the visible motion of the liquid also stopped. But angular momentum was stored in the liquid and this caused the rotation to resume when the temperature was raised, weeks later.

Entropic forces

The Lagrangian (10.1.5), in the full thermodynamic setting, is

$$\mathcal{L} = \rho(\dot{\Phi} + \dot{\vec{X}}^2/2 + \kappa\rho\dot{\vec{X}} \cdot \vec{\bigtriangledown}\Phi - \vec{\Phi}^2/2 - \phi) - f - sT,$$

where $f(\rho, T)$ is the free energy density. The Bernoulli equation becomes

$$\vec{\bigtriangledown}(\dot{\Phi} + \dot{\vec{X}}^2/2 + \kappa\rho\dot{\vec{X}} \cdot \vec{\bigtriangledown}\Phi - \vec{\Phi}^2/2 - \phi) = \vec{\bigtriangledown}p - \left(s - \rho\frac{\partial s}{\partial \rho}\right)\vec{\bigtriangledown}T - \rho T\vec{\bigtriangledown}\frac{\partial s}{\partial \rho}.$$

With the usual assumptions, $s = \rho S$, the specific entropy S uniform, both terms after $\vec{\bigtriangledown}p$ drop out. The result is the familiar hydrodynamic equation; making use of it implies that the specific entropy is assumed to be uniform. The two additional terms have been systematically ignored until now, in obedience to a popular and successful strategy: If an entropy gradient is not needed, ignore it.

In case of difficulties, one may make a different assumption about the entropy density. Allowing the specific entropy S to be non-uniform gives life to the last term. To invigorate the $\vec{\bigtriangledown}T$ term, it is necessary to expand the expression $s = \rho S$ with a term that is not linear in ρ, as in Landau and Lifshitz (1959).

In the present case, we have followed the usual protocol and it has not led to any difficulties, the experimental data having been explained to surprising accuracy. This may be less due to a correct theory than to the paucity of experimental data; it may change when the experiment is repeated with a range of different fluids. The system is not at equilibrium; some energy is being supplied; there is likely to be a temperature gradient and there may be some heat flow. Our model calculations depend only on the density and the chemical potential, only experiments can determine the pressure and the entropy distribution.

We strongly suspect that the entropy is affected by the rotation, but a different sort of data may be needed to confirm it.

10.6. Conclusions

Energy, with associated conservation laws, is the very soul of theoretical physics. It is entirely natural that, in the many contexts where an expression for it is not available, attempts are made to invoke it anyway. But a more ambitious, and potentially more profitable, approach is to insist on formulating all of physics in terms of action principles. The difficulty, of course, is to determine the action or, before that, to find the variables that make it possible.

The quest for action principles for hydrodynamics and thermodynamics was pursued most vigorously during the late 19th and early 20th centuries, by some of the most important physicists of that pioneering age: Helmholtz, Maxwell, Cartan, Einstein and others.

In a memorandum to O. Veblen, dated March 26, 1945, John von Neumann (1963) laments the fact that "hydrodynamical problems, which ought to be considered relatively simple, offer altogether disproportionate difficulties"; he says that "the true technical reason appears to be that variational methods have ... hardly been introduced in hydrodynamics". And he adds: "It is well known that they could be introduced, but what I would like to stress is that they have not been used to any practically important scale for calculations in that field".

The reason that they had "not been used" is to a large extent explained by the internal inconsistencies related to the use (or misuse) of the energy concept.

The theory developed out of a concern for natural and general physical principles with applications to many fields, how many remains to be seen, but clearly one needs to keep an open mind with regard to the physical interpretations of the new variables, a question that makes sense in each special context and that must be extracted from the results and from the experiment. Restricting ourselves to liquid dynamics we have learned, from a first physical application, that the new degree of freedom can be related to tension in the liquid. This has been reinforced by the study of other applications, studies that we hope will be published. For example:

(1) The theory provides a new degree of freedom[72] to liquids at rest; that is, in the case of configurations with the flow field $\vec{v} = 0$ but the kinetic potential K not zero. When the pressure in a van der Waals fluid is reduced, as in Fig. 10.5.11, and the pressure reaches values not allowed by the van der Waals equation of state, this tension becomes activated. It makes a contribution to the free energy of metastable states and throws entirely new light on these enigmatic states.
(2) An application to a theory of the meniscus fills a void in the theory. The explanation in terms of a surface tension is beautiful and perhaps

[72]It is "new" only in the context of variational principles.

convincing; it makes use of the fact that the surface is an isobar. But this leads inevitably to the question of calculating the other isobars, below the surface — a question that was not answered until now, nor would it seem that it was asked! The new decree of freedom of a stationary liquid under tension has been shown to fit the case admirably (Fronsdal, 2019a).

(3) An application to planetary systems, in the Newtonian approximation, is in the works. Using exactly the same Lagrangian, it was found that the theory predicts that rings must form around the planets under certain conditions. Again the Lagrangian is the same and the parameter κ has the same interpretation as a measure of compressibility (Chapter 12).

(4) In General Relativity we now have, for the first time, an action that allows for rotational flows. The cited study of rotating planets in Newtonian gravity is a first step in pursuing this application of General Relativity. This theory, with the Green-Schwarz term (the Ogievetskiĭ–Polubarinov mixing term) is a candidate for anomaly-free quantum gravity.

(5) The relation to superfluid flows offers very interesting possibility. The most economical theory would have two densities, two irrotational fields and one vector field related to a relativistic 2-form. With only six degrees of freedom, it will have much more predictive power than the theory that is current: it has eight. This leads to an interesting way to evaluate models, a rigorous criterion for deciding how many degrees of freedom are necessary and a natural preference for the most economical model.

A viable action principle for rotational Hydrodynamics, Thermodynamics and General Relativity has been proposed. It is not unique, but it is the most economical; complicated systems require more variables. It offers a first instance of an application, to a system that has resisted a complete development even in the special case of incompressible flows.

The present application of the action has led us to the interpretation of the parameters κ. In this context, it is related to the compressibility by the formula

$$\frac{\rho_i - \rho_0}{\rho_i} \approx \frac{3}{\kappa^2}.$$

See "Main results" in Section 10.5. An autonomous theory of incompressible fluids probably does not exist, certainly not in the form of an action principle. Compressibility may be small, but not zero.

Future experiments

In the present contexts of elementary hydrodynamic systems, much work has been done on incompressible flows. The more challenging problem of realistic, compressible flows require measurements of temperature and pressure profiles, which are difficult. We need more measurements of this type and we need experiments with wider horizons. Andereck and others have explored the full range of flow regimes in the Couette problem, we need to vary the fluid (even as far as using gases) as well as the density of absorbed air, ambient temperature and pressure and, especially, surface tension; that is, the surface of the cylinders. The theory presented here **predicts** that the slope of the left asymptote in Fig. 10.5.1 is sensitive to the compressibility of the fluid.

Methodology: Other applications

Other important applications of Conservative Hydrodynamics include wind tunnels, flow in pipes, ocean currents and hurricanes, and all are difficult. But even something apparently simple, such as plane Couette poses a challenge. See the Appendix to this chapter. For this reason, let us try to understand what are the essential steps that led to some understanding of cylindrical Couette flow.

(1) The geometry of the experiment was chosen to have a very high degree of symmetry, in this case rotations around the axis and translations parallel to the axis. To simplify further, we assumed that end effects can be ignored.
(2) The mass flow velocity $\vec{v}$ was required to be harmonic. This implies that the viscosity becomes irrelevant.
(3) No-slip boundary conditions. We should not be surprised if, in some applications, this turns out to be unrealistic.
(4) More experimental input may be essential. We have been able to predict the pressure profile in cylindrical Couette flow; is this prediction verified experimentally?
(5) In the case of a fluid that is nearly incompressible, it may be possible to get results that are independent of the equation of state. How far does this allow us to go?

A theory of incompressible fluids?

We set

$$\frac{\rho_i}{\rho_o} = 1+\xi, \quad c = 1-\psi, \quad 0 < \psi < \xi$$

and obtain, on the asymptotes $K = 0$, in the first quadrant,

$$\frac{\omega_i}{\omega_o} = \left(\frac{r_o}{r_i}\right)^2 \frac{2-\psi}{2-\psi+\xi} \approx 1.28$$

and in the fourth quadrant,

$$\omega_i/\omega_o = \left(\frac{r_o}{r_i}\right)^2 \frac{\psi}{\psi-\xi} = -1.28\frac{1}{\frac{\xi}{\psi}-1} < -1.28.$$

In the best fit, $\xi/\psi \approx 5$.

We get a model of an incompressible gas by taking the limit of large κ. In this case, it is better to renormalize the field $\vec{X}$ by absorbing a factor κ, to express the Lagrangian density (10.1.4) as

$$\mathcal{L} = \rho(\dot{\Phi} + \dot{\vec{X}}^2/2\kappa^2 + \rho\dot{\vec{X}} \cdot \vec{\nabla}\Phi - \vec{\Phi}^2/2 - \phi) - W[\rho] \tag{10.6.1}$$

and neglecting the second term when that gives reasonable results.

The important fields are the velocity of mass flow

$$\vec{v} = \dot{\vec{X}} - \vec{\nabla}\Phi,$$

and the renormalized vector field

$$\vec{w} = \frac{1}{\kappa^2}\dot{\vec{X}} + \vec{\nabla}\Phi.$$

In the limit $1/\kappa \to 0$, the field $\vec{X}$ becomes a non-dynamic field, as in the approach initiated by Hall and Vinen (1956).

The interpretation of this theory is difficult. In the first place, if the density is fixed, then the canonical structure is lost; it would reduce to a single canonical pair of field variables. Another difficulty is that the singularity of $\Delta\Phi$ at the center of whirls can no longer be mitigated by assuming that the density vanishes there. In an attempt to solve both problems, we may agree to retain ρ as a dynamical variable, though one that appears in the applications as essentially constant.

Since $\vec{X}$ is a non-dynamical (external) field, we can consider the case that it is a line source,

$$\dot{\vec{X}} = (0, 0, a)\delta(x)\delta(y)$$

and make contact with the string theories of Lund and Regge and of Kalb and Ramond.

10.7. The Energy–Momentum Tensor

The theory of the energy–momentum tensor was developed and applied to irrotational hydrodynamics in Chapter 3.

Irrotational case

The Fetter–Walecka Lagrangian equation (10.3.3) depends on the two scalar fields ρ and Φ and we find that

$$T_0^0 = \rho\dot{\Phi} - \mathcal{L},$$
$$T_0^i = \dot{\Phi}\rho v_i, \quad T_i^0 = \Phi_i \rho,$$
$$T_i^j = -\Phi_i \rho v_j - \delta_i^j \mathcal{L}.$$

We can verify the conservation law

$$\partial_\nu T_\mu^\nu = 0. \tag{10.7.1}$$

It is the result of a simple calculation that starts as an evaluation of the on-shell value of the quantity

$$\int d^4x \, \xi\mathcal{L}, \tag{10.7.2}$$

with $\xi = \partial_\mu$, $\mu = 0, 1, 2, 3$ (see Chapter 3). It splits into these two cases; the first,

$$\partial_0 T_0^0 + \partial_j T_0^j = 0,$$

is easy to verify, it reduces to

$$\frac{d}{dt} h + \vec{\nabla} \cdot (\rho\dot{\Phi}\vec{v}) = 0, \quad h := \rho\vec{\Phi}^2/2 + W[\rho], \tag{10.7.3}$$

which is the statement of energy conservation. The other one, with $\mu \to i = 1, 2, 3$, becomes

$$\frac{d}{dt} T_i^0 + \partial_j T_i^j = \frac{d}{dt}(\rho\Phi_i) - \partial_j(\Phi_i \rho v^j) - \partial_i \mathcal{L} = 0, \quad i = 1, 2, 3, \tag{10.7.4}$$

which is the statement of momentum "conservation". But this requires some elaboration.

It is usual to refer to T_0^0 as the energy density and to T_i^0 as (the components of) the momentum density. We shall refer to

$$\vec{P} := -\rho\vec{\bigtriangledown}\Phi$$

as the momentum density. If the system is confined within the boundaries of a domain Σ, then the total momentum is

$$\vec{P}(\Sigma) = \int_\Sigma d^3x\vec{P}.$$

Integrating (10.5.7) over Σ, we obtain

$$\frac{d}{dt}P_i(\Sigma) = -\int_{\partial\Sigma}\Phi_i\vec{v}\cdot d\sigma + \int_\Sigma d^3x\ \vec{p}.$$

The first term vanishes by the boundary conditions on the flow velocity $\vec{v}$. The other term has to be compensated by the forces that restrain the boundaries of the system. Making use of the equation of continuity, we simplify Eq. (10.7.4) to recover the Bernoulli equation.

$$\frac{d}{dt}\vec{v} + (\vec{v}\cdot\vec{\bigtriangledown})\vec{v} = -\frac{1}{\rho}\vec{\bigtriangledown}p. \tag{10.7.5}$$

We used the fact that, on shell, $\mathcal{L} = p$.

Angular momentum conservation

"The angular momentum density is the axial vector field

$$\vec{\Omega} := \vec{x}\wedge\vec{P}.\text{"} \tag{10.7.6}$$

Can we justify this remark? We have associated the operators ∂_μ with energy–momentum conservation; with the idea, especially confirmed in Quantum Mechanics, that Cartesian translations are "dual" to linear momenta. In Conservative Hydrodynamics, the usual momenta are the canonical conjugates of the Cartesian coordinates and the angular momentum (about an axis) is canonically conjugate to the corresponding angular coordinate. Thus, in (10.7.2) take $\xi = \partial_\theta = \vec{\omega}\cdot\vec{x}\wedge\vec{\bigtriangledown}$, with $\vec{\omega} = (0,0,1)$ or more generally any constant vector $\vec{\omega}$.

To derive a conservation law for angular momentum, we return to the calculation that starts with Eq. (10.7.2), but with

$$\xi = (\vec{\omega}\cdot x\wedge\vec{\bigtriangledown}), \quad \delta\phi = \xi\phi, \quad \delta\phi_\mu = \partial_\mu\delta\phi, \tag{10.7.7}$$

where ω is a constant vector. There is a complication with spherical coordinates since the kinetic energy depends not only on the fields but also on the coordinates. Cartesian coordinates are also not without complications, for if $\delta\phi = \xi\phi$, then the integration by parts gives

$$\int d^4x\ \xi\mathcal{L} = \int d^4x\ \partial_\nu\left((\vec{\omega}\cdot x \wedge \vec{\nabla}\phi)\frac{\partial \mathcal{L}}{\partial \phi_\nu}\right)$$

or

$$\epsilon^{ijk}\omega_i x_j\left(\partial_k\mathcal{L} - \partial_\nu\left(\phi_k\frac{\partial\mathcal{L}}{\partial\phi_\nu}\right)\right) + \epsilon^{ijk}\omega_i\partial_k\frac{\partial\mathcal{L}}{\partial\phi_j} = 0. \tag{10.7.8}$$

Here, most terms are space-boundary terms. If the vessel is bounded by concentric cylinders, both vanish by the appropriate boundary conditions. Integrating over the vessel leaves only the time-boundary term

$$\frac{d}{dt}\int d^3x\ \epsilon^{ijk}\omega_i x_j T_k^0 = 0.$$

This confirms the guess (10.7.6). In the irrotational theory, it is

$$\vec{\Omega} = -\rho\vec{x}\wedge\vec{\nabla}\Phi,$$

as expected.

The general case

In addition to (10.7.10), we take

$$\delta X^k = \xi X^k, \quad \delta\partial_i X^k = \partial_i(\delta X^k).$$

With the inclusion of the vortex field $\vec{X}$ in the Lagrangian density (10.1.4), we find that

$$T_0^0 = \rho\dot{\Phi} - \mathcal{L} + \vec{X}\cdot\vec{m} = \frac{\rho}{2}\big((\vec{\nabla}\Phi)^2 + \dot{\vec{X}}^2\big) + h, \tag{10.7.9}$$

$$T_0^i = \dot{\Phi}\rho\vec{v}^i, \quad T_k^0 = \Phi_k\rho + \partial_k\vec{X}\cdot\vec{m}, \quad i = 1,2,3,$$

and

$$T_k^j = \Phi_{,k}\rho v^j - \delta_k^j\mathcal{L}.$$

I have assumed that there are no modifications due to the fact that we are dealing with a gauge theory, since everything is to be evaluated in the physical gauge.

Still, we want to be sure of Eq. (10.7.1). The energy conservation law (10.7.3) retains its form, with an extra term in h, while the law of momentum conservation becomes

$$\frac{d}{dt}T_k^0 + \partial_j T_k^j$$
$$= \partial_0(\Phi_k\rho + \partial_k\vec{X}\cdot\vec{m}) + \partial_i(\Phi_k\rho v^i) - \partial_k p = 0. \qquad (10.7.10)$$

The first three terms are, respectively,

$$\Phi_k\dot{\rho} + \rho\partial_k\dot{\Phi}, \quad \rho(\partial_k\dot{\vec{X}})\cdot(\dot{\vec{X}} + \kappa\vec{\bigtriangledown}\Phi), \quad \Phi_k\partial_i(\rho v^i) + \rho v^i\partial_i\partial_k\Phi.$$

We used the vector wave equation, $\dot{\vec{m}} = 0$. The equation of continuity takes care of two terms and there remains

$$\rho\left(\partial_k\dot{\Phi} + \partial_k\dot{\vec{X}}^2/2 + \kappa(\partial_k\dot{\vec{X}})\cdot\vec{\bigtriangledown}\Phi - \frac{1}{2}\partial_k\vec{\bigtriangledown}\Phi^2 + \kappa\dot{\vec{X}}\cdot\partial_k\vec{\bigtriangledown}\Phi\right)$$
$$= \rho\partial_k(\dot{\Phi} + K) + \vec{\bigtriangledown}p = 0.$$

It is seen that Eq. (10.5.7) miraculously reduces to the Euler–Lagrange equation (10.4.5), when the equation of continuity and the wave equation are taken into account.

According to Eq. (10.5.12), the conserved quantity is the momentum density

$$P_k = -\rho\Phi_k - \partial_k\vec{X}\cdot\vec{m}.$$

The angular momentum is

$$\vec{\Omega} = \vec{x}\wedge P, \quad \Omega^i = \rho\epsilon^{ijk}x_j(-\Phi_k - \partial_k\vec{X}\cdot\vec{w}).$$

This contrasts with the expression that is normally used, namely,

$$\vec{\Omega} = \vec{x}\wedge\rho\vec{v}, \quad \Omega^i = \rho\epsilon^{ijk}x_j(-\Phi_k + \kappa\dot{X}_k),$$

where $\vec{v}$ is the mass transport velocity.

Remark. The momentum density as well as the density of angular momentum do not always vanish when there is no mass flow, when $\vec{v} = 0$. In that case, $\kappa\dot{\vec{X}} = \vec{\bigtriangledown}\Phi$ and

$$\vec{P} = -\rho\vec{\bigtriangledown}\big(\dot{\Phi} + (1+\kappa^2)\vec{\bigtriangledown}\Phi^2/2\big), \qquad (10.7.11)$$

$$\vec{\Omega} = \rho\vec{x}\wedge\vec{P}\big(1 + (1+\kappa^2)\vec{\bigtriangledown}\Phi^2/2\big). \qquad (10.7.12)$$

Such behavior has been observed in experiments with rotating superfluid. At very low temperatures the angular momentum has been observed to persist for weeks after the cylinder was stopped and the rotation of the liquid appeared to stop as well. It is a common experience that the angular momentum of rotating superfluid Helium is different from what is suggested by observation of the mass flow.

Project 7. Apply the results about angular momentum to a new analysis of experiment with rotating, superfluid Helium.

Appendix: Planar Couette Flow

When the radii of both concentric cylinders grow, with the gap held fixed, we approach a situation called planar Couette flow. Here the fluid is confined between two parallel planes. The coordinates are the standard Cartesian x, y, z. The planes are perpendicular to the y direction and the flow is in the x direction,

$$\vec{v} = (V(y), 0, 0)$$

with $v_x = V(y)$ depending only on the transverse coordinate y. The plates are at $y = y_1$, y_2. The problem is mathematically two-dimensional, but we shall treat it as a special case of three-dimensional hydrodynamics.

As in the case of cylindrical Couette flow, we shall limit our study to the most basic, laminar flow. The question that is asked is whether such a flow exists and at what point (at what speed) does it break down. There is very little experimental information about this; the simplest way to summarize the result is that this flow may be stable at all speeds. But this conclusion would be too hasty since it is based on limited evidence.

There is an unanswered theoretical question. It would seem that, once we have analyzed cylindrical Couette flow, all we have to do is to take the limit of infinite radius. It has been pointed out that this does not work; it seems that the limit does not exist. A naive attempt to calculate the limit fails, but the reader is encouraged to make an attempt.

1. Symmetry. We assume invariance with respect to translations in the z direction. 2. The function $v(y)$ must be harmonic, hence first order in y.[73] The general expression for $v(y)$ is, as always

$$\begin{aligned}\vec{v} &= \kappa\vec{w} - (1+\kappa^2)\vec{\bigtriangledown}\Phi \\ &= -\frac{\kappa}{\rho}\vec{\bigtriangledown}\tau - (1+\kappa^2)\vec{\bigtriangledown}\Phi.\end{aligned}$$

[73] During startup, the viscosity, which is small but significant in water, acts to dissipate any part of the flow that is not harmonic. Experimenters wait for this process to be completed and observations are made for the stationary, harmonic flow that remains.

We shall determine stationary, hence harmonic flows in the x direction, thus

$$\vec{v} = (\omega y, 0, 0) = -\frac{\kappa}{\rho}\vec{\nabla}\tau - (1+\kappa^2)\vec{\nabla}\Phi. \tag{A.10.1}$$

Taking the curl we obtain

$$\tau_x = \frac{\omega}{\kappa}\frac{\rho^2}{\rho'}. \tag{A.10.2}$$

Returning to (A.10.1) we have

$$(1+\kappa^2)\Phi_x = -\frac{\kappa}{\rho}\tau_x - \omega y, \quad (1+\kappa^2)\Phi_y = -\frac{\kappa}{\rho}\tau_y,$$

which can be solved for Φ by virtue of (A.10.2).

Most experiments use water, with very small compressibility, so an adequate approximation across a narrow gap is

$$\rho = \rho_0(1+\alpha y), \quad y_1 < y < y_2,$$

$$\tau_x = \frac{\omega}{\kappa}\frac{\rho^2}{\rho_0\alpha}, \quad \tau_y = \frac{2\omega}{\kappa}\rho x + f(y),$$

$$(1+\kappa^2)\vec{\nabla}\Phi = -\left(\omega y + \frac{\omega}{\alpha},\ 2\omega x + \frac{\kappa f}{\rho}\right).$$

Here f is an arbitrary function of y. Let us work out the simplest case, when $f = 0$. Then the kinetic potential is

$$K = \frac{-1}{2}\left(\frac{\vec{\nabla}\tau}{\rho}\right)^2 + \frac{\kappa^2+1}{2}\vec{\nabla}\Phi^2$$

$$= \frac{\omega^2}{2}\left(\frac{(2y+1/\alpha)^2 + (2x)^2}{1+\kappa^2} - \frac{(y+1/\alpha)^2 + (2x)^2}{\kappa^2}\right).$$

Interpretation

Note that the dependence on x cancels only in the limit of incompressible flows. This is true at all values of y and hence at the smallest velocities. It means that there is no stationary solution that is invariant under translations in the x direction, which is the same conclusion obtained by others. It is not possible to avoid this unexpected result by tweaking the form of $\rho(y)$. The same is true of the experimental setup, since in practice the walls extend only a finite distance in the x direction. There is a strong feeling among experimenters that, in a hypothetical limit of very long plates the laminar flow is stable at all velocities. In the incompressible approximation, when $\kappa \ll 1$ and α is zero, we find that

$$K = -\frac{\omega^2}{\kappa^4}(y^2 + x^2),$$

which being small and negative suggests stability for all y. To test this hypothesis further, the physical dimensions of the apparatus should be increased, along with the velocity; that is, $\omega = v/y$. An instability, if any, would be expected at the origin, at $y = x = 0$.

The possibility that the x-dependence is real, that no stationary flow is invariant under translations in the x direction, cannot be excluded. Experimenters do see a developing turbulence, but it is clearly initiated outside the experimental area. At low Reynolds number, a steady laminar flow is achieved for a short time, until this disturbance arrives. But at higher velocities, there is not enough time to approach the asymptotic, stationary distribution. And the experiment does not give much information.

Chapter 11

Rotating Planets*

There is a limit to how far one can develop **astrophysics** as an application of non-relativistic hydrodynamics; here is an attempt that pushes that limit.

Variational techniques have been used in applications of hydrodynamics in special cases, but an action that is general enough to deal with both potential flows and solid-body flows, such as cylindrical Couette flow and **rotating planets**, has been proposed only recently. This paper is one of a series that aims to test and develop the new Action Principle. We study a model of rotating planets, a compressible fluid in a stationary state of motion, under the influence of a fixed or mutual gravitational field. The main challenge is to account for the shape and the velocity fields, given the size of the **equatorial bulges**, the angular velocity at equator and the density profiles.

The theory is applied to the principal objects in the **solar system** from Earth and Mars to Saturn with fine details of its hexagonal flow and to Haumea with its odd shape. With only two parameters, the model gives a fair fit to the shapes and the angular velocity field near the surface. **Planetary rings** are an unforeseen, but a natural and inevitable feature of the dynamics; no cataclysmic event need be invoked to justify them. The simple solutions that have been studied so far are most suitable for the hard planets, and for them the predicted density profiles are reasonable. The effect of precession was not taken into account, nor were entropic forces, so far. There has not yet been a systematic search for truly realistic solutions. The intention is to test the versatility of the action principle; the indications are very encouraging.

*Originally intended as an independent article, amended.

11.1. Introduction

The ultimate aim of our work is to learn how to deal with a compressible, **rotating fluid in General Relativity**. Because much important work has been done in the past, we must emphasize the novelty of this work: We approach hydrodynamical problems within the framework of a general **Action Principle**. In the context of General Relativity this means selecting an adequate relativistic action $A_{\rm matter}$ for fluid dynamics in an arbitrary metric background and adding it to Einstein's action for the metric. The energy–momentum tensor of $A_{\rm matter}$ becomes the source of Einstein's field equations. This is the natural way to ensure that the source of Einstein's field equations satisfy the integrability conditions (the Bianchi constraint).

Our action principle incorporates traditional hydrodynamics, **including the equation of continuity**. It generalizes a well-known action principle for irrotational hydrodynamics by including rotational velocity fields. It extends to Thermodynamics where it has been extensively applied to mixtures (this volume), to Special Relativity (Ogievetskiĭ and Polubarinov, 1964), and most recently to stability of cylindrical Couette flow (this volume). It provides the first field theoretic model of rotating fluids that respects the Bianchi identity and includes the equation of continuity. Previous attempts to formulate an action principle for General Relativity (Hartle, 1967; Bardeen, 1970; Schutz, 1970; Taub, 1954) have not been developed to a stage where applications could be undertaken, but all the ideas of the older work resonate in the present one.

The principal feature of this paper is the **action principle**; in other respects it is far less complete than earlier models of planetary dynamics. See, for example, Beauvalet *et al.* (2012) or Kolb *et al.* (2013). Another feature that is not always included in planetary dynamics is the requirement that the mass flow velocity field be **harmonic**, as it needs to be for all stationary flows (see below). Most important, the action principle is a framework with a greatly improved **predictive power**, as this paper demonstrates.

Before attacking this problem in the full, Generally Relativistic setting, it is a good strategy to study the problem in the non-relativistic approximation with Newtonian Gravity. The system is a **compressible fluid** in hydrodynamics; the forces arise from a fixed, central Newtonian potential, or from the mutual, gravitational interaction between the fluid elements and the pressure. Because the new Action Principle is a new departure, we present a summary account of it, before applying it to the principal components of the solar system. Harmonic expansions of the dynamical fields will include only the first three terms, except in a study of the hexagonal flows in Saturn.

The gauged-fixed, non-relativistic action is

$$A = \int dtd^3x\mathcal{L},$$

$$\mathcal{L} = \rho(\dot{\Phi} - \vec{\bigtriangledown}\Phi^2/2 + \dot{\vec{X}}^2/2 + \kappa\dot{\vec{X}} \cdot \Phi - \Phi^2/2 - \varphi) - W[\rho]. \qquad (11.1.1)$$

The variables are the density ρ and two velocity potentials, the scalar velocity potential Φ and the vector potential $\vec{X}$ (cf. Schutz, 1970). Projected on the theory of irrotational flows, the special case $\dot{\vec{X}} = 0$, it becomes the Lagrangian that was discovered by Lagrange (1760) and brought to our attention by Lamb (1932) and by Fetter and Walecka (1980). Gravity is represented by the Newtonian potential φ; it can be a fixed, central potential or the mutual energy of interaction between fluid elements. The action (11.1.1) is the non-relativistic limit of a generally relativistic action and the potential $W[\rho]$ is the thermodynamic internal energy density for a system with fixed, uniform specific entropy density. The on-shell value of the Lagrangian density is the thermodynamic pressure; as prophesized with rare insight by Taub (1954).

The Euler–Lagrange equations are: the equation of continuity (from variation of the velocity potential Φ),

$$\dot{\rho} + \vec{\bigtriangledown} \cdot (\rho\vec{v}) = 0, \quad \vec{v} := \kappa\dot{\vec{X}}X - DD\Phi, \qquad (11.1.2)$$

the wave equation (from variation of $\vec{X}$)

$$\frac{d}{dt}(\rho\vec{w}) = \vec{\bigtriangledown} \wedge \vec{v}, \quad \vec{w} := \dot{\vec{X}} + \kappa\vec{\bigtriangledown}\Phi, \qquad (11.1.3)$$

and the Bernoulli equation (from variation of ρ),

$$\dot{\Phi} + \dot{\vec{X}}^2/2 + \kappa\dot{\vec{X}} \cdot \Phi - \Phi^2/2 - \varphi = \frac{\partial}{\partial\rho}W[\rho]. \qquad (11.1.4)$$

The incorporation of two velocity fields is an essential feature of the theory; what is both novel and effective is that they contain the correct (minimal) number of dynamical variables: Four for hydrodynamics including the density. The Lagrangian has one free parameter κ; it is related to the compressibility of the fluid.

The complete theory is a relativistic gauge theory; here we are working in the physical gauge and in the non-relativistic limit. There are two velocity fields, $\vec{v}$ represents mass flow; $\vec{w}$ allows for vorticity. A key feature of the theory is the constraint

$$\vec{\bigtriangledown} \wedge (\rho\vec{w}) = 0, \qquad (11.1.5)$$

derived by variation of the full Lagrangian with respect to a vector gauge field. (The gauge field is fixed at zero and does not appear in this chapter.) It implies that the vector field $\vec{w}$ — defined in (11.1.3) — can be expressed in terms of a scalar field,

$$\vec{w} = \frac{-1}{\rho}\vec{\nabla}\tau. \tag{11.1.6}$$

(Of course, we cannot replace $\vec{w}$ by τ in the action.) The limit of an incompressible fluid is approached as $\kappa \to \infty$. It is this constraint that connects the velocity and the density: in the context of the action principle, harmonicity of $\vec{v}$ becomes a constraint on ρ, solved by Eq. (11.2.8) below.

We have given a brief account of the theory in the physical gauge; the full gauge theory is described in two papers (Fronsdal, 2014, 2017).

What does the most to validate this theory is that it is formulated as an action principle. To account for rotational motion, it includes the velocity field $\dot{\vec{X}}$, as in classical "Lagrangian hydrodynamics", but as this would seem to add six additional degrees of freedom, we need a constraint that effectively reduces this number to 2, as is accomplished by Eq. (11.1.5). The relation of the field $\dot{\vec{X}}$ to vorticity, and to a relativistic gauge theory, was proposed, in the special case of incompressible flows, by Lund and Regge (1976). In string theory, it is the Kalb–Ramond field (Kalb and Ramond, 1974).

The action is not completely new; hints of it appeared in a classical paper by Hall and Vinen (1956) on superfluids and in a more recent review by Fetter (2009) on rotating Bose–Einstein condensates (Fetter, 2009).[74] In those papers, $\dot{\vec{X}}$ is not a local, dynamical field variable but a fixed background feature that accounts for a rigid rotation of the whole system. Although the dynamical, irrotational velocity was seen as insufficient and another degree of freedom was needed, the way to avoid an excessive number of degrees of freedom, by means of the constraint (11.1.5), was not widely known. The field components $\dot{X}^i$ appear in Bardeen (1970), they are the classical "Lagrange parameters".

A stationary flow is one that evades the dissipating effect of **viscosity**. In traditional hydrodynamics, **viscosity** is included as an additional term in the Navier–Stokes equation,

$$\dot{\vec{v}} + (\vec{v}\cdot\vec{\nabla})\vec{v} = \frac{-1}{\rho}\vec{\nabla}p + \bar{\mu}\rho\Delta\vec{v}, \tag{11.1.7}$$

[74] The equations of motion in the paper by Hall and Vinen (1956) are widely quoted; the relation of their work to an action principle, not at all.

where $\bar{\mu}$ is the kinematical viscosity, usually a constant. Viscosity cannot be accommodated within an action principle, but its effect can be acknowledged by replacing the conservation law (11.1.3) by

$$\frac{d}{dt}(\rho\vec{w}) = \bar{\mu}\rho\Delta\vec{v}.$$

In this way, a theory based on conservation laws is distorted to include a type of dissipation, precisely as is done in the familiar approach with Eq. (11.1.7). In both theories, **stationary motion is possible only when the field $\vec{v}$ is harmonic**.

The biggest surprise to emerge from this work is the natural and spontaneous appearance of **planetary rings**. Rings are predicted, not just accommodated. The simplest solutions studied in this paper predict a single ring (or none) for each planet. It is suggested that the parameter N in Eq. (11.1.9) is related to evolution, that all the planets may have had rings at one time, that the planet Mars, in particular, may have had a ring at relatively recent times and that all the planets will eventually lose their rings. The observed hexagonal rotation pattern of Saturn is useful data that reveals the presence of higher harmonics. The observation of 1–6 internal whorls is a prediction of the theory.

A brief digression on the cylindrical Couette problem

The most familiar systems treated in Newtonian gravity as well as in General Relativity have spherical symmetry. But most heavenly bodies are rotating around an axis that is more or less fixed, with approximate **cylindrical symmetry**. Much of the inspiration for our work comes from a study of cylindrical Couette flow. Following the advice of Homer Lane (1870), as is traditional in astrophysics, we apply to astrophysics what we have learned in terrestrial laboratories. The problem examined by Couette (1887; 1888a,b; 1890a,b) and Mallock (1888, 1896) is a fluid confined between two concentric cylinders that can be rotated independently around the vertical axis. In classical hydrodynamics, the balance of forces is expressed by the Navier–Stokes equation. Boundary conditions, validated by the analysis of Taylor (1923), are assumed to be non-slip and the fluid is compressible. At low speeds, any stationary motion is found to be described by the following harmonic vector field:

$$\vec{v} = \frac{a}{r^2}(-y, x, 0) + b(-y, x, 0), \quad r := \sqrt{x^2 + y^2}, \quad a, b \text{ constant}. \tag{11.1.8}$$

The first term is irrotational for $r \neq 0$ and both are harmonic.

The new action principle was used to account for the stability of basic, cylindrical Couette flow. One feature of the model is that the density profile is subjected to a strong condition that originates in the demand that $\vec{v}$ be harmonic.

11.2. A Compressible Fluid Rotating in a Fixed, Gravitational Field

Our model of a planet is an isolated system with a liquid or solid core in thermodynamic equilibrium with a gaseous atmosphere, in a stationary, rotating state and described by the action (11.1.1). In the simplest case, it consists of a single substance in two phases. The condition of thermodynamic equilibrium at the **phase boundary** is that the pressure p, the temperature and the chemical potential μ be continuous across the surface. In the case of a thin atmosphere, this implies that the pressure and the chemical potential are constant on the surface. The surface is thus a locus of the function — (see Eq. (11.1.4))

$$C(\vec{x}) := \dot{\vec{X}}^2/2 + \kappa\dot{\vec{X}} \cdot \vec{\bigtriangledown}\Phi - \vec{\bigtriangledown}\Phi^2/2 - \frac{-GM}{R},$$
$$R := \frac{r}{|\sin\theta|} = \sqrt{x^2 + y^2 + z^2}. \tag{11.2.1}$$

We have chosen the Newtonian expression for the attractive gravitational potential. It is an expedient shortcut of the present treatment, as in the simplest version of the traditional approach, and one that we hope to remove later. It is a valid approximation so long as the departure from spherical symmetry is small.

This model should be appropriate for Earth and Mars and possibly for the frozen planets Neptune and Uranus, less so for the gaseous planets. To determine the appropriate velocity fields, we begin by examining the simplest solutions.

If the velocity is irrotational, and $\dot{\vec{X}} = 0$; the shape is determined by

$$-\frac{a^2}{2}\frac{1}{r^2} + \frac{GM}{R} = \text{constant}, \quad r := R|\sin\theta| = \sqrt{x^2 + y^2}.$$

We plot the loci of this expression and vary the parameter a^2. Instead of an equatorial bulge, there is a polar hole, see Fig. 11.2.1. This attempt evidently fails.

Solid-body flow is the complementary case in which $\vec{\bigtriangledown}\Phi = 0$ and the angular velocity $\omega = b$. As in the traditional approach, the condition of

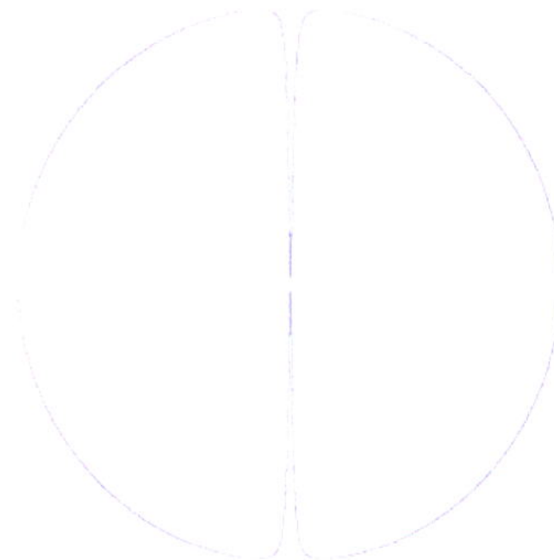

Fig. 11.2.1. Showing the "equatorial bulge", or rather the polar depression, that would be the effect of pure, irrotational flow.

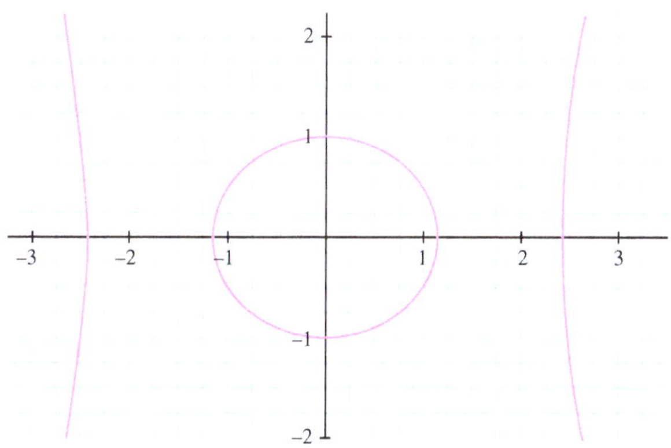

Fig. 11.2.2. The more physical locus predicted by solid-body flow. (The non-compact branch is not part of the traditional theory.)

equilibrium is

$$\frac{\omega^2}{2}r^2 + \frac{GM}{R} = \text{constant}.$$

There is a bulge, see Fig. 11.2.2. The number usually quoted is

$$\epsilon := \frac{R_{\text{eq.}}}{R_{\text{pole}}} - 1 = \frac{\omega^2}{2MG}R^3_{\text{eq.}},$$

where, to a good approximation, R_{eq} can be replaced by 1 on the right-hand side. For planet Earth, the number is $MG = R^2 g$, $g = 990\,\text{cm}/\text{sec}^2$, and the approximate value of ϵ is predicted by this model to be

$$\frac{\omega^2 R^3}{2MG} = \frac{R\omega^2}{2g} = \left(\frac{2\pi}{24 \times 3600}\right)^2 \frac{6.357 \times 10^8}{2 \times 998} = 0.0016. \tag{11.2.2}$$

Solving Eq. (11.2.1) for the azimuthal angle, we find that $(b^2/2)R^2 \sin^2\theta = C - GM/R$ and we conclude that, with the solid-body hypothesis the radius is minimal at the poles and the locus of Fig. 11.2.1 always has a non-compact branch, as in Fig. 11.2.2.

The observed value of ϵ for Earth is 0.00335, more than twice the prediction (11.2.2). The classical theory can be improved by taking into account the effect of the bulge on the potential; for example, by assuming that the shape is an ellipsoid, and that the density is uniform. That results in a value for ϵ of 0.0042, which is too large (Fitzpatrick, 2018). A further improvement results from taking the partly known density distribution into account; this has the effect of diminishing the effect of the shape on the potential.

The general case

We look for the general solution of the equations of motion. Again

$$\varphi = -\frac{GM}{R}, \quad \mathcal{R} := \sqrt{x^2+y^2+z^2}, \quad G = \text{ constant}. \tag{11.2.3}$$

The full set of equations includes the equation of continuity and

$$\vec{v} := \kappa\dot{\vec{X}} - \vec{\nabla}\dot{\vec{\Phi}}, \quad \Delta\vec{v} = 0, \quad \vec{w} := \dot{\vec{X}} + \kappa\vec{\nabla}\dot{\vec{\Phi}} = \frac{-1}{\rho}\vec{\nabla}\tau, \tag{11.2.4}$$

$$\dot{\Phi} + \dot{\vec{X}}^2/2 + \kappa\dot{\vec{X}}\cdot\vec{\nabla}\dot{\vec{\Phi}} - \vec{\nabla}\dot{\vec{\Phi}}^2/2 + \frac{GM}{R} = \mu[\rho]. \tag{11.2.5}$$

All the vector fields can be expressed in terms of the two scalar fields Φ and τ and the density. The most naive assumption is that the two scalar fields are the same as those that appear in Couette flow, proportional to the azimuthal angle, thus

$$-\vec{\nabla}\Phi = \frac{a}{r^2}(-y, x, 0), \quad -\vec{\nabla}\tau = \frac{b}{r^2}(-y, x, 0). \tag{11.2.6}$$

These are gradient-type vector fields for circular flows in the horizontal planes, with angular momentum $L_z = \pm 1$. They would not be sufficient for an ambitious attempt to construct realistic models, but they may be enough for our main purpose, which is to establish the versatility of the action principle.

Eqs. (11.2.4) and (11.2.6) give

$$\vec{v} = \kappa\vec{w} - (\kappa^2 + 1)\vec{\bigtriangledown}\Phi = \omega(-y, x, 0), \quad \omega := \frac{1}{r^2}\left(\frac{\kappa b}{\rho} + a(\kappa^2 + 1)\right). \tag{11.2.7}$$

The most general harmonic vector field of this form is a series of spherical functions with higher angular momenta. There is evidence of higher angular momenta in the flow velocities of Venus, Pluto and, most notably, Saturn. In this paper, we reduce the series to the simplest terms, the normalized inverse density taking the form

$$\frac{1}{\rho} = 1 + NR + \eta r^2 + \nu\frac{r^2}{R^3}, \quad N > 0, \quad \eta > 0, \tag{11.2.8}$$

with constant coefficients N, η, ν. The first two terms have $\ell = -1$, the others, $\ell = 1$. The introduction of a non-integrable representation of the rotation group ($\ell = -1$) should be noted. Higher harmonics are needed in the case of Saturn, since this planet shows a distinct, hexagonal flow pattern.

A non-zero value of the last term in (11.2.8) gives rise to a central hole, shaped like a donut, near the center of the planet; see Fig. 11.2.3. A very small value of ν results in a very small hole; it may serve as a regularizing device, but it is hardly relevant for the evaluation of the shape of the surface. We adopt the expression (11.2.8), with $\nu = 0$, as a plausible first approximation to the density profiles of Earth and Mars and, very tentatively, to those of the other planets.

This is our simplest model planet. It is intended, in the first place, to serve as a model for Earth and Mars and perhaps for Uranus and Neptune. A complete calculation would need a general harmonic expansion for the flow vector field $\vec{v}$.

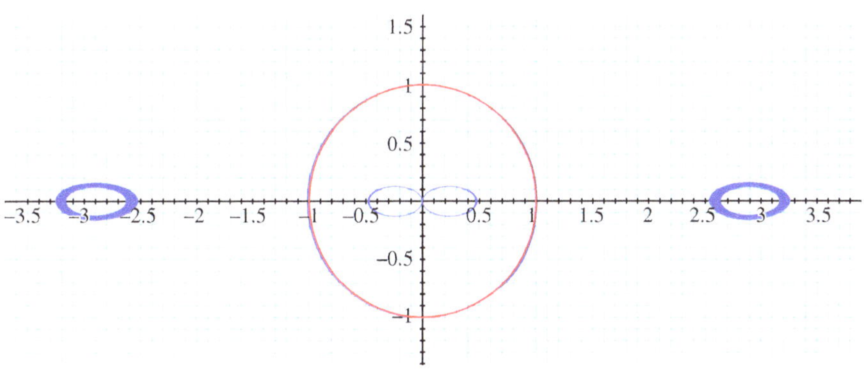

Fig. 11.2.3. The effect of including the last term in (11.2.8); a hole appears at the center, as well as a ring.

Regularity at the poles

Take the polar radius to be the unit of length and let the central density be the unit of density. Both N and η have to be positive, as we shall see.

Let us begin with a star that is spherically symmetric ($\eta = 0$) with polar radius 1 and density ratio

$$\frac{\rho_{\text{center}}}{\rho_{\text{pole}}} = N + 1;$$

then we allow for a modest violation of spherical symmetry by increasing the parameter η from zero.

The equation for the surface takes the form,

$$\frac{1}{2r^2}\left(b^2(1 + NR + \eta r^2)^2 - a^2(1+\kappa^2)\right) + \frac{GM}{R} = \text{constant}.$$

In order to avoid getting a dip at the poles (from the denominator r^2), we must have

$$a^2(1+\kappa^2) = b^2(N+1)^2. \tag{11.2.9}$$

Then the equation takes the form

$$\frac{b^2}{2r^2}((1 + NR + \eta r^2)^2 - (N+1)^2) + \frac{GM}{R} = \text{constant}. \tag{11.2.10}$$

With this constraint, the angular velocity (see Eq. (11.2.7)) is

$$\omega = \frac{\kappa b}{r^2}\left(\frac{1}{\rho} \pm \frac{N+1}{c}\right), \quad c := \frac{\kappa}{\sqrt{1+\kappa^2}}. \tag{11.2.11}$$

The visible angular velocity at the equator is

$$\omega_{\text{eq.}} = \kappa b\left(1 + N + \eta \pm \frac{N+1}{c}\right). \tag{11.2.12}$$

The mass transport velocity $\vec{v}$ can change sign within the star.

The Earth has a small equatorial bulge. To estimate the visible angular velocity, we approximate η by zero. With these approximations we obtain, using the lower sign, for the visible angular velocity of Earth, the value

$$\omega = b\frac{N+1}{2\kappa} = \frac{a}{2}. \tag{11.2.13}$$

Finally, the shape is determined by

$$f(R,r) := \frac{(1 + NR + \eta r^2)^2 - (N+1)^2}{r^2} + \frac{\xi}{R} = \text{constant}, \tag{11.2.14}$$

where ξ is the constant

$$\xi = \frac{2GM}{b^2}. \tag{11.2.15}$$

Solutions of (11.2.14) extend to very large R only if there is an effective cancellation between the terms of highest power, in $NR + \eta r^2$. If ηN is positive, there can be no cancellation, at any azimuth; hence, all the solutions are compact when we pose

$$\eta N \geq 0.$$

Overview of results

The two parameters N and ξ form a Euclidean 2-space with the latter as abscissa; it divides into a lower region (roughly $N < 1$) where the planets have a ring, and a complementary upper region where they do not, separated by a "ring-no-ring" boundary. See Fig. 11.2.4, where three versions of this dividing line are shown, for $\epsilon = 1/300$ and with $\eta = 0.01$, 0.05 and 0.1 from high to low. In the same figure, we have shown nearly vertical lines of dots, a "trajectory" for each of four planets. The coordinates of the dots on each of the planetary trajectories give a near-perfect fit to the measured ellipsoid of the respective planet.

Figure 11.2.4 shows the result of calculations as points in the plane of the parameters ξ and N. The near vertical lines connecting dots consist of points that give perfect fits to the shape of the respective planet, without rings on the upper part. Results of the calculation are tabulated in the Appendix.

It may be permissible to think of this diagram as an **evolution diagram**, each planet evolving upwards towards a state of greater compression, and losing its rings as it crosses the ring-no-ring dividing line. Earth

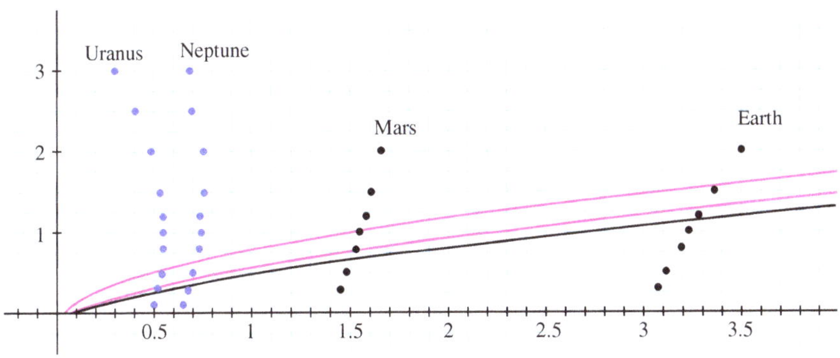

Fig. 11.2.4. The abscissa is the parameter ξ and the ordinate is N.

lost its rings long ago; it is to be placed on the upper part of its trajectory, well above the $\eta = 0.1$ line.

Increasing N means higher compression at the center; Earth may have N as high as 2 while Mars is less compressed and may have $N = 1$. Since planets are likely to become more compressed over time, we expect planets to evolve upwards. This is in accord with speculations that Mars may have had a ring in the evolutionary recent past. Uranus and Neptune still have rings and must have $N \leq 0.3$ if our model is applicable to them.

The rings will be discussed below. Other aspects of the model, including the equation of state, will be taken up in the connection with the gaseous planets, in Section 11.5.

11.3. Earth and Mars

The polar radius is our unit of length and remains fixed at unity. The equatorial radius is determined by a fourth-order algebraic equation that is obtained from (11.3.9) by setting $r = R$. We are mostly interested in shapes that are almost spherical (leaving aside the planetary rings for the moment), with a small equatorial bulge. To find surfaces that include a point on the equator with radius $s = 1 + \epsilon$, we write (11.2.14) in the form

$$B(R, r) := f(R, r) - f(s, s) = 0. \tag{11.3.1}$$

Note that $f(s, s)$ is a constant. The equatorial radius is a zero of the function $B(r) := B(r, r)$ and after division by $r - s$, this equation reduces to a cubic. For planets without rings, this cubic does not have positive roots. If a ring is about to disappear at a distance s' from the center, there will be a double zero at $r = s'$.

The measurable parameter ϵ has replaced the value of the function f. However, there are still three parameters left, N, η and ξ, and it is difficult to survey all possibilities. We shall try to find our way around this difficulty as we look at individual planets.

If Earth is an ellipsoid with eccentricity ϵ and the polar radius is normalized to unity, then the shape is

$$R - 1 = \epsilon \sin^2 \theta; \quad R_{eq} = 1 + \epsilon,$$

with $\epsilon = 0.00335$. With $s = 1.00335$, the locus $B(R, r) = 0$ passes through the equator at $R = r = s$, and through the pole. Fittings of shapes are relative to this ellipsoid, with the appropriate value of ϵ.

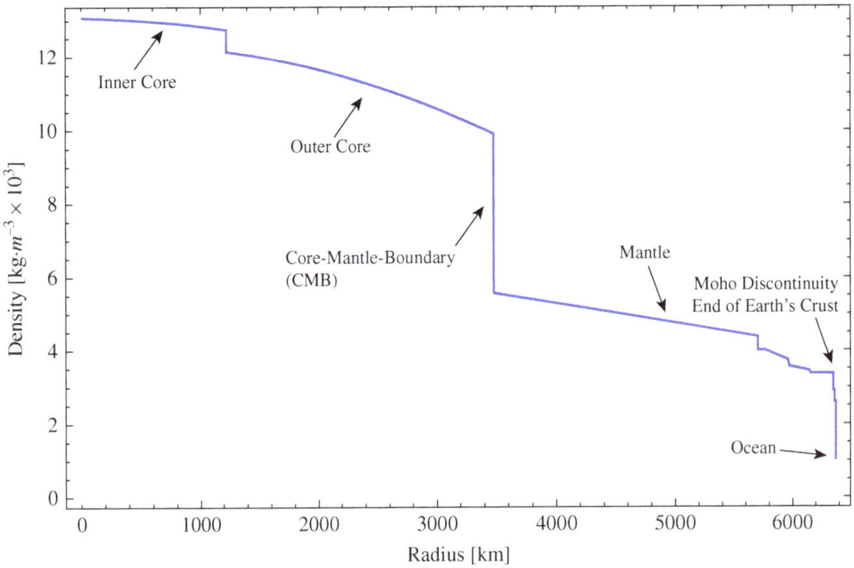

Fig. 11.3.1. Observationally estimated density profile of Earth.

The quality of the fits, examples

Very good fits to the ellipsoid are achieved with $N = 2$ and both of the following $\eta = 0.092$, $\xi = 3$, and $\eta = 0.1$, $\xi = 3.5$. The value $N = 2$ was suggested by the measured density profile shown in Fig. 11.3.1.

We tried $\eta = 0.1$ and $N = 2$, leaving only ξ to be varied. The locus is a curve that, at a small scale, resembles the geoid, the fit is perfect at the pole and at the equator. We examined the error at nine intermediate azimuths and found that a perfect fit would require ξ to vary from 0 to 3. But if we fixed $\xi = 3$, the relative error was never larger than 10^{-4}.

The conclusion is that the identification of the planetary shape with a locus of C through the pole and the equator appears to be natural and that the precise determination of the parameters applicable to each planet requires efficient use of more data. In other words, there still remains considerable flexibility to be used as more data are taken into account.

Density profile, range of N

Our own planet is unique among the planets in that the density profile has been reliably estimated, see Fig. 11.3.1. A good fit to the central core is not possible since both N and η must be positive. This can be understood since the constitution of the Earth is far from uniform; the model assumption

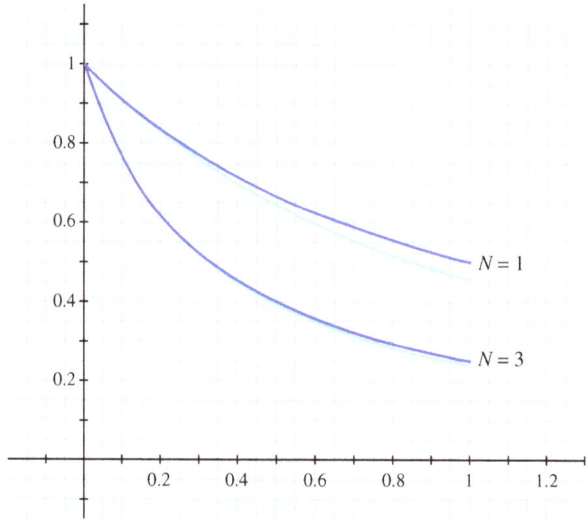

Fig. 11.3.2. Model equatorial density profile of Earth, for two values of the principal parameter N. Heavy lines $\eta = 0.01$, lighter lines, $\eta = 0.2$.

that the interior is a single phase is an over-simplification (Fig. 11.3.2). A fair approximation to the observed density suggests that N lies in the interval

$$1.5 < N < 2.5 \quad \text{(Earth)} \tag{11.3.2}$$

Rings, or not

Random sampling of the parameters of the theoretical configurations reveal that the expected, nearly spherical shape of the body is not always realized. For example, in the case that $\eta = 0.1$, $N = 1.2$ we get a good approximation to the ellipsoidal shape of the Earth with $\xi = 3.3$. But if the value of ξ is increased to 3.525, then a planetary ring appears, as shown in Fig. 11.3.3. We have crossed the ring-no-ring divider in Fig. 11.3.4, moving horizontally towards the right. For still larger values of ξ, the ring eventually dwarfs the planet.

Nothing in our model relates to the hemispheric asymmetry of Mars; that is, the depression of the northern hemisphere. But we do know the bulge ratio, $\epsilon = 1/135$, and elaborate models of the planet suggest a plausible density profile, shown in Fig. 11.3.5. We obtain excellent fits to the ellipsoid from $N = 0.7$, $\xi = 1.5$ to $N = 3$, $\xi = 1.75$.

The idea of a ring around Earth cannot be entertained, but the planet Mars is another matter. The trajectory for Mars was calculated with the equatorial bulge ratio $\epsilon = 1/135$. We fixed the value of η at 0.1 for Earth

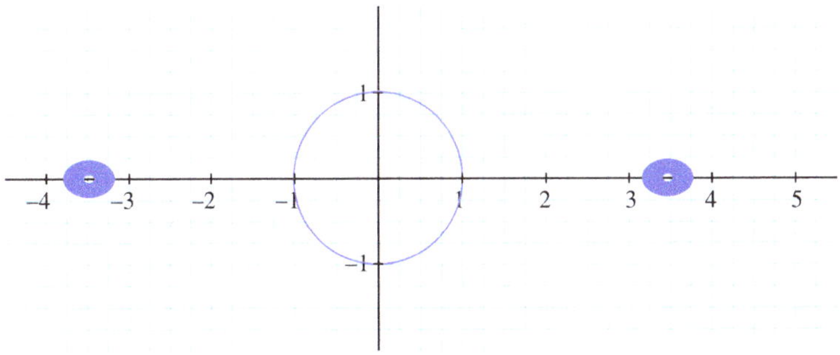

Fig. 11.3.3. Ring around the Earth; about to disappear. Parameters $N = 1.2$, $\xi = 3.525$.

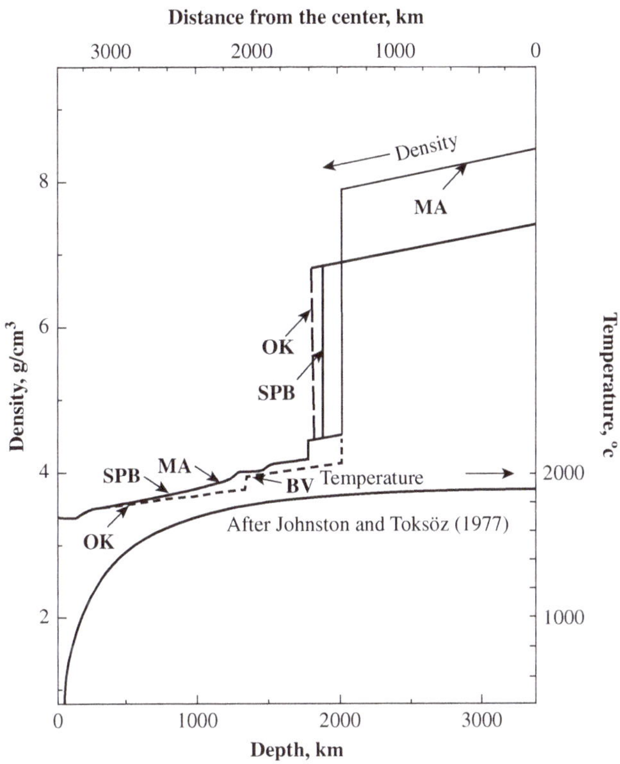

Fig. 11.3.4. Density profile of Mars, with a very thick mantle and a metallic core.

and for each of a sequence of values of N we searched for the value of ξ that gives the best approximation to the idealized shape of planet Mars. The lowest value of N for which such a fit exists is $N = 0.8$.

It is notorious that Mars shows clear evidence of having once been furrowed with large gulleys by the action of water (on the southern hemisphere).

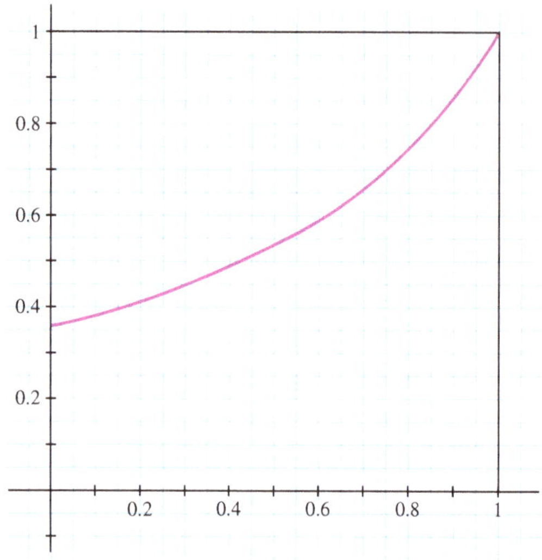

Fig. 11.3.5. Model density profile for Mars from Eq. (11.2.8), with $N = 1.7$.

It has been widely interpreted in terms of a cataclysmic event, eons past. Our calculations suggest that the value of N was once lower, that a ring actually did exist around Mars, and that the ring, consisting mostly of water or ice, disappeared as a result of the evolutionary increase in N. The alternative, that the ring may have fallen as the result of a passage very close to Earth, is less appealing, since we regard the ring as natural and expect it to resume its original shape after a shock. If it did not, then it means that the property N has evolved and that the equations of motion no longer support the ring.

11.4. Neptune and Uranus: Venus and Pluto

The shapes of Uranus and Neptune are quoted in the literature but actually they are poorly known (Bertka and Fei, 1990). If we treat them as close analogues of Earth and Mars, then they would appear far to the left as shown in Fig. 11.4.1. The progression of values of the parameter N from Neptune to Earth suggests increasing compression.

Pluto is far out among the outer planets but its small mass is a more relevant parameter (Helled *et al.*, 2010). Venus has a complicated structure and winds that are not parallel to the equatorial plane. Both are essentially spherical, but Venus has a small bulge that is believed to be induced by winds. Venus also has a very dense atmosphere with a pressure of about 92 Earth atmospheres. For all these reasons, neither Venus nor Pluto should

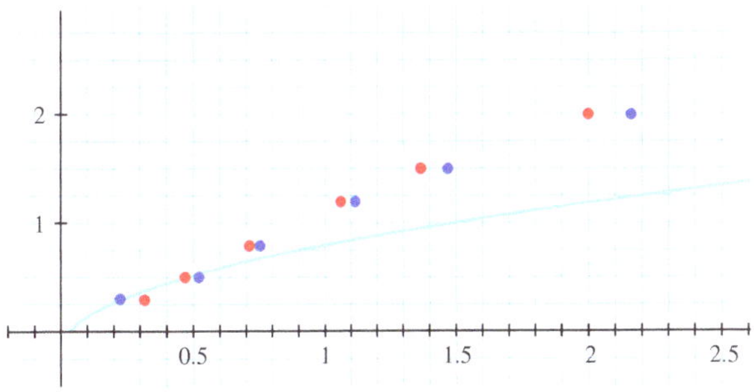

Fig. 11.4.1. This would show why Venus and Pluto do not have rings.

be expected to be suitable objects for this study. Nevertheless, we assigned very small bulges and got good fits to the ellipsoid shape.

Figure 11.4.1 shows why Venus and Pluto do not have rings. While $\eta \approx 0.1$ works well for most of the planets, Venus and Pluto require a value closer to 0.01. In Fig. 11.4.1, both are plotted with this value of η. All planets except Saturn and Jupiter have "trajectories" determined by a best fit to the ellipsoid.

11.5. Saturn, Jupiter, The Sun, Haumea, Galaxies

The search for a simple model for Saturn, initially with no expectation of accounting for anything more than the equatorial bulge, revealed that rings are a dominant feature of our model. The existence of rings gives us an additional measurable parameter, the mean radius standing in for the parameters of a complicated ring system. The radius and the width of the ring can be chosen within wide limits, but the model does not account for the flat ring system that is actually seen. The radius of the ring is closely related to the value of η.

In the case of Saturn, attempts to fit the model to ellipsoids with an equatorial bulge ratio of 1:10 failed. We did not persist in this, because: (1) Observation of the gaseous giants does not favor our model of a phase transition at the surface. (2) The surface of a gas sphere is not well defined, experimental data usually refer to isobars. See, for example, Lindal *et al.* (1985).

Saturn offers an interesting **hexagonal flow** pattern around the North pole. A photograph taken above the North pole of Saturn shows a hexagonal flow (Fig. 11.5.1).

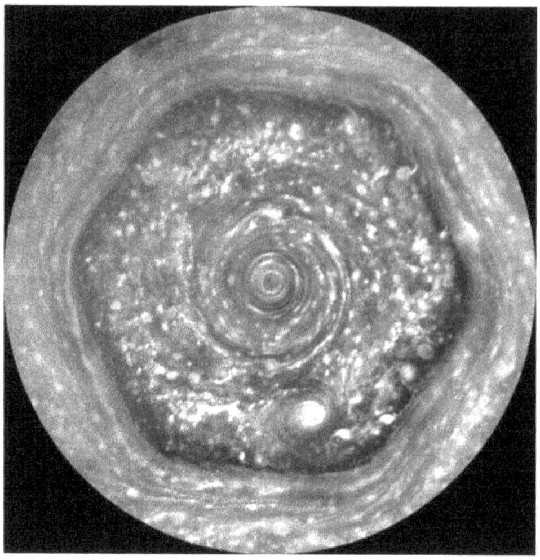

Fig. 11.5.1. Photograph taken by Voyager of the North pole of Saturn.

The photograph reveals a local, inner whorl, with a remarkable similarity to the six whorls that are present in the symmetric model.

A fair mathematical representation of the flow is

$$\Phi = \Im\left(a \ln z + \frac{A}{1+(kz)^6}\right), \quad A = \text{constant}. \tag{11.5.1}$$

The flow lines are shown in Fig. 11.5.2. The gradient of Φ is

$$\vec{\nabla}\Phi = \left(a + \frac{A(kr)^6}{D^2}((1+(kr)^{12})\cos(6\phi) + 2(kr)^6)\right) d\phi$$
$$+ \frac{A\sin(6\phi)}{D^2}(1-(kr)^{12})(kr)^5 k dr.$$

For the square, we get the surprisingly simple formula

$$(\vec{\nabla}\Phi)^2 = \frac{a^2}{r^2} + \frac{2a}{r^2}\frac{A(kr)^6}{D^2}((1+(kr)^{12})\cos(6\phi) + 2(kr)^{12}) + \frac{(kA)^2(kr)^{10}}{D^2}.$$

We have calculated only the first-order perturbation, then the modified expression for the function $\bar{f}$ in Eq. (11.2.14) becomes, for some constant α,

$$f(R,r) := \frac{(1+NR+\eta r^2)^2 - (N+1)^2}{r^2}$$
$$+ \alpha\frac{r^4}{D^2}((1+(kr)^6)\cos(6\phi) + 2(kr)^6) + \frac{\xi}{R}.$$

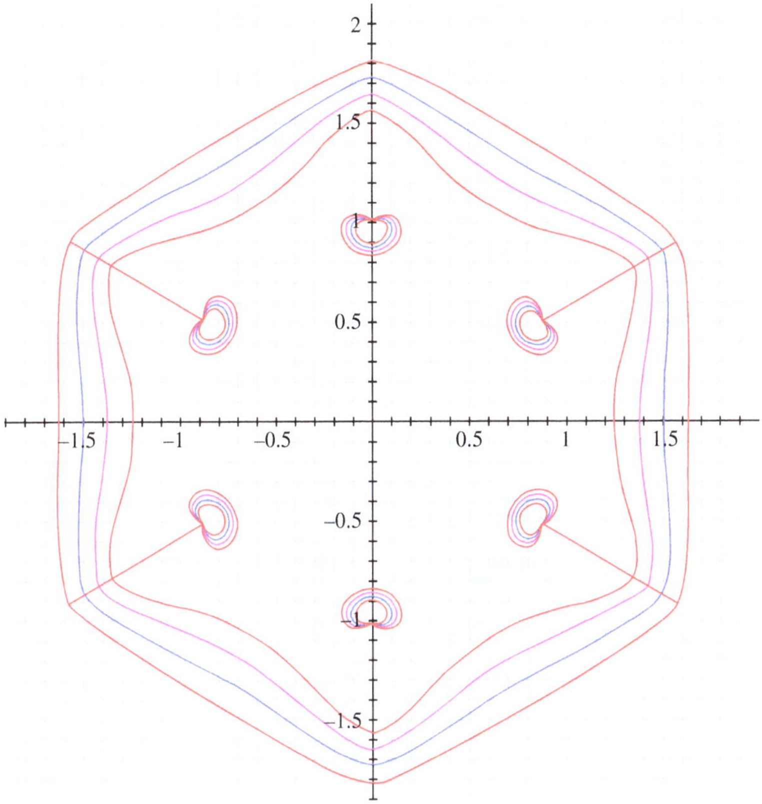

Fig. 11.5.2. Flow lines of the function (11.5.2) with $a = k = 1$, $A = 0.5$.

The result has been disappointing. It turns out to be possible to produce hexagonal rings, but no further contact with observation was discovered, so far. A more positive result: With the extra term it becomes possible to imitate the ellipsoidal shape.

Jupiter presents some of the same difficulties for analysis, but lacks the interesting hexagonal feature of Saturn. We have not constructed a model for the largest of all the planets.

The Sun is still further from our present objective, and so are galaxies. We present, however, in Fig. 11.5.3, an object that recalls, by its flatness, the shape of some galaxies.

Finally, Fig. 11.5.4 is a portrait of one of the smallest objects in the solar system, Haumea is a small moon or mini-planet in the outer Kuiper belt, remarkable for its odd shape.

Figure 11.5.5 is the real Haumea.

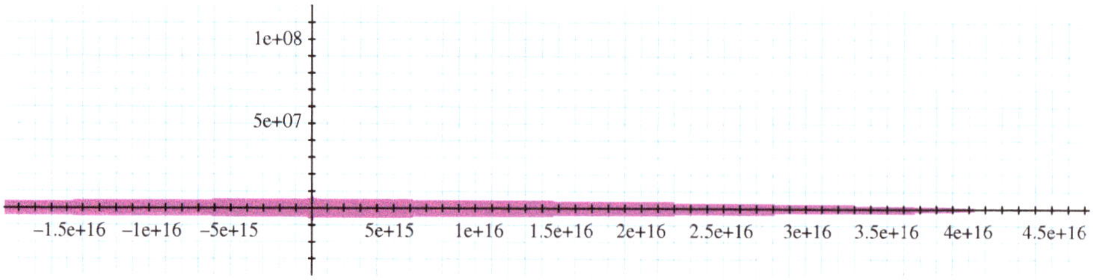

Fig. 11.5.3. A shape produced by the model, suggesting an application to galaxies. (The full figure extends equally in both directions.)

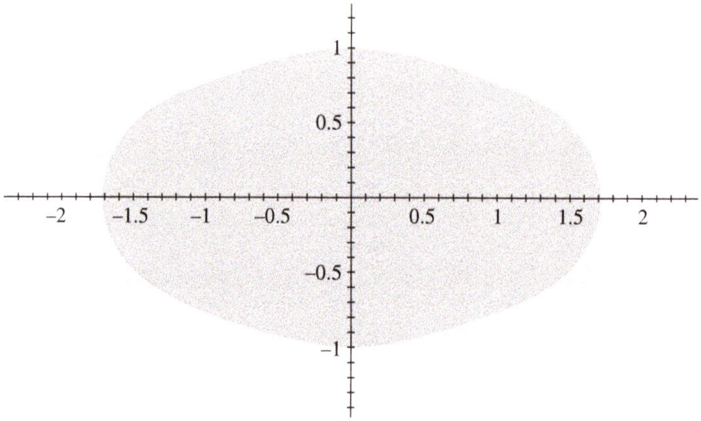

Fig. 11.5.4. A model likeness to Haumea.

11.6. Summary and Conclusions

The most significant result of this paper is that an action for hydrodynamics actually provides an effective approach to real problems. The discovery that planetary rings are natural within the formalism is a real surprise and source of encouragement. Rather than a need to invent a special historical event sequence for each ring system, we can now expect that they will emerge naturally from solutions of the equations of motion. That goes a long way towards explaining their beautiful shapes and their amazing stability.

This paper is to be regarded as a first step, to be followed by much more detailed calculations. The equatorial bulge of a planet changes the gravitational field and this affects the calculations, though often to a minor degree. For each of the objects in the solar system, it is important to consider other complications that have hardly been mentioned in this chapter, including the following.

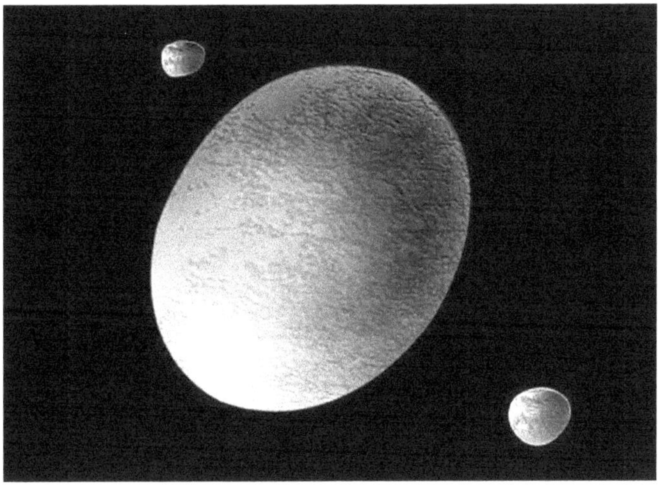

Fig. 11.5.5. A photograph of Haumea taken by Voyager.

Replace the fixed gravitational field by the solution of Poisson's equation. Include more terms in the harmonic expansion of the density, in the hope of creating more elaborate ring systems. The atmosphere was treated as empty that leaves room for improvement. For the gaseous planets, another model, without a surface discontinuity, is indicated. The effect of magnetic fields and radiation must be included.

All these complications have already been taken into account in the literature; we think that the calculations should be repeated within the framework of the action principle, where the energy is conserved and where the number of free parameters is very much reduced.

That should result in a greatly increased predictive power and, perhaps, a better understanding of the fascinating planetary rings. A separate problem can be approached in the same spirit: the shape of galaxies.

Chapter 12

Special Relativity

Einstein's theory of **General Relativity** is a theory of a dynamical metric field interacting with particles. This book exists because there is (or was) no satisfactory theory of the dynamical metric interacting with extended distributions of matter. It will be shown in Chapter 13 that any theory of interacting fields that includes the Einstein metric must be based on an **action principle**.

The non-relativistic limit of such a theory, if it can be found, is a form of hydrodynamics necessarily formulated as an action principle. One approach to the problem is therefore to study action principles for hydrodynamics, then step up, first, to a field theory that is Lorentz invariant and thereafter to one that is invariant under all coordinate transformations. We already have the theory of Lagrange, discovered by Lagrange and brought to us by Fetter and Walecka. The relativistic version was introduced in Section 10.1, where it was important to show that the appearance of the Newtonian potential in the non-relativistic Hamiltonian originates in a weak metric field. This result implies that the theory has a high degree of uniqueness. The Lorentz invariant version of irrotational hydrodynamics is based on the action

$$A_1 = \int d^4x\sqrt{-g}\,\Big(\frac{\rho}{2}(g^{\mu\nu}\psi_{,\mu}\psi_{,\nu} - c^2) - W[\rho]\Big), \tag{12.0.1}$$

with g the Lorentzian metric. The fully covariant form is exactly the same, with g the dynamical four-dimensional, pseudo-Riemannian metric. The two-step promotion of potential theory to general covariance is straightforward. This chapter is based on the paper "Relativistic thermodynamics, ..." (Fronsdal, 2016).

In Section 12.1, we shall do the same for the more general form of conservative hydrodynamics that was introduced in Chapter 10. Actually, this will be done in reverse order, for the existing literature on relativistic field theories allows us to present the relativistic version in its fully developed form; then taking the non-relativistic limit, we gain a much more complete understanding of the vector field $\vec{X}$ and its dynamical role in hydrodynamics.

To relate non-relativistic hydrodynamics to General Relativity we must pass through **Special Relativity**. All the successful relativistic field theories are based on action principles; they are theories of elementary particles, more especially, gauge theories. The very high degree of developments of this field explains the paradoxical fact that relativistic gauge theories are more familiar than non-relativistic ones.

12.1. The Relativistic, Antisymmetric Tensor Field

What is needed is a chapter from the theory of relativistic quantum gauge theories. Quantum? Yes, because a full understanding of gauge theories is reached only by quantization. That is too much for a book on thermodynamics, and for this reason, what follows can only be a sketch. The reader is encouraged to read this short summary and follow up by a study of the literature, beginning with the paper by Ogievetskiĭ and Polubarinov.

A dramatic effect of relativization is that the fields become **propagating**. In particular, propagation of the fields into empty space may be unavoidable, therefore it is prudent to keep the number of propagating modes at a minimum. And this too leads us to the antisymmetric tensor gauge field and its well-known properties. This field has at most one scalar, propagating mode. It was introduced by Ogievetskiĭ and Polubarinov (1964). In another context it is known as the B-field, or the Kalb–Ramond field. The connection to vorticity was foreseen by Lund and Regge (1976).

Let $Y = (Y_{\mu\nu})$ be an antisymmetric tensor field (a 2-form) and consider the Lagrangian density $\rho_0 dY^2$, with ρ_0 constant and[75]

$$dY^2 = \frac{c^2}{4} g^{\mu\mu'} g^{\nu\nu'} g^{\lambda\lambda'} Y_{\mu\nu,\lambda} \sum_{\text{cyclic}} Y_{\mu'\nu',\lambda'}. \tag{12.1.1}$$

This involves the 2-form field alone, there is no matter present, so far. Greek indices run over 1, 2, 3, 0, Latin indices over 1, 2, 3. The (inverse) metric tensor is Lorentzian, diagonal with $g^{11} = g^{22} = g^{33} = 1$, $g^{00} = -1/c^2$. It is

[75]Units are Gaussian, slightly modified as explained later. All components of 4-tensors have the same dimension, for example $\partial_\mu = (\vec{x}, ct)$. The fields X, Y, η are lengths and the fields $\dot{\vec{X}}, \vec{E}, \vec{B}$ are velocities.

invariant under the **gauge transformation**

$$\delta Y = d\xi.$$

In a three-dimensional notation,

$$X^i = \frac{1}{2}\epsilon^{ijk}Y_{jk}, \quad \eta_i = Y_{0i}.$$

The equation $\eta_i = \partial_0\xi_i - \partial_i\xi_0$ can always be solved for the vector field $\vec{\xi}$; there is a family of gauges in which the field $\vec{\eta}$ vanishes. In addition, the vector field $\vec{X}$ can be reduced to a gradient field. This spinless field is the only propagating mode.[76]

In flat space, with Cartesian coordinates, in terms of $\vec{X}$ and $\vec{\eta}$,

$$dY^2 = \frac{1}{2}(\dot{\vec{X}} + c\vec{\nabla}\wedge\vec{\eta})^2 - \frac{c^2}{2}(\vec{\nabla}\cdot\vec{X})^2. \qquad (12.1.2)$$

The free field equations associated with this expression for the Lagrangian density are

$$\frac{d}{dt}(\dot{\vec{X}} + c\vec{\nabla}\wedge\vec{\eta}) - c^2\vec{\nabla}\,(\vec{\nabla}\cdot\vec{X}) = 0, \quad \vec{\nabla}\wedge(\dot{\vec{X}} + c\vec{\nabla}\wedge\vec{\eta}) = 0. \qquad (12.1.3)$$

The only mode that propagates *in vacuo* is a spinless mode, but we can add sources,

$$\frac{d}{dt}(\dot{\vec{X}} + \vec{\nabla}\wedge\vec{\eta}) - c^2\vec{\nabla}(\vec{\nabla}\cdot\vec{X}) = \vec{K}, \quad \vec{\nabla}\wedge(\dot{\vec{X}} + c\vec{\nabla}\wedge\vec{\eta}) = \vec{K}', \qquad (12.1.4)$$

so that, in their presence, the field $\dot{\vec{X}}$ need not be irrotational.

The term $\rho\dot{\vec{X}}\cdot\vec{\nabla}\Phi$ in our non-relativistic Lagrangian can be made invariant under Lorentz transformations by expanding it to

$$\rho\frac{c}{2}\epsilon^{\mu\nu\lambda\rho}Y_{\mu\nu,\lambda}\psi_{,\rho}. \qquad (12.1.5)$$

Here, ψ is a Lorentz scalar. With $\rho = \rho_0$, uniform, this term is "topological"; that is, it is a boundary term and can be ignored.

Introduce the dual, $\tilde{Y}^{\mu\nu} = \epsilon^{\mu\nu\sigma\tau}Y_{\sigma\tau}$, then

$$\frac{c}{2}\epsilon^{\mu\nu\lambda\rho}Y_{\mu\nu,\lambda}\psi_{,\rho} = c^2\tilde{Y}^{\lambda\rho}_{,\lambda}\psi_{,\rho} = c(\tilde{Y}_{0i,0}\psi_{,i} + \tilde{Y}_{i0,i}\psi_{,0} - \tilde{Y}_{ij,i}\psi_{,j}),$$

$$\tilde{Y}_{i0,i} = \vec{\nabla}\cdot\vec{X}, \quad \tilde{Y}_{ij,i} = (\vec{\nabla}\wedge\vec{\eta})_j,$$

[76]For dimensional reasons, a type of density factor is required in any field theory Lagrangian. In hydrodynamics, the mass density serves naturally, in electromagnetism, the role is taken by the susceptibility of the vacuum. The densities ρ, ϵ, γt all have the same dimension. We shall take up this question when it becomes important to do so.

and

$$\frac{c}{2}\epsilon^{\mu\nu\lambda\rho}Y_{\mu\nu,\lambda}\psi_{,\rho} = \dot{\vec{X}}\cdot\vec{\nabla}\psi - (\vec{\nabla}\cdot\vec{X})\dot{\psi} + c(\vec{\nabla}\wedge\vec{\eta})\cdot\vec{\nabla}\psi. \tag{12.1.6}$$

The first term on the right is the one that appears in the non-relativistic theory.

The Lagrangian density investigated by Ogievetskiĭ and Polubarinov (1964) includes electromagnetic fields and a **mixing term**,

$$\rho_0 dY^2 + \epsilon_0 F^2 + \gamma Y F. \tag{12.1.7}$$

The last term is a 4-form, an outer product of two 2-forms,

$$\gamma Y F := \gamma(\vec{B}\cdot\vec{\eta} + \vec{E}\cdot\vec{X}), \tag{12.1.8}$$

and γ is uniform. This is not an interaction, but a mixing term; the effect is to give **mass to the photon**. The easiest way to see this is to use the transverse gauge, in which

$$\partial_\mu Y_{\mu\nu} = \partial_\mu A_\mu = 0.$$

In that case, the field equations are

$$\begin{aligned}
&\delta\vec{X}\cdot\left(\frac{d}{dt}\rho_0\big((\dot{\vec{X}} + c\vec{\nabla}\wedge\vec{\eta})\big) - \gamma\vec{E}\right)\\
&\quad + \delta\vec{\eta}\cdot\left(\vec{\nabla}\wedge\big(\rho_0(\dot{\vec{X}} + c\vec{\nabla}\wedge\vec{\eta}) - \frac{\gamma}{c}\vec{B}\right)\\
&\quad + \delta\vec{E}'(\epsilon_0\vec{E} + \gamma\vec{X}) + \delta\vec{B}\cdot(\epsilon_0\vec{B} + \gamma\vec{\eta}) + \delta AJ = 0.
\end{aligned}$$

Here, $\vec{E} = \vec{\nabla}A_0 - \dot{\vec{A}}/c$, $\vec{B} = c\vec{\nabla}\wedge\vec{A}$ and the last line transforms by an integration by parts to

$$-\delta A_0(\epsilon_0\vec{\nabla}\cdot\vec{E} + \gamma\vec{\nabla}\cdot\vec{X} - J_0) + \delta\vec{A}\cdot(\epsilon_0\dot{\vec{E}} + \gamma\dot{\vec{X}} - \vec{J}) - c\delta\vec{A}\cdot\vec{\nabla}\wedge(\epsilon_0\vec{B} + \gamma\vec{\eta}).$$

Photon mass

A formal integration over the tensor field Y shows that this theory is dual to massive electrodynamics. We wish to reach this conclusion by more elementary means. To this end, we must eliminate the tensor field from the field equations, the goal being to relate the electromagnetic potential directly to the sources. This requires that we fix the gauge. The gauge of the tensor theory has already been fixed by setting $\eta_i = Y_{i0} = 0$, $i = 1, 2, 3$.

The relevant part of the Lagrangian density is

$$\frac{\epsilon_0}{2}(\vec{E}^2 - \vec{B}^2) + \gamma(\vec{E}\cdot\vec{X} + \vec{B}\cdot\vec{\eta}).$$

Here, $\vec{E} = \dot{\vec{A}} - c\vec{\bigtriangledown}A_0$, $\vec{B} = c\vec{\bigtriangledown}\wedge\vec{A}$ and the variation is

$$\begin{aligned}(\epsilon_0\vec{E} + \gamma\vec{X})\cdot(\delta\dot{\vec{A}} - c\vec{\bigtriangledown}\delta A_0) + (-\epsilon_0\vec{B} + \gamma\vec{\eta})\cdot(c\vec{\bigtriangledown}\wedge\delta\vec{A})\\ = (-\epsilon_0\dot{\vec{E}} - \gamma\dot{\vec{X}} + c\epsilon_0\vec{\bigtriangledown}\wedge\vec{B})\cdot\delta\vec{A} + c(\epsilon_0\vec{\bigtriangledown}\cdot\vec{E} + \gamma\vec{\bigtriangledown}\cdot\vec{X})\delta A_0.\end{aligned}$$

It follows that

$$\vec{\bigtriangledown}\cdot\vec{E} + \gamma\vec{\bigtriangledown}\cdot\vec{X} = 0 \tag{12.1.9}$$

and that

$$\epsilon_0\big(-\ddot{\vec{A}} + c\vec{\bigtriangledown}\dot{A}_0 + c^2\vec{\bigtriangledown}(\vec{\bigtriangledown}\cdot\vec{A}) - c^2\Delta\vec{A}\big) - \gamma\dot{\vec{X}} = 0. \tag{12.1.10}$$

This last equation can be written

$$\epsilon_0(-\Box\vec{A} + c^2\vec{\bigtriangledown}\Lambda) = \gamma\dot{\vec{X}}. \tag{12.1.11}$$

The 4-divergence $\Lambda = \vec{\bigtriangledown}\cdot\vec{A}$ can be chosen arbitrarily, thus fixing the electromagnetic gauge. We must use the field equations, from variation of $\vec{X}$ and $\vec{\eta}$,

$$\rho_0\ddot{\vec{X}} = \gamma\vec{E}, \quad \rho_0\vec{\bigtriangledown}\wedge\dot{\vec{X}} = \gamma\vec{B},$$

to eliminate the field $\vec{X}$ from Eqs. (12.1.9)–(12.1.10). These equations tell us that, in empty space, $\dot{\vec{X}} = (\gamma/\rho_0)\vec{A}$, *modulo* a gradient. Finally, there is a gauge in which (12.1.11) reduces to

$$-\Box\vec{A} = \frac{\gamma^2}{\epsilon_0\rho_0}\dot{\vec{A}}.$$

The **photon** is thus massive, with "mass"

$$m_{ph}c^2 = \frac{\gamma\hbar}{\sqrt{\epsilon_0\rho_0}}.$$

Planck's constant has the value $\hbar = 1.05510^{-34}\,\mathrm{J\,s} = 6.5810^{-16}\,\mathrm{eV\,s}$.

In this formula, the "constants" are: Only γ is a true constant; it must be so by gauge invariance. The electromagnetic density ϵ is a constant in vacuum, but it will have a different value in a homogeneous environment with matter in it. Finally, ρ_0 is the local density of matter. All are taken to be uniform in the calculation of the photon mass.

The covariant gauge

The equations are much easier to deal with in the covariant gauge,

$$\partial^\mu Y_{\mu\nu} = 0, \quad \partial^\mu A_\mu = 0.$$

In empty space, when ρ is uniform, the equations of motion take the form

$$\rho_0 p^2 Y_{\mu\nu} + \gamma\epsilon_0 \tilde{F}_{\mu\nu} = 0, \quad \epsilon p^2 A_\nu + \rho_0\gamma p^\mu \tilde{Y}_{\mu\nu} = 0.$$

If the momentum vector is timelike, we can go to the rest system, where $p = (m, 0, 0, 0)$ and the two equations take the form

$$\rho_0 m\tilde{Y}_{ij} + \gamma_0\epsilon A_{ij} = 0, \quad \epsilon m A_{ij} + \rho_0\gamma\tilde{Y}_{ij} = 0.$$

Here, we can solve the first equation for the Y field and substitute in the second to get

$$Y_{ij} = \frac{-\gamma}{m^2}\tilde{F}_{ij},$$

and eliminate it to get

$$m^2 = \frac{\hbar^2\gamma^2}{\epsilon_0\rho_0}.$$

The effect of a photon mass is spectacular, unless it is extremely small. Many studies have been undertaken to establish an upper limit, one as low as 10^{-38} eV/c. But in the presence of important, rotating masses, the estimate may have to be revised.

Conservative hydrodynamics

Now we consider the total Lagrangian density

$$\mathcal{L} = \frac{\rho}{2}(g^{\mu\nu}\psi_{,\mu}\psi_{,\nu} - c^2) + \rho dY^2 + \kappa\rho\frac{c^2}{2}\epsilon^{\mu\nu\lambda\rho}Y_{\mu\nu,\lambda}\psi_{,\rho} + \gamma YF - f - sT \tag{12.1.12}$$

and the action $\mathcal{A} = \int d^3x dt\, \mathcal{L}$. The first term is the Lorentz invariant contribution that appears in the non-relativistic approximation as the Galilei covariant field $\rho(\dot{\Phi} - \vec{\nabla}\Phi^2/2 - \varphi)$.

The variation of the action with respect to ψ is $-\delta\psi$ times

$$\frac{d}{dt}\left(\rho\left(\frac{\dot\psi}{c^2}-\kappa\vec\nabla\cdot\vec X\right)\right)+\vec\nabla\cdot(\rho(\kappa\dot{\vec X}-\vec\nabla\Phi+\kappa\vec\nabla\wedge\vec\eta)) =: \frac{d}{dt}J^0+\vec\nabla\cdot\vec J. \tag{12.1.13}$$

In the non-relativistic limit, $\dot\psi/c^2$ is unity. The boundary conditions require that the current $\vec J$ be normal to the boundary. The last equation reduces to the equation of continuity. In a physical gauge

$$\dot\rho+\vec\nabla\cdot(\rho(\kappa\dot{\vec X}-\vec\nabla\Phi))=0.$$

This confirms that the conserved flow is

$$\rho\vec v=\rho(\kappa\dot{\vec X}-\vec\nabla\Phi).$$

The variation of the action with respect to the field $\vec X$ is

$$\int d^3x dt\,\delta\vec X\cdot\left(\frac{d}{dt}(\rho(\dot{\vec X}+\kappa\vec\nabla\psi+\vec\nabla\wedge\vec\eta))+\vec\nabla(\rho(\kappa-c^2\vec\nabla\cdot\vec X))-\gamma\vec E\right). \tag{12.1.14}$$

Setting this to zero gives the field equation

$$\frac{d}{dt}(\rho(\dot{\vec X}+\kappa\vec\nabla\Phi+\vec\nabla\wedge\vec\eta))+\vec\nabla(\rho(\kappa-c^2\vec\nabla\cdot\vec X))=\gamma\vec E. \tag{12.1.15}$$

An electric field is expected on the basis of an experiment by Tolman (1910). It may also have something to do with the anomalous Seebeck effect. But at this time, we are far from understanding all the ramifications of this interaction with the electromagnetic field. That an unexpected role may be played by the electric field in connection with vortex motion, or that an electromagnetic analogy may be glimpsed here, was suggested by Feynman in connection with liquid helium (Feynman, 1954, p. 273).

The variation of the action with respect to the field $\vec\eta$ is

$$\int d^3x dt\,\delta\vec\eta\cdot\left(\vec\nabla\wedge(\rho(\dot{\vec X}+\kappa\vec\nabla\Phi+\vec\nabla\wedge\vec\eta)-\frac{\gamma}{c}\vec B\right). \tag{12.1.16}$$

Setting this to zero gives the constraint

$$\vec\nabla\wedge(\rho(\dot{\vec X}+\kappa\vec\nabla\Phi+\vec\nabla\wedge\vec\eta)=\frac{\gamma}{c}\vec B. \tag{12.1.17}$$

In the non-relativistic context, the gauge is fixed and the field $\vec\eta$ is absent; but the constraint must be taken into account. It is this constraint, somewhat mysterious in the non-relativistic context, that reduces the number of degrees of freedom from 3 to 1.

Equations (12.2.11) and (12.2.13) are mutually consistent by virtue of Maxwell's second equation, $dF = 0$ or $\dot{\vec{B}} = c\vec{\nabla} \wedge \vec{E}$, provided that γ is constant as is required by the gauge invariance of (12.1.7). The new term is nevertheless ignorable, since it can be absorbed into $\dot{\Phi}$.

In the Lorentz invariant gauge theory, the metric is fixed, Lorentzian; the action becomes generally coordinate invariant when the fixed metric is replaced by a dynamical, pseudo-Riemannian metric field.

12.2. Non-relativistic Limit and Galilei Transformations

A concept of a **non-relativistic limit** of a relativistic field theory can be envisaged if each of the dynamical variables can be represented as a power series in $1/c$, beginning with a term of order zero; that is, $(1/c)^0$, or higher, $(1/c)^1, (1/c)^2, \ldots$. In the case of the model considered, we must assume that this is the case for the variables ρ, Y, ψ. The non-relativistic limit of the Lagrangian exists if every term is of positive or zero order. Dropping all terms of positive order, we may ask about the physical meaning of the remainder, including transformation properties.

Taking the basic variables to be ρ, Y and ψ, we find that there is one term that is visibly of order c^2, in the kinetic term (12.1.2). To overcome this obstruction, we must postulate a boundary condition

$$\vec{\nabla} \cdot \vec{X} = c^{-2}\Theta + o(c^{-3}), \tag{12.2.1}$$

with Θ of order 0 in $1/c$.

The κ term introduces another term of nominal order c^2 as is seen in Eq. (12.1.7) since $\dot{\psi} = c^2 + \dot{\Phi}$,

$$-(\vec{\nabla} \cdot \vec{X})\dot{\psi} = -\Theta + o(1/c).$$

We conclude that the existence of a non-relativistic limit of the Lagrangian depends on the validity of this equation.

We can now ask about the Galilei invariance of the bi-vector theory. The subgroup of "proper" Galilei transformations derives from Lorentz transformations of the form

$$\delta\vec{x} = t\vec{u}\gamma, \quad \delta t = (\vec{u} \cdot \vec{x}/c^2)\gamma, \quad \gamma = \frac{1}{\sqrt{1-(u/c)^2}}.$$

Infinitesimal Galilei transformations are related to first-order Lorentz transformations. It is enough to retain terms linear in $\vec{u}$, replacing γ by unity. Infinitesimal Galilei transformations are defined as the "contraction" that consists of taking the limit $c \to \infty$. But it would be imprudent to take that limit already at this stage, as we shall see. So the transformations to be considered are first-order or infinitesimal Lorentz transformations,

$$\delta\vec{x} = t\vec{u}, \quad \delta t = \vec{u}\cdot\vec{x}/c^2.$$

In what we shall call a physical gauge, the field $\vec{\eta}$ vanishes. The Lorentz group acts on the antisymmetric field in the manner that is indicated by the indices, in particular

$$\delta Y_{ij} = t\vec{u}\cdot\vec{\bigtriangledown}Y_{ij} + tu_iY_{0j} + tu_jY_{i0}, \quad \delta Y_{0j} = t\vec{u}\cdot\vec{\bigtriangledown}Y_{0j} + u_iY_{ij}/c^2.$$

Except for the first, the terms that are linear in t vanish in a physical gauge, so each component of Y_{ij} and $\vec{X}$ transforms as a three-dimensional scalar field under Galilei transformations. The last term is another matter, its presence shows that the transformed field is not in the physical gauge, since $\delta Y_{0j} \neq 0$. So we have to make a gauge transformation

$$\delta Y_{ij} = \partial_i\xi_j - \partial_j\xi_j, \quad \delta Y_{0j} = \partial_0\xi_j - \partial_j\xi_0,$$

such that

$$\partial_0\xi_j - \partial_j\xi_0 = -u_iY_{ij}/c^2.$$

The very existence of a non-relativistic limit implies that the basic variables can be represented as power series in $1/c$. So even if this transformation implies a change in Y_{ij}, that will be a change of order $1/c^2$ and we can be fairly confident that this is enough, that this change can be ignored. We shall see below that great caution is necessary and it is worthwhile to point out that if $\vec{\bigtriangledown}\cdot\vec{X}$ is of order $1/c^2$, then the change δY_{ij} will be of order $1/c^4$. To see this, take the curl of the last equation to get

$$\partial_0\vec{\bigtriangledown}\wedge\vec{\xi} = \vec{u}(\vec{\bigtriangledown}\cdot\vec{X})/c^2.$$

If, as we suppose, $\vec{\bigtriangledown}\cdot\vec{X}$ is of order $1/c^2$, then this makes ξ of order $1/c^4$. We conclude that Galilei transformations affect the field $\vec{X}$ only in the way that they affects scalar fields, by the argument shift $\delta\vec{x} = \vec{u}t$.

The reason why we are proceeding cautiously is that there are terms in the Lagrangian of apparent order c^2, see (12.1.2),

$$\rho\left(-\frac{c^2}{2}(\vec{\bigtriangledown}\cdot\vec{X})^2\right) \tag{12.2.2}$$

and (12.1.7)

$$\rho\kappa\big(-(\vec{\bigtriangledown}\cdot\vec{X})\dot{\psi}\big), \quad \dot{\psi} = c^2 + \dot{\Phi}. \tag{12.2.3}$$

Therefore, in order for the non-relativistic limit to exist, it is required that $\vec{\bigtriangledown} \cdot \vec{X}$ be a free field.

There remains a single dangerous term

$$-\rho\kappa\Theta.$$

To sum up, for the existence of the non-relativistic limit, it is necessary and sufficient that $\vec{\bigtriangledown} \cdot \vec{X}$ is of order ≥ 2 in $1/c$, and the only correction needed in the non-relativistic Lagrangian discussed earlier is the addition of $-\rho\kappa\Theta$, the field Θ defined as

$$\Theta := \lim_{c \to \infty} c^2 \vec{\bigtriangledown} \cdot \vec{X}.$$

This restores the invariance under Galilei transformations that was lost with the introduction of the κ term in the Lagrangian.

Unfortunately, this quantity Θ cannot be calculated within the non-relativistic theory. The relativistic equation of continuity, obtained by variation of ψ, yields

$$\dot{\rho}\vec{\bigtriangledown} \cdot \vec{X} = \dot{\rho} + \vec{\bigtriangledown} \cdot (\rho(\kappa\dot{\vec{X}} - \vec{\bigtriangledown}\Phi) = 0.$$

The non-relativistic conservation law must be valid up to corrections of order $1/c^2$. These corrections are encoded in the value of the field Θ and this introduces a degree of arbitrariness into the theory. The field Θ is an independent field, although its variation under Galilei transformations is related to that of $\vec{X}$, making

$$\Theta + \dot{\vec{X}} \cdot \vec{\bigtriangledown}\Phi$$

invariant. Invariance under Galilei transformations is thus formal; the choice $\Theta = 0$ characterizes a preferred Galilei frame of reference. This is an unsatisfactory result of including the κ term. We shall retain it nevertheless, for it has turned out to be essential. More work is required to settle questions that remain.

The total, non-relativistic Lagrangian density is thus

$$\rho\big(\dot{\Phi} - \Theta(1+\kappa) + \dot{\vec{X}}^2/2 + \kappa\dot{\vec{X}} \cdot \vec{\bigtriangledown}\Phi - (\vec{\bigtriangledown}\Phi)^2/2\big) - f - sT.$$

The field θ plays the same role as the B field in electromagnetism; the theory, in the non-relativistic approximation, is not manifestly **Galilei invariant**.

12.3. Dark Matter

In order that the new Conservative Hydrodynamics be generally useful, it is important that the vorticity field $\vec{X}$ have a universal coupling to matter. If that includes electromagnetism, then one effect will be to regularize the infrared. The Higgs boson will no longer be needed.

Another immediate effect will be another source of matter to the population of the universe; the new field will clearly be a contributor to Dark Matter. It would seem reasonable to expect that the greatest field strengths will be found in the neighborhood of clusters of galaxies and that they may be sufficiently strong to be recorded in experiments on gravitational lensing. A still more promising direction is the study of the Milky Way (and other galaxies) and the curious, non-Newtonian stellar orbits that are being observed.

Chapter 13

General Relativity

13.1. Astrophysics and Hydrodynamics

The original impetus for developing an action principle for hydrodynamics and thermodynamics was to prepare for a study of astrophysics, the structure of **stars and galaxies**. The relation of these fields of science to each other is less than evident. However, already in 1870 the physicist Homer Lane published a paper with the following title:

> "On the Theoretical Temperature of the Sun, under the Hypothesis of a Gaseous Mass Maintaining its Volume by its Internal Heat and Depending on the Laws of Gases as Known to **Terrestrial Experiments**" (Lane, 1870).

Half a century later, Eddington marveled at the success of an approach that treated the interior of the Sun as if it consists of an ideal gas. And even today, the best that can be done to describe the interior structure of stars is to apply the experience that has been gained by terrestrial experiments.

We are going to consider the hydrodynamics of self-gravitating fluids by means of Einstein's equations for the metric, in the context of the General Theory of Relativity. Stars and galaxies tend to rotate and of some it may be said that **rotation** is their dominant characteristic; this has been the principal difficulty for some time. It is a problem that the preceding chapter has already prepared us for, but more preparation is needed: we can approach General Relativity only after promoting our theory to a relativistic, Lagrangian field theory in the sense of the Special Theory of Relativity. This is what we are going to try to do.

This sketch of an introduction to General Relativity divides naturally into several parts:

A very short summary of the first stage of General Relativity. The elevation of the Lorentzian metric of Special Relativity to a dynamical field with matter as its source. The development of a dynamical action principle for the metric. The Bianchi identity. The search for a hydrodynamical source for the metric field. The phenomenological approach. Gravity waves. Action principle.

13.2. Introduction to General Relativity: The Metric

Special relativity was created in 1905. It arose from the demand that the laws of mechanics have the same invariance group as Maxwell's theory of electromagnetism, the Poincaré group or the inhomogeneous Lorentz group. It became necessary to formulate physical laws in a language that makes Lorentz invariance manifest, in order that the attention could be directed to physics without the need to worry about consistency with the new principles at each stage. This had the effect of introducing the **Lorentzian metric** into all the equations of fundamental physics. For example, Maxwell's field equations now take the form

$$g^{\mu\nu}\partial_\mu F_{\nu\lambda} = J_\lambda. \tag{13.2.1}$$

The prominence of the metric led Einstein to seek a larger role for it, and this resulted in interest in an **geodesic motion**. The great leap forward was the realization that Newton's equations for the motion of a particle in a gravitational field could be interpreted as geodesic motion in the special metric

$$ds^2 = \left(1 - \frac{2\varphi}{c^2}\right)(cdt)^2 - d\vec{x}^2, \tag{13.2.2}$$

where φ is the Newtonian gravitational potential. The distance between two points in space-time is

$$\int_a^b ds = \int_a^b \sqrt{-g_{\mu\nu}dx^\mu dx^\nu} = \int_a^b \sqrt{-g_{\mu\nu}\dot{x}^\mu\dot{x}^\nu}d\tau, \quad \dot{x}^\mu := dx^\mu/d\tau, \tag{13.2.3}$$

and the equation that minimizes the value of this expression is[77]

$$\ddot{x}^\mu - \Gamma^{\mu}_{\nu\lambda}{}^{\mu}\dot{x}^\nu\dot{x}^\lambda = 0, \tag{13.2.4}$$

[77]This is an oversimplification. The formula (13.2.4) is valid if the parameter τ is chosen so as to make $g_{\mu\nu}\dot{x}^\mu\dot{x}^\nu = 1$.

where the dots stand for derivatives with respect to s and Γ is the **metric connection**,

$$\Gamma^{\mu}_{\nu\lambda} = \frac{1}{2} g^{\mu\rho}(g_{\nu\rho,\lambda} + g_{\rho\lambda,\nu} - g_{\rho\lambda,\rho}). \qquad (13.2.5)$$

The proliferation of indices is frightening, but most of the coefficients of Γ are zero and a little work leads to the revelation that Eq. (13.2.4), in the non-relativistic limit with (13.2.2), is precisely the Newtonian equation of motion for a particle with unit mass in the field φ. This "geometric" interpretation of gravity has had a great appeal, but we shall take the attitude, by now prevalent, that the metric is a dynamical field, analogous to the Maxwell potential. As we shall see, this was Einstein's standpoint as well.

13.3. The Dynamical Metric

The recognition of geodesic motion was not just an interpretation of Newton's equations of motion; it was a major discovery with important experimental consequences. But the theory needed further development. In Newtonian theory, the potential was determined by solving the **Poisson equation** with the mass distribution as a source,

$$\Delta\phi = -4\pi G\rho, \qquad (13.3.1)$$

with G a universal constant. It relates the field φ to the source ρ, the matter density. The search for a suitable generalization would take 10 years of labor by Einstein, Hilbert, Poincaré and others.

The answer was found by insisting (1) that the main equations of the theory must have the same form in all space time coordinate systems and (2) that the theory must be governed by an action principle. Here is how it was done.

In practical terms, it is obvious that equations that look the same in all coordinate systems must be relations among **tensor fields**. Besides, for a physicist it is clear that the field components $g_{\mu\nu}$ must satisfy **differential equations**. The difficulty is that the derivatives of the metric do not form a tensor.

However, a "covariant derivative" is defined with the help of a connection,

$$D_\mu := \partial_\mu - \Gamma_\mu,$$

where each Γ_μ is a linear operator on the tensor algbera; the **covariant derivative** of a tensor is a tensor. This merely tells us that a connection is needed, but the remarks about geodesics strongly suggests that the metric connection (13.2.4) is to be used. This makes the metric tensor "covariantly constant",

$$D_\mu g_{\nu\lambda} = 0.$$

It is the most economical choice and it simplifies everything.

The Gaussian **curvature tensor** is

$$[D_\mu, D_\nu] = R_{\mu\nu} = R_{\mu\nu}{}^b{}_a L^a_b,$$

where the (L^b_a) are matrices, the generators of "**local Lorentz transformations**". Alternatively, if A is any co-vector field,

$$[D_\mu, D_\nu]A_\rho = (R_{\mu\nu})^\lambda_\rho A_\lambda.$$

Two contractions lead to the **curvature scalar** $R = g^{\mu\rho}(R_{\mu\nu})^\nu_\rho$ and this scalar field, by virtue of being the only candidate with required characteristics, is the Lagrangian density for the metric field.

Variations of the action

$$A_{\text{metric}} = \int d^4x\sqrt{-g}R \tag{13.3.2}$$

with respect to the metric,

$$\delta A_{\text{metric}} = \frac{1}{2}\int d^4x\sqrt{-g}\,\delta g^{\mu\nu}G_{\mu\nu}, \tag{13.3.3}$$

define the **Einstein tensor**

$$G_{\mu\nu} = 2\frac{\delta R}{\delta g^{\mu\nu}} - g_{\mu\nu}R$$

and give a unique field equation

$$G_{\mu\nu} = 0. \quad \textbf{Einstein's equation}\ \textit{in vacuo}. \tag{13.3.4}$$

It is a set of second-order differential equations for the components of the metric, an elaborate generalization of Eq. (13.3.1). It makes no reference to any other fields or sources; it characterizes the metric field in a space-time that is empty (we shall refer to it as a "**vacuum**") except for the metric itself.

The discovery of field equations for the gravitational metric ***in empty space*** was a milestone in the development of a General Theory of Gravitation. It has received a direct confirmation only recently, with the discovery of traveling **gravitational waves**. But it is not yet a theory of gravitation. Just as the mass density ρ appears as a source for the potential in Newton's theory, we need to add sources to the right-hand side of (13.3.4). This hydrodynamical source has remained unknown for 100 years.

We shall take great pains to justify this last statement. The difficulty is related to the **Bianchi identity**.

13.4. The Bianchi Identity

Theorem. *The Einstein tensor, defined by (13.3.3), satisfies the following equation:*

$$D_\nu G_\mu{}^\nu = 0, \quad \textit{Bianchi identity} \tag{13.4.1}$$

identically.

Proof. The action (13.3.2) is invariant under infinitesimal coordinate transformations,

$$\delta x^\mu = \xi^\mu,$$

under which

$$\delta g_{\mu\nu} = D_\mu \xi_\nu + D_\nu \xi_\mu.$$

Hence,

$$\delta A_{\text{metric}} = 2\int d^4x \sqrt{-g}(D_\mu \xi_\nu) G^{\mu\nu} \tag{13.4.2}$$

is identically zero. An integration by parts gives

$$\int d^4x \sqrt{-g}` \, \xi_\nu D_\mu G^{\mu\nu} = 0.$$

Since the vector field ξ is arbitrary, this implies that the statement (13.4.1) is true.

Implications of the Bianchi identity

Let us add a source to Einstein's equation (13.3.4):

$$G_{\mu\nu} = T_{\mu\nu}. \quad \textit{Einstein's Equation cum fontibus.} \tag{13.4.3}$$

Applying the covariant derivative, we find

$$D_\nu G^{\mu\nu} = D_\nu T^{\mu\nu} = 0. \tag{13.4.4}$$

The left side is identically zero; therefore an inevitable consequence is that *there can be no solution of Einstein's equation for any source unless*

$$D_\mu T^{\mu\nu} = 0. \quad \textit{The } \textbf{Bianchi constraint}. \tag{13.4.5}$$

Equation (13.4.2) is a condition of integrability. This condition was received with great enthusiasm and led to the first examples of interactive gravitation theory, among the most important, a theory of the metric tensor field interacting with the vector field of electromagnetism, Einstein–Maxwell theory. We shall see how it came about.

This is where Emmy Noether entered history. We have seen, in Sections 3.8 that, if an action is constructed from scalar fields $\phi, \psi, \ldots$ alone, then there is a tensor field

$$T_\mu{}^\nu = \sum_\phi \phi_\mu \frac{\partial \mathcal{L}}{\partial \phi_\nu} - \delta_\mu^\nu \mathcal{L} \tag{13.4.6}$$

that satisfies a "conservation law" that resembles (13.4.2), with ∂_μ instead of D_μ.

Theorem (Noether). *Consider a theory of scalar fields defined by an action*

$$A_{\text{matter}}[g, \phi, \psi, \ldots] = \int d^4x \sqrt{-g}\mathcal{L},$$

in which the tensor $g = (g_{\mu\nu})$ *is a* **`Riemannian metric tensor`**. *Suppose further that the action is invariant under infinitesimal coordinate transformations; that is, under infinitesimal variations with*

$$\delta g_{\mu\nu} = D_\mu \xi_\nu + D_\nu \xi_\mu, \quad \delta\phi = \xi\phi, \ldots. \tag{13.4.7}$$

Finally, suppose that the fields $\phi, \psi, \ldots$ *satisfy the Euler–Lagrange equations of the action* A_{matter}. *Then the tensor field with components*

$$T_\mu{}^\nu = \sum_\phi \phi_\mu \frac{\partial \mathcal{L}}{\partial \phi_\mu} - \delta_\mu^\nu \mathcal{L}$$

satisfies the divergence condition (13.4.5).[78]

Sketch of proof. The proof of Noether's second theorem in Section 3.8 applies with some changes. The components of the metric may depend on the coordinates, but this can be allowed for by interpreting the variations (13.4.4) of the fields in terms of covariant derivatives. Partial integration then brings in the covariant derivative of the metric that, as we have seen, vanishes. The essential points are that the covariant derivative acts as a derivation of the tensor algebra and that it allows to carry through the partial integrations as usual.

We omit details of the proof, but verify the statement in the special case of the relativistic action associated with Lagrange's theory, when

$$A_{\text{matter}} = \int d^4x \sqrt{-g} \left(\frac{\rho}{2} (g^{\mu\nu} \psi_\mu \psi_\nu - c^2) - f - sT \right). \tag{13.4.8}$$

In this case,

$$2 \frac{\partial \sqrt{-g} A_{\text{matter}}}{\partial g_{\mu\nu}} = \rho \psi_\mu \psi_\nu - g_{\mu\nu} \mathcal{L}, \tag{13.4.9}$$

[78]Essentially, this tensor is the same as the tensor defined by (13.4.2). This is a slight oversimplification, the two tensors are not always identical, but both are conserved; the symmetric one is the one that must be used.

which is the Noetherian energy–momentum tensor. In the general case of the complete, conservative hydrodynamics, the calculation is more complicated due to the appearance of tensor field variables.

This result points to a theory that includes Einstein's metric and Maxwell's electromagnetic fields with mutual interaction. Later the inclusion of Dirac's electronic field was worked out by the same principles. The question that inevitably arose was this: How can we add hydrodynamics to this family of interacting theories? A preliminary answer was soon found. □

Particles as sources: Gravity waves

Formulating an action principle for a metric in interaction with a number of particles has been attempted in connection with the recent observation of Gravity Waves (LIGO). A finite number of point particles, intended as models of "Black Holes", are moving in a metric field. The matter action is a sum of contributions, one from each particle,

$$\begin{aligned} A &= \int d^4x\sqrt{-g}\mathcal{L}_{\text{particles}} \\ &= \sum_a \int d\tau_a \int d^4x\sqrt{-g}\sqrt{g_{\mu\nu}\dot{x}_a^\mu(\tau_a)\dot{x}_a^\nu(\tau_a)}\delta^4(x-x_a). \end{aligned}$$

Each term is the familiar **geodesic action**. The inclusion of the factor $\sqrt{-g}$ may be unexpected. The dynamical variables are the four-dimensional positions x_a^μ of the objects, following paths in 4-space, each parameterized by its own "proper time" τ_a. So far, when this action is considered on its own, the metric is fixed, not yet a dynamical variable.

We have two ways to obtain the energy–momentum tensor. The result is, fortunately, exactly the same. If we make the usual choice of the parameters, the proper times, then each of the big square roots is equal to 1 and Noether's formula gives

$$T^{\mu\nu} = \sum_a \left(\dot{x}_a^\nu\dot{x}_a^\mu - \frac{1}{2}g^{\mu\nu}\mathcal{L}_{\text{particles}}\right)\delta^4(x-x_a). \tag{13.4.10}$$

This is very similar to (13.4.7). The second term comes from the factor $\sqrt{-g}$. The same result is obtained by variation of the metric.

13.5. Sources for G.R. Tolman's Formula

Following the main ideas, a source for hydrodynamics is provided by the action (13.4.8); it was proposed some time ago and it was used for a preliminary study of the internal constitution of stationary stars (Fronsdal, 2007).

But to stay closer to the historical sequence, we return to 1934. By this time it must have been clear that the success of Einstein–Maxwell theory would have little impact on the problem of the effect of gravity on **extended mass distributions**. No action principle that could serve as a basis for relativistic hydrodynamics had been found, and in 1934 Richard Tolman proposed, *faute de mieux*, a phenomenological formula for the right-hand side of Einstein's equation,

$$T_{\mu\nu} = (\rho + p/c^2)U_\mu U_\nu - g_{\mu\nu}.$$

It was based on Tolman's review of **Cauchy's approach** to hydrodynamics.[79] The vector field U was interpreted as 4-velocity, reflecting the usual practice of upgrading all three-dimensional velocity fields to 4-vectors. To agree with experience in particle theory, the length was restricted,

$$g_{\mu\nu}U^\mu U^\nu = c^2,$$

imposed as a constraint. One effect is to make this theory incompatible with the **equation of continuity**.

But this critique of Tolman's formula is actually irrelevant; What makes us determined to eschew Tolman's approach is that it fails to take into account the Bianchi constraint.

The question still remains: what is the source that would define relativistic hydrodynamics?

13.6. Conservative Sources for Relativity

We have already stressed the importance of action principles in physics. In the context under discussion, an action principle has been found that gives a highly satisfactory answer to a very important problem: to find a source for Einstein's equation that respects the **integrability condition** (13.4.2).

[79]The term p/c^2 appears to be an error, made possible because Tolman followed the practice of setting the speed of light equal to unity. The combination $\rho + p/c^2$ is unnatural and certainly not justifiable in a formula that was wholly inspired by non-relativistic physics; the natural combination $h + p$, where h is the Hamiltonian density, is the enthalpy density. How Tolman was able to make this capital mistake can be observed in his book, comparing with Section III.8 of this one. But how it is possible that it was never found out, that is incomprehensible; he confused an energy density with a mass density. One physicist who did not is Chandrasekhar. The confusion between energy density and mass density and ensuing conflicts are evident in a famous paper by Oppenheimer and Volkov (1939).

Unfortunately, for more than 80 years a consensus has been maintained, to the effect that the problem has no solution. For a large part of this period, countless papers have been published that have acted upon Tolman's suggestion. In the early part of this period, there were those who hoped for, and believed in, a better solution to the problem. Today, the idea has lost all support, being almost universally regarded as impossible. But this is a defeatist position and it has left General Relativity in a deplorable state.

What is it that we are trying to do? We have in hand a theory of the metric field and several theories of matter, more specifically hydrodynamics. We wish to couple them to each other. That is our problem, but many do not see it that way. The attitude that seems to prevail is that Einstein's equation is a point of departure, subject to be developed and improved on by adding additional terms to it. But that ignores the fact that General Relativity has to coexist with particle theory, the Standard Model, as we know after the discovery of gravitational waves (LIGO, 2016). Note that the "variables" ρ, p and U are just letters, with no *a priori* properties, they are not the variables of a theory or model. All that we can know about them is the information that comes from the gravitational field equation $G = T$ itself. This information is very unlikely to constitute a theory.

To demonstrate this, suppose that we have a theory that passes for hydrodynamics, a theory with variables ρ, p and U, and interpret the corresponding letters in Tolman's formula accordingly. If the Bianchi constraint is not satisfied on shell, then the gravitational field equation selects a subset of the solutions of that theory. We know from experience, most clearly from quantum theory, that this is likely to make the theory untenable. In quantum mechanics, it leads to a violation of **`unitarity`**; in our present situation, to a loss of **`mass conservation`**.

To end this protest against the *status quo*, let us raise a point that is the most compelling reason for persisting in respecting the imperative conditions of integrability; it is this: when we compromise, we lose the hope of making predictions. It is often pointed out, in defense of Tolman's formula, that it can easily be generalized, and when carefully considered, this is the strongest objection that can be raised against it.

Bibliography

Allen, J.F. and Misener, A.D. (1939). The properties of flow of liquid He II, *Proc. Roy. Soc. Lond. Math. Phys. Sci. A* **172**, 467.

Amontons, G. (1699). Moyen de substituer commodement l'action du feu a la force des hommes et des chevaux, *Mém. Acad. Royale Sci.* **112**; De la resistance causé dans les machines, *Mém. Acad. Royale Sci.* **206**.

Andereck, C.D., Dickmann, R. and Swinney, H.L. (1983). New flows in a circular Couette system with co-rotating cylinders, *Phys. Fluids* **26**, 1395.

Andereck, C.D., Liu, S.S. and Swinney, H.L. (1986). Flow regimes in a circular Couette system with independently rotating cylinders, *J. Fluid Mech.* **164**, 155.

Anderson, P.W. (1966). Considerations of the flow of superfluid helium, *Rev. Mod. Phys.* **38**, 298.

Andronikashvili, E.L. (1946). Direct observation of two types of motion in Helium II, *Zh. Eksp. Teor. Fiz.* **16**, 780.

Andronikashvili, É.L. and Mamaladze, Yu.G. (1966). Quantization of macroscopic motions and hydrodynamics of rotating Helium II, *Rev. Mod. Phys.* **38**, 567.

Arnowit, C., Deser, S. and Misner, C.W. (1962). The dynamics of general relativity. In: Witten, L. (ed.), *Gravitation: An Introduction to Current Research.* Wiley: New York. Reprinted from arXiv:gr-qc/0405109.

Baez, J.C. (2012). https:johncarlobaez.wordpress.com.

Bardeen, J.M. (1970). A variational principle for rotating stars in general relativity, *Astrophys. J.* **162**, 71.

Barnett, S.J. (1915). Magnetization by rotation, *Phys. Rev.* **6**, 239.

Batchelor, G.K. (1953). *The Theory of Homogeneous Turbulence.* Cambridge University Press.

Beauvalet, L., Lainey, V., Arlot, J.-E. and Binzel, R.P. (2012). Dynamical parameter determinations in Pluto's system? Expected constraints from the New Horizons mission to Pluto, *Astron. Astrophys.* **540**, A65.

Berdichevsky, V.L. (1970). On variational principles, *Astrophys. J.* **162**, 7.

Bernoulli, D. (1738). *Hydrodynamica*, Dulsecker, Argentoratum.

Biot, J.B. (1802). Sur la théorie du son, *J. Phys. Chim. Wist. Nat.* **55**, 173.

Birkhoff, G. (1950). *Hydrodynamics*, Dover, New York.

Bois, P.-A. and Kubicki, A. (2000). Double diffusive aspects of the convection in moist-saturated air. In: *Continuum Thermomechanics. Solid Mechanics and Its Applications*, Vol. 76, Springer.

Boltzmann, L. (1896). *Gastheorie*, Leipzig, Barth.

Boltzmann, L. (1909). *Wissenschaftliche Abhandlungen*, Vol. 2, Hasenoehrl, F. (ed.), Leipzig.

Bordoni, S. (2014). J.J. Thompson and Duhem's Lagrangian approaches to thermodynamics, *Entropy* **16**, 5876.

Bowler, J.R. and Johnson, E.A. (1985). Anomalous dispersion in disparate-mass gas mixtures, *Phys. Rev. Lett.* **54**, 329–332.

Boyle, R. (1660). *Collected Works*, London.

Brenner, H. (2013). Proposal of a critical test of the Navier–Stokes–Fourier paradigm for compressible fluid continua, *Phys. Rev. E* **87**, 013014.

Brooks, J.S. and Donnelly, R.J. (1977). The calculated thermodynamic properties of superfluid helium-4? *J. Phys. Chem. Ref. Data* **6**, 51.

Brush, S.G. (1976). *The Kind of Motion We Call Heat*, North-Holland.

Callen, H.B. (1960). *Thermodynamics*, Wiley, New York.

Campbell, L.J. and Krasnov, Yu. K. (1982). Transient behavior of rotating superfluid helium, *J. Low Temp. Phys.* **49**, 377.

Carathéodory, C. (1909). Untersuchungen über die Grundlagen der Thermodynamik, *Math. Ann.* **67**, 355.

Carnot, S. (1824). *Réflections sur la Puissance Motrice du Feu et sur les Machines Propres á développer cette Puissance*, Bachelier, Paris.

Chandrasekhar, S. (1939). *An Introduction to the Study of Stellar Structure*, University of Chicago Press.

Chueh, P.I. and Prausnitz, J.M. (1967). Vapor–liquid equilibria at high pressures: Calculation of critical temperatures, volumes, and pressures of nonpolar mixtures, *AICHE J.* **13**, 1107.

Clark, A.M., Din, F. and Robb, J. (1954). The liquid–vapour equilibrium of the binary system argon/oxygen, *Proc. R. Soc. of Lond. A Math. Phys. Sci.* **221**, 517.

Clausius, R. (1864). *Abhandlungen Über die Mechanische Warmetheorie*, Friedrich Vieweg und Sohn, Braunschweig.

Cox, J.P. and Giuli, R.T. (1968). *Principles of Stellar Structure*, Gordon and Breach.

Couette, M. (1887). Oscillations tournantes d'un solide de révolution en contact avec un fluide visqueux, *C. R. Acad. Sci. Paris* **105**, 1064.

Couette, M. (1888a). Sur un nouvel appareil pour l'étude du frottement des fluides, *C. R. Acad. Sci. Paris* **107**, 388.

Couette M. (1888b). La viscosité des liquides, *Bull. Sci. Phys.* **4**, 40, 123, 262.

Couette, M. (1890a). Distinction de deux régimes dans le mouvement des fluides, *J. Phys. [Ser. 2]* **IX**, 414.

Couette, M. (1890b). Etudes sur le frottement des liquides, *Ann. Chim. Phys. [Ser. 6]* **21**, 433.

Dalton, J. (1802). Experimental essays on the constitution of mixed gases: on the force of steam or vapor from water and other liquids, both in a Torricellian vacuum and in air; on everporation; and on expansion of gases by heart, *Mem. Proc. Manchester Mem. Lit. Phil. Soc.* **5**, 535.

Dalton, J. (1808). *A New System of Chemical Philosophy*, Manchester.

Donnelly, R.J., Herman, R. and Prigogine, I. (1965). *Non-equilibrium Thermodynamics, Variational Techniques and Stability*, University of Chicago.

De Groot, R. and Mazur, P. (1962). *Non-equilibrium Thermodynamics*, North-Holland, Amsterdam.

Drazin, P.G. and Reid, W.H. (1981). *Hydrodynamic Stability*, Cambridge University Press.

Eddington, A.S. (1926). *The Internal Constitution of Stars*, Dover, New York.

Emden, R. (1907). *Gaskugeln*, Leipzig, B.G. Teubner.

Euler, H. (1936). Über die Streuung von Licht an Licht nach der Diracschen Theorie, *Ann. Phys.* **26**, 398.

Fernández de la Mora, J. and Puri, A. (1986). Two-fluid Euler theory of sound dispersion in gas mixtures of disparate masses, *J. Fluid Mech.* **168**, 369–382.

Fetter, A.L. (2009). Rotating trapped Bose–Einstein condensates, *Rev. Mod. Phys.* **81**, 647.

Fetter, A.L. and Walecka, J.D. (1980). *Theoretical Mechanics of Particles and Continua*, McGraw-Hill.

Fitts, D.D. (1962). *Nonequilibrium Thermodynamics*, McGraw-Hill, New York.

Fitzpatrick, R. (2018). Newtonian mechanics, *Lecture Notes*, University of Texas, Available on Prof. Fitzpatrick's webpage.

Fomenko, A.T. (1989). *The Plateau Problem: Historical Survey.* Gordon & Breach, Williston, VT.

Fourier, J.B.J. (1822). *Théorie Analytique de la Chaleur*, Didot, Paris.

Fronsdal, C. (2007). Ideal stars in General Relativity, *Gen. Relativ. Grav.* **39**, 1971–2000.

Fronsdal, C. (2011). Heat and gravitation. The action principle, *Entropy* **16**(3), 1515–1546.

Fronsdal, C. (2014). Action principle for hydrodynamics and thermodynamics including general, rotational flows, preprint, arXiv:1405.7138v3[physics.gen-ph].

Fronsdal, C. (2016). Relativistic thermodynamics, a Lagrangian field theory for general flows including rotation, *Int. J. Geom. Math. Methods Phys.* **14**, 1750017.

Fronsdal, C. (2017a). Action principles for thermodynamics, review and prospects, *Phys. Elem. Part. and Nucl.* (JINR).

Fronsdal, C. (2017b). *Memorial Volume on Abdus Salam's 90th Birthday*, Lars Brink, Michael Du and Kok Khoo Phua (eds.), (World Scientific Publishing), ISBN: 978-981-3144-86-6.

Fronsdal, C. (2018a). A classic mistake and what it tells us, preprint.

Fronsdal, C. (2018b). Analysis of cylindrical Couette flow, revision of previous entry, see fronsdal.physics.ucla edu.

Fronsdal, C. (2019a). Rotating planets in Newtonian gravity, preprint, arXiv:1803.09625 [physics.gen-ph].

Fronsdal,C. (2019b). Noetherian field theories for General Relativity, arXive.

Fronsdal,C. (2020a). Inside the Menususcus (in progress).

Fronsdal,C. (2020b) Metastable liquids, capillary action and superuids, see fronsdal.physics.ucla edu.

Galilei G. (1638). *Discorsi e Dimostrazioni Mathimatiche.* Leidon, English translation by Dover Publication in 1914 and 1952.

Gay-Lussac, J.L. (1802). The expansion of gases by heat, *Ann. Chim.* **43**, 137.

Gibbs, J.W. (1878). On the equilibrium of heterogeneous substances, *Trans. Conn. Acad.*

Glavatskiy, K. (2009). Multicomponent interfacial transport as described by the square gradient model; evaporation and condensation, Thesis, NTNU, Trondheim.

Graeff, R.W. (2007). Viewing the controversy Loschmidt–Boltzmann/Maxwell through macroscopic measurements of the temperature gradients in vertical columnsof water, preprint. Additional results are on the webpage: "firstgravitymachine.com".

Hall, H.E. (1954). The inertia of heat flow in liquid helium II, *Proc. Phys. Soc. Sect. A* **67**, 485.

Hall, M.B. (1965). *Robert Boyle on Natural Philosophy: An Essay with Sections from this Writings*, Indiana University Press, Bloomington.

Hall, H.E. and Vinen, W.F. (1956). The rotation of liquid helium II. The theory of mutual friction in uniformly rotating helium II, *Proc. R. Soc. Lond. A Math. Phys. Sci.* **238**, 215.

Hartle, J.B. (1967). Slowly rotating relativistic stars I. Equations of structure, *Astrophys. J.* **150**, 1005.
Helled, R., Anderson, J.D. and Schubert, G. (2010). Uranus and Neptune: Shape and rotation, *Icarus* **210**, 446.
Helmholtz, H. (1883). Die Thermodynamik chemischer Vorgänte, In: *Wissenschaftliche Abhandlungen*, Vol. 2, Barth, Germany, pp. 958–978.
Heyerdahl, T. (1987). *The Maldive Mystery*, Ballantine Books.
Hicks, C.P. and Young, C.L. (1975). The gas–liquid critical properties of binary mixtures, *Chem. Rev.* **75**, 119.
Holman, J.P. (1974). *Thermodynamics*, McGraw-Hill.
Hooke, R. (1678). *Lectures De Potentia Restitutiva, or, of Spring*, John Martyn.
Johnston, D.H. and Toksöz, M.N. (1977). Internal structure and properties of Mars, *Icarus* **32**, 73.
Joseph, D.D. (1976). *Stability of Fluid Motions*, Vols. 1 and 2, Springer-Verlag.
Joule, J.P. (1857). Remarks on the heat and constitution of elastic fluids, *Phil. Mag.* **4**, 211.
Kalb, M. and Ramond, P. (1974). Classical direct interstring action, *Phys. Rev. D* **9**, 2273.
Kapitza, P. (1938). Viscocity of liquid helium below the λ point, *Nature* **141**.
Kaufman, A.N. (1984). Dissipative Hamiltonian systems: A unifying principle, *Phys. Lett. A* **100**, 419.
Kell, G.S. (1983). Early observations of negative pressures in liquids, *Am. J. Phys.* **51**, 1038.
Khalatnikov, I.M. (1965). *An Introduction to the Theory of Superfluidity*, Benjamin.
Knobler, C.M.P., Van Heijningen, R.J.J. and Beenakker, J.J.M. (1961). *Physica*, **27**, 296.
Kolb, S.M., Stute, M., Kley, W. and Mignone, A. (2013). Radiation hydrodynamics integrated in the PLUTO code, *Astron. Astrophys.* **559**, A80.
Lagrange, J.M. (1760–1761). *Miscellanea Taurinensia, II.* Oeuvres, Paris, 1867, 1892.
Lamb, H. (1916). *Hydrodynamics*, Cambridge University Press.
Landau, L. and Lifshitz, E. (1959). *Fluid Mechanics*, Pergamon Books.
Landau, L. and Lifshitz, E. (1994). *Physique théorique. Tome 6: Mécanique des fluides*, Editions Mir.
Lane, H.J. (1870). On the theoretical temperature of the Sun, under the hypothesis of a gaseous mass maintaining its volume by its internal heat, and depending on the laws of gases as known to terrestrial experiment, *Amer. J. Sci. Arts Ser.* 2, **4**, 57.
Laplace, P.S. (1825). *Traité de Mécanique Céleste*, Tome V, Duprat, Paris.
Lemmon, E.W. and Jacobsen, R.T. (1999). A generalized model for the thermodynamic properties of mixtures, *Int. J. Thermophys.* **20**, 825–835.
Lemmon, E.W., Jacobsen, R.T., Penoncello, S.G. and Friend, D.G. (2000). Thermodynamic properties of air and mixtures of nitrogen, argon, and oxygen from 60 to 2000 K at pressures to 200 MPa, *J. Phys. Chem. Ref. Data* **29**, 331.
Lewis, G.N. and Randall, M. (1936). *Thermodynamics*, McGraw-Hill, New York.
Liao, C. (2009). Temperature gradient caused by gravitation, *Int. J. Mod. Phys. B* **23**, 4685.
Lieb, E.H. and Yngvason, J. (2013). The entropy concept for non-equilibrium states, preprint, arXiv: 1305.3912.
Lin, C.C. (1955/1981). *The Theory of Hydrodynamic Stability*, Cambridge University Press; Academic Press.
Lindal, G.F., Sweetnam, D.N. and Esleman, V.R. (1985). The atmosphere of Saturn: An analysis of the voyager radio occultation measurements, *Astronom. J.* **90**, 1136.
Linus, F. (1661). *Tractatus de Corporum Inseparabilitate*, London.

Löfqvist, T. and Delsing, J. (2003). Speed of sound measurements in gas-mixtures at varying composition using an ultrasonic gas flow meter with silicon based transducers, *Eleventh IMEKO TC9 Conf. FLOMEKO-2000.*

London, F. (1938). The λ-phenomenon of liquid helium and the Bose–Einstein degeneracy, *Nature* **141**, 643. Reprinted in *Superfluids I, II*, (1964) Dover, New York.

Loschmidt, J. (1865). Zur Größe der Luftmoleküle, *Sitzungsber. Kais. Akad. Wiss. Wien* **52**, 395.

Lund, F. and Regge, T. (1976). Unified approach to strings and vortices with soliton solutions, *Phys. Rev. D.* **14**, 1524.

Mallock, A. (1888). Determination of the viscosity of water, *Proc. R. Soc. A*, **45**, 126.

Mallock, A. (1896). III. Experiments on fluid viscosity, *Philos. Trans. R. Soc.* **187**, 41.

Massieu, F. (1876). Thermodynamique: Mémoire sur les fonctions caractéristiques des divers fluides et sur la théorie des vapeurs, Académie des Sciences de L'Institut National de France, 92 pp.

Maxwell, J.C. (1868). XXII. On the dynamical theory of gases, *Lond. Edinb. Dubl. Philos. Mag. J. Sci.* **35**, 185–217.

McKenna, J. and Platzman, P.M. (1962). Nonlinear interaction of light in vacuum, *Phys. Rev.* **129**, 2354.

Medeov, M. and Skyba, P. (2002). A NMR spectrometer for superfluid helium-3 study at very low temperature, *Czech. J. Phys.* **52**, 149.

Meixner, J. (1970). On the foundation of thermodynamics of processes. In: *A Critical Review of Thermodynamics*, Mono Book Corp., Baltimore, pp. 37–47.

Milone, E.F. and Wilson, J.F. (2008). *Solar System Astrophysics*, Springer.

Misner, W., Thorne, K.S. and Wheeler, J.A. (1972). *Gravitation*, W.H. Freeman.

Modell, M. and Reid, R.C. (1974). *Thermodynamics and Its Applications*, Prentice-Hall.

Müller, I. (2007). *A History of Thermodynamics*, Springer, Berlin.

Müller, I. and Mueller W.H. (2009). *Fundamentals of Thermodynamics and Applications*, Springer.

Nernst, W. (1864–1941). *Die theoretischen und experimentellen grundlagen des neuen wärmesatzes.* Translation by Dover Publications, New York, 1969.

Newton, I. (1730). *Opticks: Or, A Treatise of the Reflections, Refractions, Inflections and Colours of Light*, William Innys at the West-End of St. Paul's, London.

Noether, E. (1971). Invariant variation problems, *Transport Theory Statist. Phys.* **1**, 186, arXiv:physics/0503066.

Ogievetskiĭ, V.I. and Polubarinov, I.V. (1963a). Interacting spin 1 fields and symmetry properties, *J. Exptl. Theoret. Phys.* (*U.S.S.R.*) **45**, 966.

Ogievetskiĭ, V.I. and Polubarinov, I.V. (1963b). On interacting fields with definite spin, *J. Exptl. Theoret. Phys.* (*U.S.S.R.*) **45**, 237.

Ogievetskiĭ, V.I. and Polubarinov, I.V. (1964). Minimal interactions between spin 0, 1/2 and 1 fields, *J. Exptl. Theoret. Phys.* (*U.S.S.R.*) **46**, 1048.

Palatnik, L.S. and Landau, A.I. (1964). *Phase Equilibria of Multicomponent Systems*, Holt, Rinehart and Winston, New York.

Peshkov, V.P. (1946). *J. Phys. USSR* **10**, 389; (1948). *J. E. T. P. USSR* **18**, 951.

Pierrehumbert, R.T. (2010). *Principles of Planetary Climate*, Cambridge University Press.

Poincaré, H. (1908). *Thermodynamique*, Gauthier-Villars, Paris.

Poisson, S.D. (1835). *Théorie mathématique de la chaleur*, Bachelier, Paris.

Pool, R.A.H., Saville, G., Herrington, T.M., Shields, B.D.C. and Stavely, L.A.K. (1962). Some excess thermodynamic functions for the liquid systems argon+oxygen, argon+nitrogen, nitrogen+oxygen, nitrogen+carbon monoxide, and argon+carbon monoxide, *Trans. Faraday Soc.* **58**, 1692.

Prigogine, I. (1949). *Etude thermodynamique des phénoménes irréversibles*, Desoer, Liège.

Prigogine, I. (1965). Evolution criteria, variational properties and fluctuations, In: *Non-equilibrium Thermodynamics, Variational Techniques and Stability*, University of Chicago Press.

Prigogine, I. (1970). Dynamic foundations of thermodynamics and statistical mechanics, In: *A Critical Review of Thermodynamics*, Mono Book, Baltimore.

Putterman, S. (1974). *Superfield Hydrodynamics*, American Elsevier Pub. Co.

Ramsey, A.S. (1936). *Hydrostatics*, Cambridge University Press.

Ramsey, A.S. (1956). Thermodynamics and statistical mechanics at negative absolute temperature, *Phys. Rev.* **103**, 20.

Rayleigh, L. (1880). On the stability or instability of certain fluid motions, *Proc. London Math. Soc.* **11**, 57.

Reif, F. (1965). *Statistical and Thermal Physics*, McGraw Hill, New York.

Reisman, A. (1971). *Phase Equilibria*, Academic Press.

Ritter, A. (1938). A series of papers in Wiedemann Annalen, *Ann. Phys.* 214–256.

Rowlinson, J.S. (1959). *Liquids and Liquid Mixtures*, Butterworths, London.

Rukeyser, M. (1942). *Willard Gibbs*, Ox Bow Press, Woodbridge.

Saha, M.N. and Srivastava, B.N. (1935). *A Textbook on Heat*, Indian Press Limited, Calcutta.

Schmid, P. and Henningson, D.S. (2000). *Stability and Transition in Linear Flows*, Springer-Verlag.

Scott, R.L. and van Konynenburg, P.H. (1970). Static properties of solutions. Van der Waals and related models for hydrocarbon mixtures, *Discuss. Faraday Soc.* **49**, 87.

Scott, R.L. and van Konynenburg, P.H. (1980). Critical lines and phase equilibria in binary Van Der Waals mixtures, *Philos. Trans. Royal Soc. A: Math. Phys. Eng. Sci.* **298**, 495.

Schutz, B.F. Jr. (1970). Perfect fluids in General Relativity: Velocity potentials and a variational principle, *Phys. Rev. D* **2**, 2762.

Schwarzschild, K. (1906). Ueber das Gleighgewicht der Sonnenatmosphäre, *Nach. von der Gesellschaft Wissen. Göttinger Nachrichten.* **1906**, 41.

Stanyukovich, K.P. (1960). *Unsteady Motion of Continuous Media*, Pergamon Press.

Sychev, V.V. and Selover, T.B. (1987). *Thermodynamic Properties of Helium*, Hemisphere Publishing Corporation.

Taub, A.H. (1954). General relativistic variational principle for perfect fluids, *Phys. Rev.* **94**, 1468.

Taylor, G.I. (1923). Stability of a viscous fluid contained between two rotating cylinders, *Philos. Trans. R. Soc. London A* **223**, 289.

Thomson, W. (Lord Kelvin) (1848). On an absolute thermometric scale, *Philos. Mag.* **33**, 313.

Thomson, W. (Lord Kelvin) (1862). On the convective equilibrium of temperature in the atmosphere, *Manchester Phil. Soc.* **70**, 176.

Thomson, W. (Lord Kelvin) (1890). *Mathematical and Physical Papers*, Vol. 3, Cambridge University Press, p. 255.

Thomson, W. (Lord Kelvin) (1904). *Baltimore Lectures*, C.J. Clay and Sons, London.

Thomson, W. (Lord Kelvin) (1907). On Homer Lane's problem of a spherical gaseous nebula, *Nature* **75**, 232.

Thomson, W. (Lord Kelvin) (1911). *Mathematical and Physical Papers*, Vol. 5, Cambridge University Press, p. 232.

Tilley, D.R. and Tilley J. (1986). *Superfluidity and Superconductivity*, Adam Hilger Ltd, Bristol and Boston.

Tisza, L. (1938). Transport phenomena in helium II, *Nature* **141**, 913.
Tisza, L. (1961). The thermodynamics of phase equilibrium, *Ann. Phys.* **13**, 1.
Tisza, L. (1969). *Thermodynamics in a State of Flux. A Search for New Foundations*, University of Pittsburgh, PA.
Tolman, R.C. (1910). The electromotive force produced in solutions by centrifugal action, *Phys. Chem.* **59**, 121.
Tolman, R.C. (1934). *Relativity, Thermodynamics and Cosmology*, Clarendon Press.
Torricelli, E. (1644). *Experienzia dell'Argento vivo*, Correspondence with Ricci.
Truesdell, C. (1980). *The Tragicomical History of Thermodynamics 1822–1854*, Springer-Verlag.
Trupp, A. (1999). Energy, entropy: On the occasion of the 100th anniversary of Josef Loschmidt's death in 1895: Is Loschmidt's greatest discovery still waiting for its discovery? *Phys. Essays*, **12**, 614.
Turner, R.S. (1977). The Ohm–Seebeck dispute, Hermann von Helmholtz, and the origins of physiological acoustics, *Br. J. Hist. Sci.* **10**, 1.
Uhlenbeck, G.E. (1964). Van der Waals revisited, A talk given at the Creation of the van der Waals chair at the University of Amsterdam.
Van der Waals, J.D. (1873/1899). Die continuitat des Gasformige und flussige Zustandes, Ph.D. thesis, Leipzig.
Van der Waals, J.D. (1893). *Square gradient model, Verhandel. Koninkl. Akad., Wetenschap., Amsterdam* **1**, 8.
von Neumann, J. (1963). Memorandum addressed to O. Veblen. In: *Collected Works*, A.H. Taub (ed.), Pergamon Press.
Waldram, J.R. (1985). *The Theory of Electrodynamics*, Cambridge University Press.
Weinberg, S. (1972). *Gravitation and Cosmology: Principles and Applications of the General Theory of Relativity*, Wiley.
Wissner-Gross, A.D. and Free, C.E. (2013). Causal entropic forces, *Phys. Rev. Lett.* **110**, 168702.
Zheltukhin, A. (2014). On brane symmetries, preprint arXiv:1409.6655.

Index

A

action functional, 36
action principle, 7, 119, 250, 274, 295
adiabatic, 82
adiabatic condition, 13, 45, 83, 179, 222
adiabatic derivative, 45, 83
adiabatic dynamics, 51
adiabatic index, 31
adiabatic invariants, 51, 128
adiabatic process, 34
adiabatic relation, 95
adiabatic thermodynamics, 19
aerodynamic lift, 27
air and Argon, 225
air and water, 227
angular momentum, 248
antisymmetric tensor field, 296
assigned entropy, 37, 79, 122, 126
astrophysics, 273
atmospheres, 220
attractive force, 86

B

Bernoulli equation, 49, 67, 122, 180
Bianchi constraint, 311
Bianchi identity, 310
Boltzmann's formula, 140
Bose–Einstein, 210
boundary conditions, 48, 223
boundary terms, 10
boundary velocity, 201
bubble point, 90, 92

C

capillary action, 21
Carnot cycle, 15
Cauchy's approach, 314
chemical interactions, 137
chemical potential, 45, 63, 93
chemical reactions, 143
Clapeyron equation, 63
closed system, 63
coexistence, 158, 163
common tangent, 96, 154–155, 228
compressible fluid, 274, 278
conduction, 53
conservation, 115
conservation laws, 34, 66
conservative hydrodynamics, 300
constant lapse rate, 226
constraint, 83, 242
convection, 61, 131
Couette, 277
Couette flow, 243
covariant derivative, 309
critical, 167
critical behavior, 202
critical point, 100, 132
critical temperature, 87, 162
curvature scalar, 310
curvature tensor, 310

cyclic, 14
cylindrical Couette flow, 244
cylindrical symmetry, 277

D

Dalton's law, 124
degrees of freedom, 241
densities, 74, 63, 116
density field, 2
density space, 160
dew point, 90, 92, 97, 160
differential equations, 309
dissipation, 20, 221
dissipative process, 34
dissociation, 137
dividing wall, 124
dustribution of entropy, vii

E

Earth and Mars, 284
Earthly atmosphere, 220
Einstein tensor, 310
Einstein's equation, 310
empty space, 310
energy, 4
energy conservation, 67
energy–momentum tensor, 66
entropic force, 46, 154
entropic rubber, 60
entropy, vi, 14, 37, 54, 61, 195, 204
entropy functiona, 196
entropy of mixing, 118
entropy parameters, 120
entropy path, 122
entropy production, 141
equation of continuity, 5, 180, 242, 314
equation of state, 49
equatorial bulge, 273
equilibrium, 3, 69
equilibrium concentration, 195
Euler–Lagrange equations, 64, 82, 141
Eulerian version, 25
evaporation energy, 92
evolution diagram, 283
excess free energy, 150
excluded volume, 85
extended mas distributions, 314
external forces, 50

F

first law, 16
fluid, 2
fountain effect, 193
free energy, 17, 30
free energy density, 44
free expansion, 132
fundamental relations, 35

G

galaxies, 289
Galilei invariance, 73, 201, 304
gas in pipes, 27
gauge field, 242
gauge transformation, 297
Gauss' theorem, 5
General Relativity, 219, 240, 274, 295
geodesic action, 313
geodesic motion, 308
Gibbs, 76, 77
Gibbs function, 39
Gibbs potential, 116
Gibbs–Dalton, 127, 133, 180
Gibbs–Dalton hypothesis, 222
Gibbs–Duhem, 118
global thermodynamic system, 29
gravitation, viii, 219
gravitational potential, 219
gravitational waves, 310
greenhouse effect, 221

H

Hamiltonian, 11
Hamiltonian density, 9
harmonic, 274
harmonic perturbation, 81
Haumea, 289
heat capacity, 107, 209
heat content, 14
hexagonal flow, 289
hydrodynamics, 1

hydrogen, 137
hydrostatic condition, 46

I

ideal gas, 31
ideal gas law, 12, 229
incompressible fluids, 265
independent variables, 240
infrared radiation, 220
integrability condition, 314
integration factor, 14
interaction, 113, 133, 186
interaction energy, 150
internal disspation, 56
internal energy, 13, 29
irrotational, 9
isentropic atmosphere, 221
isobars, 161

J

Joule, 2

K

kinetic energy, 48, 248
kinetic energy density, 9
kinetic potential, ix, 18

L

Lagrange multiplier, 8
Lagrangian, 114, 138
Lagrangian density, 9, 57
Lagrangian version, 25
laminar Couette flow, 244
least action, 7
Legendre transformation, 18, 30, 52, 63, 68, 116
lifting force, 27
linear constraint, 144
local Lorentz transformations, 310
local variation, 48
locally irrotational, 196
Lorentzian metric, 308

M

main relations, 64
mass conservation, 315
mass current, 5
mass density, 2
maximal entropy, 40, 93
maximal temperature, 166
Maxwell's rule, 90
mechanical energy, 13
metastable, 86
metric, 219, 240
metric connection, 309
minimal energy, 40, 140
miscible, 149
mixing, 124
mixing term, 298
mixture, 113, 149
molar fractions, 116
molecular weight, 12
momentum conservation, 67
multiplier, 14

N

Navier–Stokes equation, 21, 27, 66, 245
Neptune and Uranus, 288
neutron scattering, 200
Newtonian fluid, 6
Newtonian potential, 219, 240
nitrogen and oxygen, 222
Noether's theorem, 64
non-euilibrium thermodynamics, v
non-relativistic limit, 241, 302
non-slip boundary condition, 243
normal conditions, 184, 224
normalized free energy, 97

O

off shell, 38
open system, 63
organic fluids, 164

P

parameters, 105
partial pressures, 152
Pascal, 2
perturbation, 81
phase boundary, 278
phase change, 88

phase rule, 139
phases, 229
phonons, 196
photon mass, 298–299
piston, 3
planetary rings, 273, 277
Poisson equation, 309
Poisson bracket, 11
polytropic atmosphere, 226
polytropic relation, 32
potential type, 9
potentials, 63
predictive power, vi, 120, 274
pressure, 3, 17, 48, 54
pressure wave, 81
Prigogine, 77
propagating, 296
propagation matrix, 179, 181

Q

quasistatic change, 55

R

$\mathcal{R}$, 12
radiation, 53
rates, 147
reaction constant, 146
real gases, 133
reference state, 120, 125
relativistic generalization, 242
Riemannian metric, 312
rotating fluid, 274
rotating planets, 273
rotation, 307
rotons, 196

S

saturation, 89
Saturn and Jupiter, 289
scalar fields, 73
scalar velocity potential, 240
second law, 52
second sound, 198
simple system, 2, 81
solar system, 273
sono-luminescence, 85
sound, 81
sound at critical point, 102
sound propagation, 81
space of variations, 76
Special Relativity, 295–296
specific entropy, 83
specific entropy density, 44, 71
spinodial, 159
spinodial of binaries, 159
square gradient model, 92
square root interaction, 171
stability, 27, 84, 248
stars and galxies, 307
state variable, 16
statistical mechanics, 54
stress tensor, 66
Sun, 289
superfluid component, 194
superfluid helium, 193
surface tension, 92

T

temperature, 11
temperature field, 124
temperature gradient, 220
tensor fields, 309
terrestrial experiments, 307
terrestrial gravity, 220
thermally isolated, 16
thermodynamic pressure, 45
thermodynamics, 1
timescales, 33
total variation, 6
two velocities, 200
types of mixtures, 181

U

unitarity, 6, 315
units, 2

V

vacuum, 310
Van der Waals atmosphere, 227
Van der Waals fluid, 85

vapor pressure, 91, 97, 164
variables of hydrodynamics, 240
variation of the volume, 48
variational principles, vi
velocity, 4
velocity potential, 10, 50, 91
Venus and Pluto, 288
viscosity, 251, 276
vorticity, 19, 251

W

wind tunnel, 27

www.ingramcontent.com/pod-product-compliance
Ingram Content Group UK Ltd.
Pitfield, Milton Keynes, MK11 3LW, UK
UKHW052114170926
8624IPUK00026B/383

9 789811 200670